# Strategies for Sustainability

Spatial Planning and Sustainable Development

## Strategies for Sustainability

The series, will focus on "implementation strategies and responses" to environmental problems – at the local, national, and global levels.

Our objective is to encourage policy proposals and prescriptive thinking on topics such as: the management of sustainability (i.e. environment-development trade-offs), pollution prevention, clean technologies, multilateral treaty-making, harmonization of environmental standards, the role of scientific analysis in decision-making, the implementation of public-private partnerships for resource management, regulatory enforcement, and approaches to meeting inter-generational obligations regarding the management of common resources. We will favor trans-disciplinary perspectives and analyses grounded in careful, comparative studies of practice, demonstrations, or policy reforms. We will not be interested in further documentation of problems, prescriptive pieces that are not grounded in practice, or environmental studies. Philosophically, we will adopt an open-minded pragmatism – "show us what works and why" – rather than a particular bias toward a theory of the liberal state (i.e. "command-and-control") or a theory of markets.

We invite Authors to submit manuscripts that:

– Prescribe how to do better at incorporating concerns about sustainability into public policy and private action.
– Document what has and has not worked in practice.
– Describe what should be tried next to promote greater sustainability in natural resource management, energy production, housing design and development, industrial reorganization, infrastructure planning, land use, and business strategy.
– Develop implementation strategies and examine the effectiveness of specific sustainability strategies. – Focus on trans-disciplinary analyses grounded in careful, comparative studies of practice or policy reform. – Provide an approach "...to meeting the needs of the present without compromising the ability of future generations to meet their own needs," and do this in a way that balances the goal of economic development with due consideration for environmental protection, social progress, and individual rights.

Submit your Proposal!

Authors interested in submitting a book proposal under these series should contact the Publishing Editor for a Book Proposal Form:

Fritz Schmuhl, Senior Editor: Environmental Economics, Management, and Policy, Springer Dordrecht, Van Godewijckstraat 30, P.O. Box 17, 3300 AA Dordrecht, The Netherlands. Phone: +31 78 657-6281, e-mail: Fritz.Schmuhl@springer.com

More information about this series at http://www.springer.com/series/8584

Zhenjiang Shen • Ling Huang
KuangHui Peng • Jente Pai
Editors

# Green City Planning and Practices in Asian Cities

## Sustainable Development and Smart Growth in Urban Environments

Springer

*Editors*
Zhenjiang Shen
School of Natural Science and Technology,
Environmental Design Division
Kanazawa University
Kanazawa City, Japan

KuangHui Peng
Institute of Design
Taipei University of Technology
Taipei City, Taipei, Taiwan

Ling Huang
Faculty of Architecture and Urban Planning
Chongqing University
Chongqing, Chong Qing, China

Jente Pai
Chengchi University
Taipei, Taiwan

ISSN 2212-5450 ISSN 2452-1582 (electronic)
Strategies for Sustainability
ISSN 2522-8463 ISSN 2522-8471 (electronic)
Spatial Planning and Sustainable Development
ISBN 978-3-030-09926-8 ISBN 978-3-319-70025-0 (eBook)
https://doi.org/10.1007/978-3-319-70025-0


Softcover re-print of the Hardcover 1st edition 2018


Printed on acid-free paper

This Springer imprint is published by the registered company Springer International Publishing AG part of Springer Nature.
The registered company address is: Gewerbestrasse 11, 6330 Cham, Switzerland

# Foreword by Zhi Yin

With the rapid development of cities in Asian countries, awareness of environmental degradation and climate change such as global warming continues to rise. Many Asian researchers and practitioners such as urban planners, designers, and policy makers make efforts in engaging a sustainable approach to reducing the environmental footprint caused by urbanisation. The development brings a dynamic and complicated impact in many aspects of cities such as social, economic, and environmental issues. The provision of people's health, safety, and security needs induces many challenges, such as exploitation of natural resources; increasing energy demand; pollution impact on air, water, and soil; and involvement of sophisticated technologies. These aspects are interconnected and considered in a sustainable approach for a green city concept.

To attain a utopian city that is smart and sustainable, the approach of green city concept offers an ideal thinking of city environment to safeguard the equilibrium of natural environment, social-culture effect, and technology. However, implementing the whole concept of green city is a very difficult task that requires complex strategies within different regions of the world. As the original concept of a green city comes from European countries, modification is needed in terms of strategies to create a green city in Asian countries. This is due to differences in history, development, natural and social-culture situation, governance systems, and available technology among cities in the European region and those in the Asian region. Although there are many challenges and obstacles pertaining to green city construction in Asian countries, these do not discourage the urge to accomplish sustainable development in the region.

The most concrete principles and strategies of Asian green cities are manifested in planning and practice by utilising local potential and strength to overcome the disadvantage and weakness of each city. Hence, the most appropriate technique can be adapted and addressed for the local context. Integration of up-bottom and bottom-up approach becomes particularly interesting to accomplish sustainability in all levels of the green city at building, community, and urban scales. The feasibility of implementing an integrative green city concept can be perceived from a higher level of the city by determining the general guidelines, followed by performing detailed strategies in a lower level of unit scales such as building code for indicators assessment and performance evaluation.

In Asian countries, the core of green city practice appertains with three-dimensional city greening and green and blue space network interlink. Green spaces accommodate multiple functions that include natural restoration and preservation, oxygen supplier, public spaces, and city aesthetics. Cities are often divided into cells formed by unique natural and/or artificial barriers as a frame to place city elements. Preservation of characteristics of natural ecology and land resources is an essential consideration in planning and design of Asian cities. Conforming to the natural ecological process for balancing the urban construction and natural system is usually recommended in green planning and designing of urban form, space layout, and urban expansion.

Green building is the most favourite element in applying the green concept in most of Asian cities that is conformed in this book's chapters. This book describes the current developments of green city concept, research, and application in Asian cities through a multidisciplinary approach that includes planning perspective, design practice, and green city assessment. We underline several necessary action stages to attain a green city by defining the goals and suitable concepts, preparing appropriate evaluation tools, determining strategies and techniques, and implementing them into practice.

The sustainable approaches presented in this book are derived from the most concrete method of green city practised in different areas, which is responsive to Asian cities' environment and demographic situation. Although the ideal green city is difficult to achieve in recent situation, this book offers feasibilities in constructing a green city in Asia by presenting several case studies that demonstrate strategies and techniques in different contexts by overcoming obstacles and barriers and maximising local resources. Finally, this book offers tantalising ideas to encourage innovative design that can be practised in different parts of Asian cities.

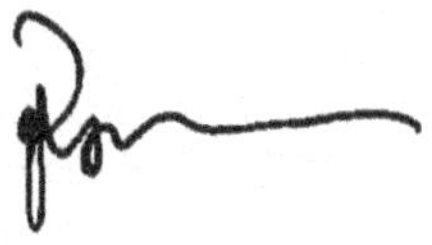

Deputy Head, Institute of China Sustainable Urbanization Zhi Yin
Tsinghua University, Beijing, China
Chair, Scientific Committee, School of Architecture
Fuzhou University, Fuzhou, Fujian, China
Vice President, Urban Planning Society of China

# Foreword by Youguo Qin

Up till now, there has been no universally accepted definition of the green city. By general thinking, the green city is an ideal city that balances the construction and natural ecology of the city by encompassing high environmental quality, liveable design, humanistic spaces, graceful landscape, convenient city service, and technology support.

The green city concept strives to lessen the environmental burden resulting from urban development by lowering emissions, supplying oxygen, reducing-reusing-recycling waste, increasing compact urban services, promoting historical and cultural heritage preservation, and expanding green space. Constructing a green city is more complicated than just putting in place a perfect urban planning and design, certain policies, and codes. With dynamic and complex urban problems, integrative strategies combined with best practices of city government and supportive attitude from every citizen are indeed needed to achieve a true green city. The possibility of a green city to be practiced, maintained, and managed is one of the important keys in achieving a green city.

In planning a green city for a sustainable development, the urban setting is essential as it influences the system layout and allows for crucial environmental change. To tackle many urban problems, it is imperative to formulate a comprehensive and integrative planning interrelated across all levels from a single building, to the community, to regional and urban scale. Every city in the world has a unique characteristic and encounters different challenges that need to be resolved. Numerous kinds of solutions and strategies with refreshing ideas and novel thinking which are genuine and ingenious may appear in various places. The differences in natural conditions, history, development, social-culture situations, governance systems, and available technology among the cities may lead to different preferences of green city principles.

Even though the origin of the green city concept is not in Asia, various ancient and traditional wisdoms from Asian cultures deliver many indigenous ideas, strategies, and techniques that complement the green city concept which is adapted to their context. The most accurate strategies and appropriate techniques are employed to achieve a green city adapted to particular local conditions that are derived from the utilization of local potential to overcome specific urban problems. As a result, many compelling green solutions allow for the creation of unique characteristics of Asian green cities which are different from green cities in Europe and North America where the original green city concept comes from. The newest idea of the green city concept in Asia is the integration with the smart city concept aimed at providing a comfortable, healthy, safe, and secure living environment while maintaining carrying capacity of natural ecology and accelerating efficient technology. The rapid development of information and communication technology (ICT) in recent years triggers this concept integration. The role of ICT in supporting the realization of the green city is placed on smart technology in data gathering, data processing, and action execution. This technology is helpful in several stages of the green city, from city planning and designing to city management processes.

Green city practice in Asian cities mostly focuses on city greening and green building. This practice also encourages technology development on smart-green building and smart-green gardening in Asian cities. Certain sensors and smart management system are already embedded in many green buildings and green houses in Asia to support energy efficiency. Energy efficiency mostly gained from integration of passive-active techniques in lighting, air conditioning, and energy is balanced with renewable energy supply and energy consumption. Green spaces which accommodate multi-functions that include natural restoration and preservation are supported by smart-green watering involving renewable energy supply to generate pumps, advanced technology of water efficiency, and smart treatment for planting.

This book presents the current state of green city development in Asia that includes the idea, concept, research, and implementation from a multidisciplinary approach in regard to planning, design practice, and green city evaluation. Distinctive methods to construct a green city are sketched in this book by associating the most possible strategies and techniques of the green city as practiced in different areas responsive to Asian city environment. The integration of the green city concept with the smart concept is introduced in this book. Although the idea of the smart-green city is still in process in Asian countries, it is still worth mentioning that one of the book's chief contributions is on how to construct the green city. Finally, innovative smart-green design ideas offered by the book are expected to encourage green city practice in distinct parts of Asian cities.

Youguo Qin

Former Dean, School of Architecture
Tsinghua University, Beijing, China
Honor Dean, School of Architecture
Fuzhou University, Fuzhou, Fujian, China
Director, Green Building Committee,
Architectural Society of China, Beijing, China

# Acknowledgment

We would like to express our deep appreciation to all the authors for their outstanding contributions to this book. The series editor of Strategies for Sustainability, Dr. Fritz Schmuhl, and his editorial assistant, Catalina Sava, of Springer are greatly appreciated for their kind encouragement, support, and editorial work. With their kind help, we have such diverse topics on "Green City Planning and Practices" published as a book in the subseries Spatial Planning and Sustainable Development of the series Strategies for Sustainability.

We are deeply indebted to our colleagues in the international community of Spatial Planning and Sustainable Development (SPSD, https://www.spsdcommunity.org/) from Asian countries and territories. In our SPSD workshops or symposiums, we have shared a wonderful time with our members, and all their stimulating suggestions and arguments have helped us throughout the time of considering and editing this book project with Springer.

We also appreciate the support of Mrs. Puteri Fitriaty, who edited this book and has worked as a lecturer at Tadulako University. Now she is studying as a PhD candidate instructed by Prof. Zhenjiang Shen at Kanazawa University. We enjoy cooperation with all members of SPSD and thank them sincerely for their assistance and contributions to this book.

# Contents

# Chapter 1
# Overview: Green City Planning and Practices in Asian Cities

**Zhenjiang Shen and Puteri Fitriaty**

**Abstract** This chapter is provided as an introduction and summary of the whole book's contents. From the viewpoint of planners and architects, the summary presents the key contributions of the book on a three-dimensional scale of the green concept, including buildings, the community, and the city. This book draws extensively on multidisciplinary research and practice to convey implementation of green concepts in Asian cities in achieving sustainable development. The book aims to present current developments in the green city concept, research, and applications in Asian cities. Most of the works focus on building scale as an essential unit of the green city, which can be functionalized through ecology restoration as well as user comfort where the building envelope can serve as a three-dimensional landscape for greening the city and modifying indoor environments. There are several indications that the green city concept may shift to absorb and merge with the smart concept in Asian cities to accomplish sustainable and smart living environments by technology innovation and embedding of information and communications technology (ICT) into city units and systems.

**Keywords** Sustainable development · Climate change · Green building · Green infrastructure · Community development · Smart city and urban design

Z. Shen (✉)
School of Natural Science and Technology, Environmental Design Division, Kanazawa University, Kanazawa City, Japan
e-mail: shenzhe@se.kanazawa-u.ac.jp

P. Fitriaty
Urban Planning Laboratory, Environmental Design Division, Graduate School of Natural Science and Technology, Kanazawa University, Kanazawa, Japan

Architectural Department, Engineering Faculty, Tadulako University, Palu, Indonesia



Z. Shen et al. (eds.), *Green City Planning and Practices in Asian Cities*,
Strategies for Sustainability, https://doi.org/10.1007/978-3-319-70025-0_1

## 1.1 Introduction

Environmental degradation caused by urban development has prompted many city planners, architects, and scholars to promote ideal concepts for city development. One concept among others that can be implemented to achieve sustainable development is the green city concept. European countries have been a pioneer in the area of the green city and have contributed to important sources of ideas for sustainable development (Beatley 2012). In the USA, this concept was driven when some large cities started to use advanced drinking water, sewerage, and sanitary systems, with public open spaces, in the late 1800s (Kahn 2006; Karlenzig 2007). The concept was aimed at achieving an environmental balance in supporting human activities by considering the carrying capacities of the natural environment. Several aspects are included in this approach, such as climate and context, green water and energy, a green built environment, green transportation, a green community and government, and green waste management. Unfortunately, to implement the whole concept of the green city is a difficult task and requires different strategies in different regions of the world (El Ghorab and Shalaby 2016).

This book is expected to contribute by describing the recent evolution of the green city concept, standards, and assessment in Asian cities that are absorbing the smart concept to achieve sustainability. The book also makes every effort to highlight the most feasible approaches to planning and design to implement the green city concept in Asian city planning and practice. The diversity of the green city approach, which is covered by different case studies at different locations in Asia, can provide many ideas for constructing a green city in a wide range of different regions. Finally, the book offers contemporary innovative planning and design that can be implemented in preserving and restoring the ecology of Asian cities.

The role of Asian cities is very crucial in reducing negative impacts of development on the earth's carrying capacities. Asian countries accommodate nearly two thirds of the world's human population, where 54% of this population are concentrated in urban areas (Haase et al. 2017). Thus, with the biggest population out of the five continents, Asian cities are responsible for stability of the natural equilibrium and depletion of the ozone layers of the earth's atmosphere, which leads to global warming. With the awareness of global warming, Asian cities have made much effort in ecology preservation and restoration. Several concrete principles and strategies of the green city that can be manifested and implemented in Asian city planning and practice in different local and regional contexts are described in this book. Several cities have been chosen as case studies, which cover East Asia and the Central East Asia region. The case study analyses are selected from the top ten most developed countries in Asia, namely China, Japan, South Korea, Saudi Arabia, and the United Arab Emirates.

## 1.2 The Green City Concept and Sustainable Development

The term *sustainable development* was first used in 1972 by Donella Meadows and other authors of *The Limits to Growth* and by Edward Goldsmith and the other British authors of *Blueprint for Survival* in that same year (Downton 2009). Generally, the terms addressed a development process that learns from past wisdom to meet present needs and considers future generations for healthy living, restoring the ecology and preserving the environment. The green city concept is addressed for sustainable development to minimize exploitation of energy, water, and materials at each stage of the city, community, and building life cycle (Lehmann 2011). The stages include the embodied energy in the extraction and transportation of materials, fabrication and assembly of unit elements, and operation of units or buildings, until the recycling stage, when an individual building or city element is demolished, which adheres to the biomimetic approaches of "cradle to cradle" and "cradle to grave."

The green concept is an environmentally friendly concept for city planning and development. The concept covers different layers of planning and design, and encompasses several stages. The underlying action for successful sustainable city development is to ensure that eco-cities are managed carefully on every level. No less important, all city design components need to work interactively and inseparably (Lehmann 2011). Although there are many ideas related to the green city, here the author can represent the idea of the green city as covering six elements and five principles based on a triple-zero framework of zero fossil-fuel energy use, zero waste, and zero emissions.

As to our opinion on the challenges in achieving the goal of the green city concept, a city is considered by planners and architects as a living organism, which is built from many cell units. The cell units that compose a green city are classified on the basis of the seven elements of the green city, which can work independently and cooperate with each other to form the city's metabolism and shape a solid–void green city form. The seven elements of the green city include:

1. Green resources (water, energy, and materials)
2. A green social system
3. Green open space
4. Green waste
5. Green transportation
6. Green building

The principles of the green city already proposed by Lehmann cover the 15 principles of green urbanism (Lehmann 2011). Here the authors present principles included in the green concept at each level of the green city, which comprise five main principles:

1. Adaptation and responsiveness to climate, ecology, and context
2. People's quality of life
3. Reduction, reuse, and recycling of resources
4. Cradle to cradle and cradle to grave
5. Clean technology and innovation

The six elements of the green city and the five principles of the green city, as a pilot for achieving sustainable development, thus became a basis for organizing the chapters in this book. The discussion content in the case studies in the book covers the domain of the green city principles that have been implemented in Asian cities. Most of the contributors to this book are planners and architects, who are mainly focusing on adaptive planning and design, according to local and regional contexts, which accentuate the identity and image of the city, and restore and preserve the ecology of the city. A minority of the cases discuss people's quality of life, renewable resources, and clean technology.

The discussions in the book point to several elements of the green city, such as a green social system, green open space, and green building. As a unit of the green city, these elements definitely can work independently and contribute to restoring the ecology of the city through minimization of carbon emissions, even though the action can be greater when all of the elements work together.

### *1.2.1 Green City Planning and Design Practice in Asia*

The concept of the green city as mentioned above originally came from Europe and the North American countries, where it is very different from the concept in Asian countries because of climate, the people's backgrounds, and the culture, goals, and history of city development. These distinctions lead the implementation of the green city concept in Asian countries to be a bit different from the concept in those pioneer countries. For better understanding of the background to this book, here we compare the practice of the green city concept in European countries, North American countries, and Asian countries.

In European countries, the approach to green cities in achieving sustainability is the "creative blending" of social systems. New and old construction are blended and balanced, and sustainability is embedded in a deeper span of history, commitment to place, and treasuring of high quality of life (Beatley 2012). The achievement of European cities in practicing the green city concept essentially comes from a long-term sustainability plan, which is supported by their green social system and the history of their cities, which have been developed as shrinking cities and thus will be more compact in serving people's needs. The compact form of cities can support green transportation by shortening the time and distance, thus many places can be easily accessed by pedestrians and bicycle users, with less energy expenditure. Additionally, the infrastructure of European cities was well planned long before the green concept officially emerged, which therefore accelerated the construction of green cities. The identity of the city is preserved by maintaining historical buildings and sites while blending them with new construction.

Green open spaces are multifunctional spaces serving as community places, preserving the city's ecology and oxygen supply, and balancing the proportion of hard built surfaces with soft surfaces of vegetation and water. Apart from the urban form's history, the essential key to European green cities lies in conscious policies aimed at strengthening a tight urban core, supported by community awareness. Cities such as Freiburg and Copenhagen are supported by strict policies to maintain the population living in the centers of these cities; therefore the cities will not succumb to urban sprawl. The new growth areas have been placed near to existing developed areas with high densities and are supported by natural and built environments balanced by utilizing wind power and solar energy, community food production, and natural drainage, such as in the Vauban in Freiburg and the Thames Gateway in London.

The green city concept in North America can be seen in San Francisco, New York, and Vancouver. San Francisco is recognized as the greenest city in North America. The keys to its achievement in implementing the green concept are a strong economy and regulation to enhance the green building market and clean technology. Economic incentives for green businesses have been used as strategies to actualize a green city. In municipal building codes, the processing of permits for highly ranked green buildings is prioritized, with tax incentives for distributed solar generation (Zhou and Williams 2013). In addition, waste within the city has been controlled by aggressive waste policies.

The main challenge for New York, as a mega city, has been population forces, which have required more energy and land. Hence its strategies for a green city are primarily concerned with energy security and land development. Because the city government does not control or regulate energy consumption, policies have been directed toward reducing energy demands in public and private buildings. This action is approved by the Green Code Task Force (http://urbangreencouncil.org/GreenCodes), which requires regular energy audits, retrocommissioning, and data sharing for government buildings larger than 10,000 ft2 and private buildings larger than 50,000 ft2 (Zhou and Williams 2013). The long-term plan for the green city aims to decrease the city's environmental loads by reducing material and energy use, improving natural spaces, mitigating the effects of climate change, and creating a more equitable, engaged society.

In Canada, the city of Vancouver, through the Greenest City Action Plan, identified ten goals based mainly on screening and evaluation of measurements, indicators, and best practices used in leading green cities around the world. The goals refer to three overarching areas: (1) zero carbon (climate leadership, green transportation, and green building); (2) zero waste; and (3) healthy ecosystems (accessibility to nature, clean water, local food, and clean air) (Affolderbach and Schulz 2017). The action has a good foundation with three important keys, which are legalizing the policy of a green city, promoted by municipality, and public participation in decision making. Public opinion on a green city in Vancouver identified several goals:

economic development, green jobs, and investment in infrastructure, with an emphasis on quality of life that resonated with the broader public.

The approach of North American cities in complying with the green city concept tends to be focused on strengthening the economy, reducing energy demand, and enhancing clean energy production. In the case of urban form, North American cities may still learn from European cities with regard to maintaining city compactness for sustainability (Beatley 2012). The success of both European and North American green cities in pursuing the green city concept apparently closely involves the roles of local and municipal authorities in terms of policy, action plans, and cooperation of the private sector. The effectiveness in achieving the goals of a green infrastructure and green transportation appear to be related to the cities' compactness, with benefits in savings of time, cost, and energy.

Conversely, most Asian cities have sprawled during their development. Thus, services will be scattered and the impacts of this will be high fiscal and infrastructural costs (Beatley 2012). This fact has made it more difficult for Asian cities to achieve comprehensive and integrative green city practices, especially in the elements of infrastructure and transportation. To avoid high infrastructural costs, green city practice in Asia is more focused on unit elements involved in green open space, green building, and green technology. With this solution, it is expected that a lot of green city costs can be divided into smaller portions, and this is true of the green city practices in Asia described in this book.

City greening is one of the ways to pursue a green city. It is aimed at replacing the loss of green spaces that have disappeared in ongoing urban land conversion for housing and transportation. The objective is to provide multifunctionality in improving people's health and well-being, maintaining ecosystem services, enhancing the attractiveness of open spaces in cities (Haase et al. 2017), and providing passive cooling of buildings and open spaces (Zeng et al. 2017). In this book, we describe several strategies that have been used in city greening practices in Asian cities. The practices of city greening described in this book involve several techniques, from a conservative approach of two-dimensional horizontal green spaces to three-dimensional green spaces embedded in walls and roofs of buildings. Furthermore, the creative idea of converting waste land, abandoned land, and idle land into green spaces for revitalization and urban renewal is also described in this book.

Green building is the most favored aspect practiced in Asian countries among the six elements of green cities that are described in this book. Some green building practices in Asia utilize green roofs to increase the green coverage of the city in limited urban space, which has the effect of retention, collection, and filtration of urban water resources. Some green buildings in Asian cities may fulfill the criteria for climatically responsive building while maintaining energy efficiency and retaining local identity connected to the city's history and image. Conversely, some of the buildings demonstrate a deliberate departure from the past identity of the city and project a futuristic image reflecting the technology and age of the buildings.

### 1.2.2 Future Development of Green Cities in Asia

Recent progress to achieve sustainable development has introduced a smart concept into city planning and design. The concept itself started from the development of technology, especially information and communication technology (ICT). Through this technology, many devices can be connected by certain protocols with the main control system. This technology enables management of devices or even systems and sustains data sharing from the server unit. With recent development of ICT, it makes it possible to manage the system using the internet of things (IoT) with a remote controller.

With the arrival of smart technology, there has been a shift in the sustainability goals approached using smart concepts at the levels of building design, community management and sharing, and city planning. Effective, sustainable smart cities have emerged as a result of dynamic processes in which the public and private sectors coordinate their activities and resources on an open innovation platform (Lee et al. 2014). Smart cities base their strategy on the use of information and communication technologies in several fields such as the economy, the environment, mobility, and governance to transform the city infrastructure and services (Bakici et al. 2013).

At present, many Asian cities still struggle to actualize the green city concept. With the incoming smart city concept, some developed cities in Asia—for example, in Japan, China, Korea, and Singapore—are trying to integrate the green concept and the smart concept into their planning and policies to attain sustainability as well as reducing $CO_2$ emissions. While the green concept is heavily focused on reducing the environmental footprint, maintaining the city's carrying capacity, and employing passive strategies, the smart city concept is responsible for the system management of green elements by embedding ICT to achieve energy conservation, balance, and efficiency, and quality of life for sustainable smart living. Even though the smart concept nowadays can be seen only in developed cities in Asia, the possibilities of sustainable, smarter, and greener cities can be achieved in developing Asian countries.

## 1.3 Part I: Green Concepts, Standards, and Assessment

In pursuing and constructing the green city, several documents must first be prepared by the authority, to specify the goals, concepts, and approach, with strategies to facilitate planning and design processes. The documents serve as a guideline to ensure that the development of the green city will follow the aims. Assessment tools are one of the documents that should be well prepared in pursuing a green city. They can be used as a support for decision making in urban development as they provide assessment methodologies for cities to show progress toward defined targets. Comprehensive documents and assessment tools must be provided on different scales, from the building code to community evaluation, and from the district level up to the city level. In this way, quality can be monitored in the planning and designing processes, construction process, and maintenance process.

The four chapters in this part of the book focus on green concepts, parameters, evaluation, and assessment tools. In Chap. 2, Chang analyzes the vision, strategy, and planning principles integrated into a green city. The purpose of this study is to propose an ideal green city based on an ecological approach to determine a policy suitable for the city of Taipei in Taiwan. Chapter 2 indicates that publicity and promotion of green city benefits are important for stimulating people's environmental consciousness so they can participate in creating, maintaining, and preserving the quality of the living environment. Furthermore, it is recommended that implementation of the green city concept is done gradually in several stages: the nature resource conservation stage, the renewable green stage, and the restorative organic stage. In the last stage, ICT is introduced, with the scale of implementation growing bigger from the building scale to the community scale to the global scale.

Recently, ICT has been gaining popularity, especially for its relationship with smart concepts for accomplishing sustainable smart living. The smart concept and green concept have been blended to pursue sustainability in Asian cities. Fitriaty, Shen, and Sugihara discuss the smart house concept and its position in relation to the green building concept in Chap. 3. This concept is devoted to creating a high quality of life based on technologies of functional automation within residential buildings. Some literature has conveyed that the smart house aims to ensure energy efficiency in the house, which is included in green concepts (Kofler et al. 2012; Fensel et al. 2013; Solaimani et al. 2015; GhaffarianHoseini et al. 2013), and provides a platform to monitor and control certain appliances in the house environment. In this chapter, parameters of assessment have been derived by combining literature study, online materials, and smart house pilot projects in Japan. The results indicate that the smart house concept and the green building concept have a relationship as intersecting sets. The intersection set parameters are parameters included in the principles of respect for users, conservation of energy and water, and working with the climate. An interesting finding in this study is that the concept of the smart house evolved originally from the green concept and developed into a greener concept where energy efficiency was also included in smart house considerations.

Assessment tools are very important to evaluate the degree of green concept implementation in practice. In China, the Three-Star evaluation standard is semi-mandatory and provides certification of green buildings, contributing to maximum resource conservation (including energy, land, water, and materials), protecting the natural environment, and minimizing pollution. In Chap. 4, the Three-Star evaluation is employed by Liu and Lin to assess energy efficiency and user satisfaction in three typical green buildings located in Shenzhen (a hot summer–warm winter zone), Shanghai (a hot summer–cold winter zone), and Tianjin (a cold zone). The chapter presents the green buildings' significance in reducing building energy consumption and the implications for occupant satisfaction, as well as comparing them with conventional buildings. The evaluation results indicate that mixed-mode green buildings (naturally and mechanically air conditioned) are significantly more energy efficient than conventional buildings, especially in the hot summer–cold winter zone, whereas in all climate zones, mechanical air conditioning does not have a significant impact on energy performance in green buildings versus conventional buildings. Additionally, performance on several indicators of indoor environmental quality and building service performance—such as thermal comfort, indoor air quality, facilities, operating and maintenance indicators, and user satisfaction—is marked higher in green buildings than in conventional buildings. The results confirm the role of passive strategies in energy efficiency for green building practice.

Ecological assessment systems in the UK, the USA, Germany, Japan, Taiwan, and China are reviewed in Chap. 5 by Shu and Huang. The ecological community terms used in Chap. 5 cover various aspects such as people, buildings, and environment. A contribution is made to ecological community assessment in subtropical regions by pointing out important indicators. The results may be used to design a model for weighting ecological community assessment by displaying the degree of importance of each assessment category for sustainable development. It is recommended that the eco-community assessment system take into consideration the weighting structure. Index weighting and evaluation criteria should recognize the actual needs of different regions, different communities, and different living standards so that humanistic indicators can be strengthened. The indicators for eco-community assessment in Taiwan give more emphasis to ecology and to living and natural environments than those in other countries, where the assessment systems cover a full range of health, finance, and the economy.

## 1.4 Part II: Ecological Preservation and Restoration Planning by Greening the City

In the context of urban expansion and the corresponding impact on the natural ecology, several concepts are closely related to the natural ecology and environmental balance in a city environment—such as a green city, eco-city, decarbonized city, or smart city—for preserving, conserving, and restoring the natural ecology. This part includes five chapters, which describe several techniques related to ecological restoration and preservation.

In Chap. 6, Lu and Jiang perform a challenging task related to boundary expansion problems in mountainous cities in China because of natural land resource limitations and the unrecyclable natural ecology. Their study proposes a new developmental model for resource-conserving cities, which focuses on internal renewal of the cities by utilizing existing resource advantages. Chongqing is taken as a case study, as one of the largest mountainous cities in western China. The city is facing a series of grave problems in its development with regard to economic remodeling, functional updating, and resource reorganization. The model presented in this chapter could become a good example for other mountainous cities in the same region. In evaluating the characteristics of the natural ecology and land resources, Lu and Jiang emphasize three design strategies: (1) defining the ecological carrying capacities as the boundary of urban development by centering on natural resources, and creating an ecological green network and public open space inside the boundary; (2) strengthening the accessibility of riverside areas and landscape diversity, and connecting public spaces in different districts; and (3) constructing a three-dimensional public space system that gives priority to pedestrians and public traffic. The street network is optimized to highlight the cultural and historical characteristics of the high-density urban public space.

It is generally known that inherited vernacular strategies are responsive to the local climate and environment. In Chap. 7, Yan, Yang, Xu, and Ren present vernacular wisdom from ancient times, which can be implemented in contemporary urban planning and design. The lesson is taken from the ancient Chinese urban civilization in the era of the Qin and Han Dynasty city system (200 BC). The cities were proved to have strong vitality in terms of site selection, spatial patterns, and urban forms, and were adapted to geomorphological agents, hydrological features, and climate states. The chapter analyzes the relationship between site selection, spatial patterns, and building clusters in typical capitals (Chang'an in the Sui and Tang Dynasties, Lin'an in the Song Dynasty, and Beijing in the Ming and Qing Dynasties) and typical county cities (Lijiang, Suzhou, and Baoding) in ancient China. An interesting finding reveals that the ancient cities of China had a good relationship with and adaptability to nature. The coordination between urban space and ecological wisdom was skillfully utilized to support human needs and activities in the process of planning and design of the ancient cities. The study conveys that the ancient Chinese cities were embodied with several concrete manifestations of planning and design strategies, as follows: (1) the cities were located on the basis of defense functions, rich water bodies, sufficient land spaces, and comprehensive analysis of regional terrain patterns; (2) the spatial layout was adapted to natural terrain fluctuations; and (3) urban land was arranged to facilitate city life, and the road network skeleton was designed for to local conditions. The regional water systems were improved on the basis of natural hydrology features to provide abundant water resources and efficient water transportation. Urban drainage was used to irrigate the landscape and to control flood and drought disasters by controlling the rivers and lakes to create a comfortable living environment. The textures of the ancient Chinese cities were adapted to the regional climate. Climatic determination influenced the street directions, building orientation, and building cluster forms to create comfortable temperature and wind environments.

Adapting this ancient wisdom, contemporary planning and designing for urban forms, spatial layout, and urban expansion need to conform to natural ecological process and pattern rules to balance urban construction and natural systems. Provision of energy is preferably done using natural sources, and it is necessary to consider adaptation of natural space textures and water systems as important objects in urban planning. Thus, urban planning and design are directed toward energy efficiency through reasonable design of buildings, building clusters, and city groups, and through strengthening of the utilization of wind energy, solar energy, and other renewable energy.

In city ecology, green space is one of the essential elements for restoring the natural ecology. Conversion for green space through urban regeneration is discussed by Jin, Chen, and Jing in Chap. 8. It is done by exploiting the space stock resources to reduce costs and establish a new way to supply land for green space. An action planning strategy is proposed, based on the planning principles of dividing levels, connecting networks, and building multipaths. Stock space resources are converted into green space and connected to a network, thereby expanding three-dimensionally, with creation of an integrated urban green open space system. This contribution provides a novel planning method for greening the city, especially in high-density areas, in order to improve city space values that serve multiple functions, which are ecological conservation, public service, and community vitality. The action planning strategy is done in three steps—namely, reconstructing the green skeleton, extending the meridians of the green network, and fleshing it out by constructing three-dimensional green spaces. A multilevel system of green space is established in high-density urban areas as a planning solution to land availability shortages. With limited space resources, the diversity of the space demands is resolved by reconstructing abandoned urban industrial, waste land, and idle land. The backbone of the urban green–blue network is reshaped and moderately adjusted in order that the existing spatial city structure can be reorganized to create a new ecological network system. Hence, negative space can be transformed into positive space as an urban green growth economic belt to guide city life.

The roof garden also can be used to restore city ecology and plays an important role in reducing greenhouse effects through energy efficiency (Collins et al. 2017; Zeng et al. 2017). The ecological function of roof greening by using roof gardens is discussed in Chap. 9 by Hong, Wu, Chen, and Lin. The aim is to test the ecological benefits of roof greening in reducing energy consumption, cooling and humidifying, and controlling rainwater runoff. The chapter provides experimental evidence of the role of the green roof in modifying the indoor environment and controlling rainwater runoff for improvement of the city's ecology. The rooftop of the Fujian Agriculture and Forestry University's administration building is used as a case study, where measurements were taken to verify the effects of cooling and moisture, and lower energy consumption. The changes in environmental factors were analyzed before and after construction of the roof garden by using ecological environment monitoring data. Rainwater in the roof garden was monitored from January to May to test the rainwater collection utility of roof greening. According to the experimental results, the roof greening had significant effects on indoor cooling and humidification compared with the former state. The indoor temperature dropped by 1.8–4 °C and relative humidity increased by 5–10%. It can be concluded that roof greening has a significant insulating effect and effectively reduces building energy consumption. Additionally, the results of rainfall monitoring show that the roof garden also plays a significant role in controlling rainwater runoff.

Another method for city greening is presented by Chen, Shao, and Peng in Chap. 10. The method used for green space replacement is vertical greening, which can be useful for lowering indoor air temperatures. The concepts of green planning, such as building–human coexistence, use and reuse of rainwater, and green materials have been incorporated into a new green wall system that manifests the new thinking of sustainable greening, which is different from existing green wall systems using chemical foams with a short service lifetime. The proposed novel system is innovative and feasible, enables green walls to deliver long-lasting effects, and uses the exterior structure of buildings for energy conservation.

In Chap. 11, Son discusses green roof policies and practices in Korea and introduces a case study of a rooftop garden, Hanul Madang (which means "garden in the sky") at Seoul National University. Since the 2000s, Seoul City's policy has been increasingly concerned with qualitative growth indicators for green and livable cities. Green roofs were included in the *2030 Seoul City Park and Green Space Master Plan*, because they are a very effective way of increasing the quantity and quality of urban green spaces. Two types of green roofs are described in this study. The first type is a leisure and recreation space to enrich green culture in the urban environment, while the other type has more ecological effects and aims to enhance the environment through greening of artificial grounds. This chapter introduces the functions of the green roof in ecosystem services and cultural values. Ecosystem services enable the building envelope to reconnect disconnected natural areas at the ground level, while their cultural value provides leisure, recreational, healing, educational, and tourism values. It offers a place for new experiences—a healing space that relieves stress from everyday activities, and a space for events and welcoming visitors. A dynamic rooftop system is also introduced in this chapter, where different types of plants can be planted seasonally and can provide an aesthetic view or food security through planting of vegetables and fruit.

## 1.5 Part III: Green Building Practice and Innovation

Part III of the book presents several design strategies to achieve green city goals through green building and green technology. This part includes five chapters, which describe strategies to control wind velocities and direction in a mountainous environment, daylighting and ventilation utilization, mixed-mode building conditioning, and energy efficiency for a comfortable living environment. The strategies are executed by adaptation to the local climate, environment, and context, using vernacular and contemporary knowledge.

Vernacular architecture is a type of architecture that highlights local characteristics, emphasizing harmony with the natural environment, in line with local conditions (Zhai and Previtali 2010; Bouillot 2008). In Chap. 12, Su and Chang discuss passive strategies used in the vernacular tulou architecture of the Fuyu House in Fujian Yongding, maintaining an internal environment that is warm in winter and cool in summer. It adheres to the concept of sustainable development through ecology and energy conservation, which is a good example to implement in green buildings. This chapter contributes to the green city concept by describing passive building strategies to achieve thermal comfort in mountainous and windy winter environments, which have been proven by years of implementation of vernacular building. The key strategies lie in a round form with a courtyard in the center, with an appropriate arrangement of openings and patios, and existence of water bodies near to the front yard. The surface temperature differences created by the land and water bodies stimulates wind convection, and an indoor courtyard surrounded by two-story walls create a stack effect; hence wind can be introduced into the courtyard atrium and create indoor thermal comfort even though it has an externally closed architectural form. A good barrier is achieved by the surrounding construction of an atrium, which blocks invasion by external wind, thereby reducing the temperature difference between day and night and creating a warming effect in the winter.

A green strategy for lighting and ventilation in a museum is presented in Chap. 13 by Li and Zu. Most museum buildings presently use artificial lighting, ventilation, and heat preservation devices to maintain indoor exhibits at constant temperature and humidity. As a result, energy consumption incurs high running costs for museum buildings. Li and Zu addressed this problem by designing and renovating a museum on the basis of green building concepts. Energy savings can be achieved by employing insulation systems in the building envelope or structure (Bao 2013), such as the external wall, doors and windows, roof, and other building parts (Zhong 2015). Combining lighting and ventilation devices is believed to be the most effective methodology for improving energy saving in the building. This chapter contributes to green concepts by presenting a practical passive energy-saving strategy for museum lighting and heat preservation under special condition requirements. This chapter describes the ventilation and light-resistant energy-saving design of the Dinosaur Egg Museum. Based on phenomenological theory, the museum's overall layout is studied. The design is proposed by considering local meteorological conditions to install a natural ventilation and daylighting structure and create adaptability between the indoor thermal environment and lighting environment. The results indicate that a design based on green building concepts can achieve the purpose of an energy-saving effect. The automatic adjustment of temperature and ventilation can be improved and refined by further research works in the future.

An interesting concept of simplified green building—which means minimization of complexity, reduced use of materials and resources, and improvement of the quality of the indoor environment—is presented in Chap. 14 by Liu and Chou. In Taiwan's humid climate, the simplified green building concept puts an emphasis on the regional natural environment and climate characteristics in establishing an innovative green building protocol for design, focusing on necessity, sustainability, and passivity. The manifestation of the concept is the AGS1 experimental house, which fulfills several requirements such as environmentally friendliness, structural strength, material conservation, energy efficiency, safety, healthiness, and comfort. The chapter contributes to the literature by establishing a development model for green buildings with high functional performance in terms of safety, healthiness, comfort, energy efficiency, and earth resource conservation. The experimental house is constructed as a three-floor residential detached house, providing indoor thermal comfort by reducing envelope structure weight, controlling heat gain through windows, avoiding excessive interior decoration, and using high-resistance envelope materials. For moisture permeability, an air exchange mechanism is employed in an intermediary space, with a suitable fitted air conditioning system. The study results show that the experimental house design can yield building weight reductions of up to two thirds of the reinforced concrete used in the house, with carbon footprint and total carbon emission reductions (calculated over 40 years of usage) of up to 30.7% and 43% during occupied and nonoccupied periods, respectively. In addition, the overall thermal transfer value (OTTV) is equivalent to only 15.32% of the heat gain with the combination of the reinforced concrete building structure and window openings without thermal insulation treatment.

A new building typology, suited to the locality of Abu Dhabi and utilizing environmental design principles, is presented by El Amrousi in Chap. 15. This building typology is reshaping the urban identity of Abu Dhabi, using new forms of architecture that manifest sustainable design and development. The typologies are aimed at retaining and improving the quality of life of its large expatriate community, who in the past conceptualized it as a transit city. The new design paradigm recontextualizes the concept of the brise-soleil and advocates environmental design principles. Discussing the design precedent of buildings in Masdar City, this chapter highlights the emergence of new building typologies that are aimed at generating energy and implementing passive cooling strategies through window orientation and facade treatments to support sustainable design in Abu Dhabi. New generations of brise-soleil are being implemented to provide shading around the building and passive cooling to control the indoor environment. The brise-soleils are lighter and more flexible, and are easily applicable to building facades to create a new design form that addresses the hot, humid climate. These new forms represent contemporary looks, attractive designs, and technology innovation, and manifest as newly emerging iconic designs. Although the aspirations for a full-scale green city still have a long way to go in Abu Dhabi, individual buildings can represent a new venue of sustainable design that can be viewed and experienced by people. Therefore, collectively, they can begin to form a powerful image of sustainable design within an innovative context to stimulate the process of forming a green city.

One of the factors that support green city concepts is green technology. The implementation of green technology can be utilized to support construction of green buildings. The smart green approach has been used by Hung and Peng, who discuss vertical greening in a high-density urban area in Chap. 16. This vertical greening presents a suitable display of the level of smartness and greenness in a city. This contribution present a proposed novel prototype of a green energy water-autonomous greenhouse (GEWA) system to green the city, employing a smart watering system. The GEWA system is a sophisticated smart technology employing water resources and solar energy in a rational way, which can serve as an alternative technology approach toward sustainable and smart vertical greening in the city.

## 1.6 Conclusion

In this book, we present current developments in green city concepts, research, and applications in Asian cities from multidisciplinary approaches including planning perspectives, design practices, and assessment of development codes. We underline several necessary action stages to attain a green city—from defining the goals and suitable concepts to preparing appropriate evaluation tools, determining strategies and techniques, and implementing them in practice.

The chapters in this book indicate that the green city concept should be implemented by merging top-down and bottom-up methods to accomplish sustainability at all scales and levels of the green city—from buildings to communities and the urban scale. The possibility of integrative green city concept implementation requires action at the higher levels of the city authorities to determine the general strategies, followed by lower level implementation of detailed strategies at the unit scale, such as building codes for indicator assessment and performance evaluation of green city elements.

Various findings in this book have been generally practiced in Asia to construct green cities along the following lines:

- The recent evolution of the green city concept in Asia tries to integrate the concepts of the green city and the smart city to establish the new green–smart concept to achieve an ideal city with a sustainable, smart living environment, with a healthy ecology, low carbon emissions, and resource security in the future. Embedding of ICT, data sharing, virtual base management, and IoT have already been applied in developed cities in Asian countries such as Japan, China, Korea, and Singapore. This statement is supported by several chapters in the book, such as Chaps. 2, 3, 10, and 15.
- Green city practices in Asia tend to focus on constructing green buildings and green landscapes by producing a series of documents such as guidelines, assessment and evaluation tools, and technology upgrades involving ICT and automatic mechanisms. The evaluation tool is very important in controlling the planning and designing processes, certification needs and evaluation target achievement, and quality control of the whole process.
- Asian cities tend to retain local and regional values that represent the people in the space by adapting vernacular passive strategies, which shift the concept of green building implementation from their original concept developed in the European and North American regions. These vernacular strategies are implemented on different scales such as individual buildings, communal buildings, public spaces, and district design.
- Ecology restoration and preservation through city greening is generally the most important factor identifying the green city in Asia, which is a bit different from the approach used in Europe and North America.

## References

Affolderbach J, Schulz C (2017) Positioning Vancouver through urban sustainability strategies? The greenest city 2020 action plan. J Clean Prod 164:676–685

Bakici T, Almirall E, Wareham J (2013) A smart city initiative: the case of Barcelona. J Knowl Econ 4(2):135–148

Bao Y (2013) Discussion on the application of energy-saving technology in building external envelope structure design. Urban Archit 22:116

Beatley T (2012) Green cities of Europe. Island Press, Washington, DC

Bouillot J (2008) Climatic design of vernacular housing in different provinces of China. J Environ Manag 87(2):287–299. Available at: http://linkinghub.elsevier.com/retrieve/pii/S0301479707002514

Collins S et al (2017) Thermal behavior of green roofs under Nordic winter conditions. Build Environ 122:206–214

Downton PF (2009) Ecopolis architecture and cities for a changing climate. Springer, Adelaide

El Ghorab HK, Shalaby HA (2016) Eco and green cities as new approaches for planning and developing cities in Egypt. Alex Eng J 55(1):495–503. Available at: https://doi.org/10.1016/j.aej.2015.12.018

Fensel A, Tomic S, Kumar V (2013) Sesame-s: semantic smart home system for energy efficiency. Informatik-Spektrum 36(1):46–57. Available at: http://link.springer.com/article/10.1007/s00287-012-0665-9

GhaffarianHoseini AH et al (2013) The essence of future smart houses: from embedding ICT to adapting to sustainability principles. Renew Sust Energ Rev 24:593–607. Available at: https://doi.org/10.1016/j.rser.2013.02.032

Haase D et al (2017) Greening cities—to be socially inclusive? About the alleged paradox of society and ecology in cities. Habitat Int 64:41–48

Kahn ME (2006) Green cities: urban growth and the environment. Brooking Institution Press, Washington, DC. Available at: https://www.brookings.edu/wp-content/uploads/2016/07/greencities_chapter.pdf

Karlenzig W (2007) How green is your city? New Society Publishers, Gabriola Island

Kofler MJ, Reinisch C, Kastner W (2012) A semantic representation of energy-related information in future smart homes. Energ Buildings 47:169–179. Available at: https://doi.org/10.1016/j.enbuild.2011.11.044

Lee JH, Hancock MG, Hu MC (2014) Towards an effective framework for building smart cities: lessons from Seoul and San Francisco. Technol Forecast Soc Chang 89:80–99. Available at: https://doi.org/10.1016/j.techfore.2013.08.033

Lehmann S (2011) What is green urbanism? holistic principles to transform cities for sustainability. In: Blanco J (ed) Climate change—research and technology for adaptation and mitigation. InTech, Rijeka, pp 243–266. Available at: http://www.intechopen.com/books/climate-change-research-andtechnology-%0Afor-adaptation-and-mitigation

Solaimani S, Keijzer-Broers W, Bouwman H (2015) What we do—and don't—know about the smart home: an analysis of the smart home literature. Indoor Built Environ 24(3):370–383. Available at: http://journals.sagepub.com/doi/10.1177/1420326X13516350

Zeng C et al (2017) Optimal parameters of green roofs in representative cities of four climate zones in China: a simulation study. Energ Buildings 150:118–131. Available at: https://doi.org/10.1016/j.enbuild.2017.05.079

Zhai Z, Previtali JM (2010) Ancient vernacular architecture: characteristics categorization and energy performance evaluation. Energ Buildings 42(3):357–365. Available at: http://linkinghub.elsevier.com/retrieve/pii/S0378778809002400

Zhong A (2015) Application of thermal insulation mortar in energy-saving retrofitting on the external wall of an existing structure. ICCREM 2014: Smart Construction and Management in the Context of New Technology. ASCE, pp 140–146

Zhou N, Williams C (2013) An international review of eco-city theory indicators, and case studies. Available at: https://china.lbl.gov/sites/all/files/lc_eco-cities.pdf

# Part I
# Green Concept, Standard, and Assesment

# Chapter 2
# Green City Vision, Strategy, and Planning

**Hsiao-Tung Chang**

**Abstract** Nineteenth century, the City Beautiful Movement, known as the "white movement," contributed to the urban design system and attached importance to the cityscape. At the end of the twentieth century, the "green movement" became popular and urban development. More and more countries now attach importance to the planning issues of ecological and livable cities. Under the vision of the "green movement," a lot of similar urban development being discussed, such as sustainable development, environmental coexistence, sustainable cities, compact cities, healthy cities, quality of life, intermediate cities, eco-cities, shan-shui cities, and so on. The thinking of green city planning were integrated into these series of developmental strategies. An ideal green city is aimed to achieve via the vision, strategy and planning, based on ecological environment approach.In this chapter, the concept of green city and the theories of developmental strategies are discussed, and the policy of promoting a green city is expounded.

**Keywords** Green city · Green movement · Green infrastructure · Eco-city

## 2.1 Introduction

### *2.1.1 Background and Motivations*

The "green city" concept emerged in the 1970s, developed rapidly, and became an important concept of urban development. The philosophy of the green city existed in many planning theories, such as Chinese garden design, garden city theory, urban park theory, urban ecology, and so on. In the early twentieth century, the trend of

H.-T. Chang (✉)
Graduate Institute of Architecture and Urban Design, Chinese Culture University, Taipei, Taiwan
e-mail: changht4613@gmail.com

 
Z. Shen et al. (eds.), *Green City Planning and Practices in Asian Cities*, Strategies for Sustainability, https://doi.org/10.1007/978-3-319-70025-0_2

centralization and decentralization for city greening were all based on green objectives, implied green ideas to improve the living urban design theory including greenbelts of garden city, green cores of neighborhood units, green network city designs, etc. In the late twentieth century, many new urban design theories emerged. With the growing recognition of the environmental crisis in the 1970s and of climate change in the 1990s, environment awareness, the concepts of sustainable cities, healthy cities, compact cities, and ecological cities evolved gradually and then gave birth to the concept of the green city, revealing a new beginning in urban design theory. Green city planning and design is one solution to the global warming phenomenon (Wikantiyoso and Tutuko 2013).

Concepts such as the "garden city," "new town," and "techno-city," which occurred in the nineteenth and twentieth centuries, are some of the major representatives (Joss 2011; Zou and Li 2014). Later, "climate-neutral city," "low-carbon city," "smart city," and "sustainable city" can also be considered as sister terms of the "eco-city" (Register 1987), which covers various notions and approaches to sustainable urbanism, rather than a conceptually coherent and practically uniform phenomenon (Joss 2012; Zou and Li 2014).

"Green city" is a new term; so far, there is no universally accepted definition, and no clear concept of it is recognized. Many scholars have studied and defined the green city from different viewpoints. Some have devised their own indicators to express the green city. Various scholars have believed that the green city is essentially human civilization and the nature of the earth. Some have believed that the green city is equivalent to the garden city, ecological city, sustainable development city, or forest city, but with different language expression. Many other scholars have supposed that a balanced social, economic, and environmental system may be called a green city.

### 2.1.2 Definition of a Green City

The city is a place characterized by its social, environmental, and economic system, and a specific function or administrative region forming patterns of human settlement, population, structure, and characteristics of artificial activities (OECD and China Development Research Foundation 2010). The city is not only a man-made environment. Parks, gardens, trees, and other green spaces are integrated parts of the city environment. Green spaces represent far more than simply an effort to balance the manufactured with the natural. Greening the city is one of the most direct definitions of the green city. However, the green city is not just greening of a city; "green" contains a deeper meaning and refers to a comprehensive design method, innovation, and renovation, to minimize the impact on the environment.

The United Nations Environment Program (2011) defines the green city from the perspective of urban development and environmental quality. A green city is defined as an environmentally friendly city that is represented by environmental performance indicators (Meadows 1999). Other indicators are diversified housing, a public transportation system, human-oriented traffic, an ecological footprint, etc. (Ewing et al. 2010). The definition of green cities through environmental quality performance does not mean ignoring social justice. In fact, a greener living environment can promote more equitable treatment of a city's residents. There is no universal solution that could be applied to every city in any country. Adaptable, responsive, and innovative solutions that differ from one place to another enable green cities to emerge in various guises and recognize the variation and dynamism of cities (Asian Development Bank 2015).

A green city is a healthy city full of green and vital spaces, while also being a people-oriented and livable city. On the other hand, it is a cultural city with its own characteristics and style, and a city with sustainable development of its environment, economy, and society. Planning the future of the city with the green development concept, improving the urban ecological environment, and enhancing people's well-being will be the inevitable trend of urban development. Therefore, the green city should be planned according to local conditions, combined with local resources and climate characteristics to actively promote greening and beautification, and to build the natural beauty and ecology of the city. In addition, we should actively develop urban forests, protect rivers and wetlands, improve the natural and ecological functions of the city, and enhance the comprehensive carrying capacity of urban ecology. The green city should pay attention to city pollution control, advocate energy saving and carbon reduction, develop low-carbon industry, and encourage green transport. Cities should promote green ideas, protect the earth's ecosystem, and jointly create a better green future.

Even a city that is completely filled with green buildings cannot a green city. An ecological, low-carbon, and livable green city should facilitate green, low-carbon construction, applying reasonable patterns of urbanization and minimizing resource consumption. One of the key contents of a green city is alleviation of traffic pressure, which requires residential areas and business areas with mixed living and functions. Meanwhile, the concept gives a lot of weight to technology in the pursuit of a green, energy saving, carbon reducing, and healthy living style. To achieve a green city, overwhelming control of all pollution and highly efficient use of resources, as well as friendly communication between people and nature, are indicated.

## 2.2 Construction of a Green City

### *2.2.1 Conceptual Development*

The world now is facing climate change, while the human life system has exceeded the load capacity of the earth. A considerable number of serious challenges have been raised for the foundation that the civilization of all human kind intends to maintain. Instead of timid gradual efforts, the best strategy for each city should be construction of a new era of the green city through transformation of the global challenge into future opportunities, coupled with the efforts of governments and the public, with a macro vision.

The green city, which is not easily achieved, needs to experience a series of gradual changes, which can be divided into three stages: nature resource conservation, renewability, and organic restoration. The ultimate goal is to transform the city into a restorative, organic, and green city.

The first stage is nature resource conservation, which means that city assets are saved, maintained, and used with efficiency. Resource recycling is dependent on reducing importation of energy and resources. The city itself is an efficient machine, which maximizes the performance of each city asset and decreases the overdraft of production. The city economy is constructed based on economic strategies that can advance ecological benefit, green industrial technology, ecological conservation, carbon reduction, alternative energy, and green buildings. For instance, carbon reduction and ecological systems could contribute to ecological resources in the future. For the long-term future of the city, it is essential to conserve the forests and preserve the water environment. The development strategy of the city is to preserve the local environmental assets and strategically invest in the emerging green industry. Urban planning must carefully consider stable supply of energy and food, reduce community costs, and strengthen neighborhood service facilities.

The second stage is renewability. A green and ecological city is eco-effective. The city's energy primarily comes from renewable energy sources, and other natural resources follow a process of an economical closed cycle. Organic city gardens are vital to deploy reusable resources and make use of unused resources. All waste will become materials for other resources, realizing zero waste in the city. This solution for the twenty-first century is constructed on the systematic science of ecological industry, green buildings, green public facilities, and green urban design. The ecological strategies of open space, meanwhile, address the management of storm water and water resources. The entity of a city is a complete community comprising several regional centers and certain neighboring areas. This kind of city focuses on ecological interests and zero waste for the benefit of various city assets through gradual improvement of products and creation of additional ones in the production process. A city should decrease its carbon footprint and reuse water resources, as well as local reusable resources, to fulfill its energy demands. The financial resource of a city mainly depends on local development, economic strategies, and export of techniques and experience, demonstrating local and strategic investment. Community pays attention to sustainable learning and applies systematic science in all kinds of fields, such as government, the private sector, the academic sector, and nongovernmental organizations. The city—itself a living laboratory—serves as a learning center for eco-cities, leading the practice of green renewability.

The third stage is organic restoration. In this stage, a green city aims to recycle the organic environmental system and is dedicated to supporting itself with energy received from the sun, wind, and water, geothermal energy, and natural resources. The city becomes just as alive as a tree, which is supported by local resources. Relevant science and technology creatively offer lifestyle and organic solutions. The entity of the city is represented by a series of communities of different scales, from the local community scale to the global scale. This city must take full advantage of ecology in production processes for the ultimate achievement of improving its ecological performance and human life, as well as benefiting from city assets.

A city can gradually transform itself into an organic green city through natural resource conservation, renewability, and organic restoration.

### *2.2.2 Vision*

Amid the rush of the green movement, six themes are focused on in discussions and policies worldwide: (1) green and natural resource waste management, natural resource restoration and management, air quality, etc.; (2) water resource and river basin management, climate change, etc.; (3) public construction and green transportation, etc.; (4) human education in urban development and management, etc.; (5) energy saving, use of renewable resources, green buildings, etc.; and (6) international information and international cooperation, etc.

The author proposed strategies and actions according to the above topics, with th consideration of city characteristics and requirements to resolve problems simultaneously. (1) city greening: to conserve the nature, to improve the city coverage, and to develop city agriculture; (2) green water: for the purpose of city storm water management, a low-impact development (LID) is recommended to improve permeability and water, create a water-friendly environment, and strengthen the attraction of water storage; (3) green transportation: to replace the vehicles with advanced public transportation systems, pedestrian systems, and cycle-friendly systems; (4) green block: ecological blocks and environmental awareness. (5) green energy to reduce emission of greenhouse gases in daily life and improve air quality; (6) green information and communications technology (ICT) to promote beneficial application of the internet and digitization according to international standards.

## 2.3 Development Strategies for a Green City

### *2.3.1 City Greening*

City greening refers to green spatial networks (Benedict and McMahon 2001) with multiple functions, which are beneficial to improve the quality of natural and architectural environments. It covers connected networks constituted by natural green spaces (forests, natural resource zones, conservation zones, etc.), artificially managed green land (city parks, green land, water storage, and historical landscape design), and connecting spaces (pedestrian streets, cycle paths, green corridors, and waterways). Greenbelts have been designated to contain expansion and engender more compact growth (Ma and Jin 2014). City greening should be involved on all space scales from the city center to rural areas. The planning of green infrastructure provides, conserves, and connects these green spatial networks to further link the strategy planning for rivers, forests, natural conservation zones, city green land, and historical remains, achieving a network of green land and green corridors with a diversity of landscapes and ecology. (Chris Blandford Associates 2007).

#### 2.3.1.1 Optimize Forest Resources

The existing natural resources, green systems, and other green infrastructure are optimized to avoid loss of the original ecological function of the existing green resources, caused by urbanization and expansion. From the city-forming perspective, the substance of urban design actually involves three main elements: natural environmental factors, artificial environmental factors, and nonphysical environmental factors (Wikantiyoso and Tutuko 2013). Regulations, action plans, and management strategies are therefore established to further solidify, strengthen, and maintain the vitality of urban nature plots and to optimize natural resources and ecological performance. The specific points are: (1) maintenance of natural green resources: to define and control the division and management of natural green resources, green resource conservation, and sensitive areas, with increased city afforestation and protection of original forests, and establishment of forests, rivers, coastal areas, and windbreak forests; (2) establishment of urban growth boundaries (UGBs): UGBs, setting up city development restrictions, avoiding violation of environmentally sensitive mountains, and green resources and changes of land use; (3) division and setting-up of green corridors and buffer green belts: to divide, set up, and manage the buffer zones at the city margins, as well as the buffer green belts between incompatible land use, to recognize and manage important production and ecological agricultural land, with a series of green corridors and communities and extension of ecological green, ensuring the quality of natural green resource; and (4) conservation of existing ecological corridors and vegetation diversity, to identify and manage key rivers and green corridor buffer zones, ecological corridors, wetlands, and network nodes, and minimum widths of ecological corridors and their connectivity (Chang 2010).

#### 2.3.1.2 Replenish Green Resources

Vertical and horizontal types of blue and green corridors are coordinated by creation, improvement, and consolidation to achieve artificial development of a green infrastructure. A comprehensive matrix space is created, based on points and lines. Review and transformation of land use further adjustment of idle space and changes of land use. The strategies are as follows: (1) optimization of existing green land in parks to improve the quality and quantity of green land; (2) advancement of green systems development of leveled green spaces in parks and management of leisure, recreational and nature-oriented parks, creation and management, and facilitation of green reserve land for parks; (3) addition of potential green space in the city: addition of green space to schools and to official and public facilities, prioritizing green suggestions for division of agricultural areas into urban–rural development areas; offering to set up citizen farms and community-experience organic farming areas, standardizing landscape management and buffer zones around agricultural areas; (4) transformation of idle land: conducting strategic plans to cooperate idle ground with the green infrastructure system, prioritizing green infrastructure establishment before development, ensuring green development of new land after construction of underground railways; and (5) review and remediation of land use with regard to gray infrastructure: reviewing land use by existing gray infrastructure, incorporating green infrastructure planning into major construction projects; setting up buffer green belts around gray infrastructure, and improving the green coverage and quality of gray infrastructure.

#### 2.3.1.3 Enlarge the Area of Green-Covered Resources

In the existing city space, gray infrastructure is relocated to enlarge green-covered areas. Green-covered areas are enhanced with regard to the current infrastructure to improve the environmental effectiveness (Chang 2010). The development strategies are as follows: (1) construction of green corridors on city roads: defining roads over 30 m in width as avenues, greening of city freeways without interfering with traffic visibility, and conservation and maintenance of the city's old trees and characteristic vegetation; (2) urban microclimate environmental regulation: planning of vegetation in the city's wind-guiding channels, mitigation of heat islands in the community, use of green land plans, and planning of freshwater resources, windbreaks, and vegetation planting; (3) strengthening of the greenery in the surrounding areas of green infrastructure: greening of undeveloped reserve land for public facilities, improvement and establishment of green coverage, ensuring high permeability ratios in open space, and promotion of corner space and community greens; and (4) greening advancement of city buildings: use of green roofs and vertical greening of buildings (platform and walls, etc.), encouraging cooperation between green land plans and water retention (the water cycle), maintenance and management of urban habitats for living things, and improvement of the urban green coverage ratio and green visibility ratio.

#### 2.3.1.4 Develop City Organic Agriculture

With limitations on the availability of agricultural land in cities, high-rise agriculture is proposed to incorporate green ideas and agricultural production into the city. City leisure agriculture promotes organic farming in the city and healthy organic farming in the surrounding areas. Vertical leisure agricultural building centers can be introduced regionally to allow the experience of farming for fun and achievement. On the community scale, more empty space in schools, which can be transformed into cultivation space for leisure agriculture (Chang 2010). The development of organic agriculture, meanwhile, improves the health and the quality of life and air quality (Figs. 2.1 and 2.2).

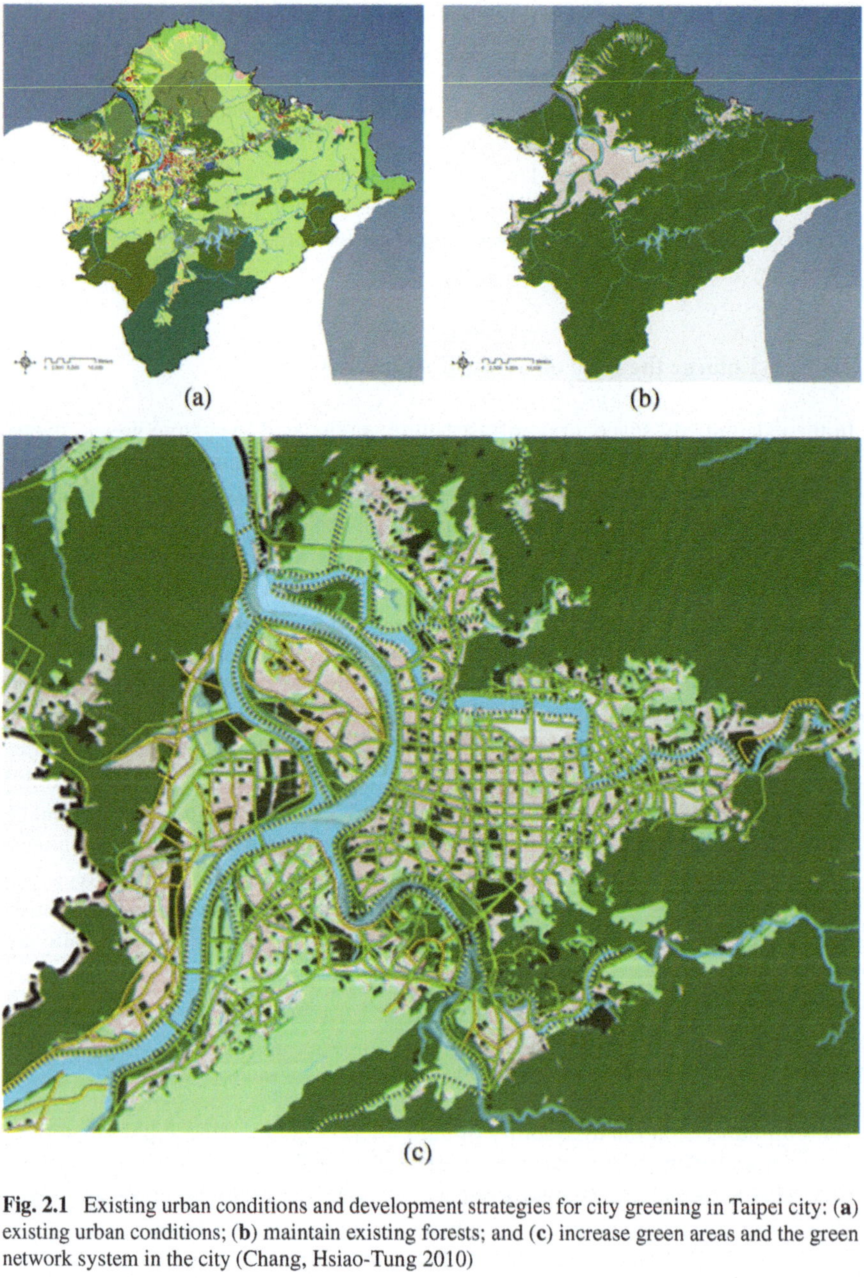

(a) (b)

(c)

**Fig. 2.1** Existing urban conditions and development strategies for city greening in Taipei city: (**a**) existing urban conditions; (**b**) maintain existing forests; and (**c**) increase green areas and the green network system in the city (Chang, Hsiao-Tung 2010)

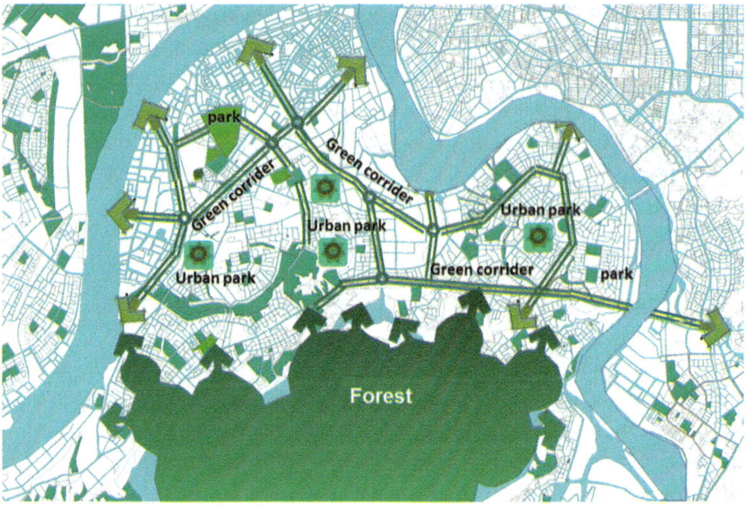


**Fig. 2.2** City greening strategies to increase green areas and green corridors in the city (Chang, Hsiao-Tung 2010)

## *2.3.2 Green Water*

Green water is defined as methods to reduce rainwater runoff and pollutants using LID. LID contributes to water quality with the techniques of permeability, evaporation, and rainwater reuse, avoiding and reducing pollution in rivers, streams, lakes, coastal waters, and groundwater.

During the urbanization process, impervious pavement caused effects on the city environment. Ground runoff and peak flow after pouing rain results urban flooding and waterlogging. The impervious ground coverage ratio has direct influences on ground runoff and changes city flooding and the macroclimate. A decrease in the impervious ground coverage ratio can effectively contribute to reducing city flooding, increasing the groundwater supply, restoring the normal water cycle, and mitigating urban heat island effects (Dietz 2007).

### 2.3.2.1 Improve Water Quality and Conserve Water

To improve water utilization in the city, an effective system for the city water cycle, ecological purification methods, and the popularization of sanitary sewers are required.

A high-efficacy city water cycling system is constructed within the city cope with water use and transform city rivers into a hydrophilic space for fishing and fun. With climate change, the reuse of water and rainwater is improved, and water leakage is reduced. Sanitary sewer construction all discharged sewage is treated. In water environments, water quality is purified ecologically with constructed wetlands by natural treatment to achieve double purification, vitality of living things in the water, and activation of the water environment.

#### 2.3.2.2 Reduce Storm Water and LID of Public Infrastructure

A set of plans is drawn up for ground permeability, water storage, and reuse of storm water, such as rainwater gardens, green streets, green roofs, green highways, permeable pavements, and recycling of rainwater and reused water, etc.

LID development of a permeable city calls for construction of green streets to collect rainwater from roads, bridges, and pedestrian streets, with connection to rainwater gardens (Origin Ltd., Team Orgon 2009) (Figs. 2.3 and 2.4). In particular, ecological tanks are to prevent flooding in low-lying areas, on hillsides, and in marginal and waterlogged areas. On hillsides and in areas, each building site endeavors to improve the permeable area and the water conservation area, as well as the utilities for rainwater, cooperating with water cycle strategies. The effects of rainwater are decreased, and it is discharged into a freshwater river system after purification, which also reduces pollution in the river. In contrast to traditional storm water treat-

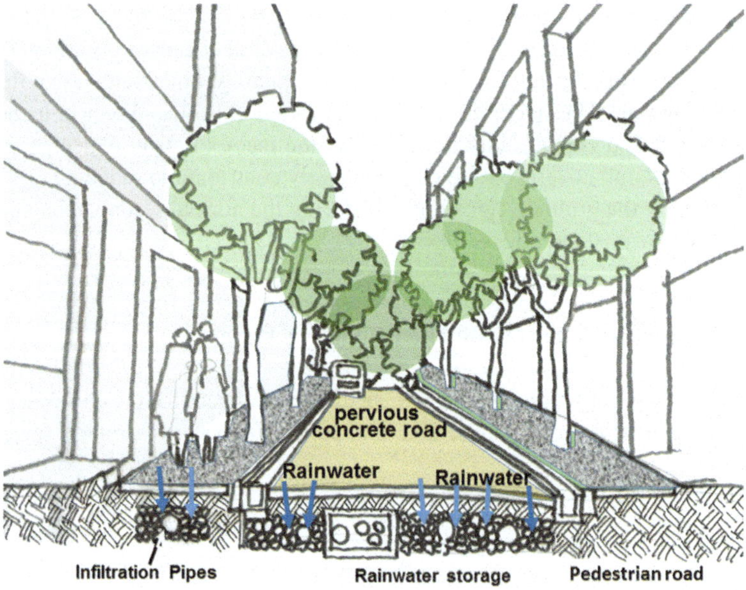


**Fig. 2.3** Development strategies for city green roads

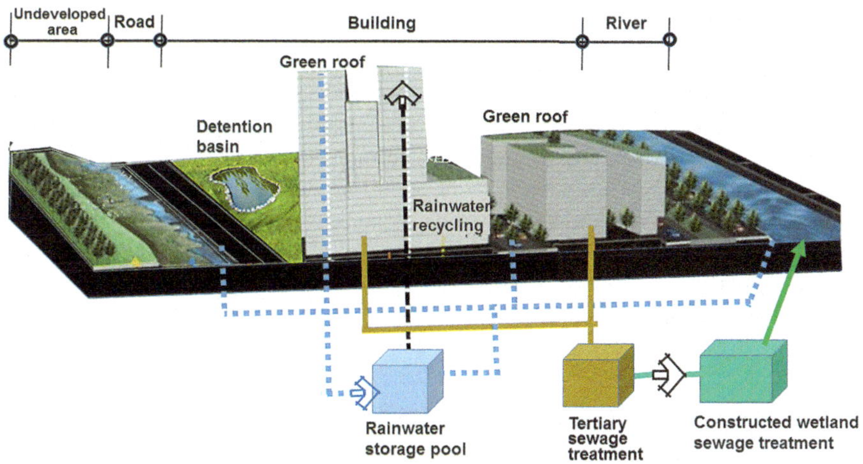


**Fig. 2.4** Concept for storm water permeability and storage and reuse facilities

ment, which typically only mitigates peak flow rates, the use of LID also helps to maintain the predevelopment runoff volume. Cluster layouts, grass swales, rain gardens or bioretention areas, and pervious pavements all reduce the "effective impervious area" (Booth and Jackson 1997) of the watershed, or the area that is directly connected to the storm water system (Dietz 2007). Implementation of LID principles is a shift (in storm water management) toward volume-based hydrology (VBH), a storm water control approach that focuses on management of storm water volumes. VBH is founded on the premise that reduction of storm water volumes will automatically help to solve other related problems such as pollutant loading, water velocity, peak flow rate, erosion, and sedimentation (Reese 2009).

#### 2.3.2.3 Construct a Hydrophilic Environment

Initially, waterways such as rivers in the city were managed as a resource for human benefit, including water supply, flood mitigation, disposal of wastewater, and minimization of disease (Walsh 2000; Paul and Meyer 2001; Morley and Karr 2002). However, this has led to degradation of their ecological functions—an issue that was initially ignored (Paul and Meyer 2001). Rivers and waterways could be rehabilitated for natural control of city temperatures and implementation of hydrophilic planning.

## 2.3.3 *Green Transportation*

The goal of green transportation is to replace transportation based on fossil fuels.

Energy consumption is dominated by transportation to meet the basic needs of citizens. Transportation could effectively promote the construction of ecological cities with low energy consumption and pollution, which are the key technologies required for transportation in this century. A sophisticated system of public transportation pedestrian and cycling environments to improving quality of life and are also symbols of an advanced city. Green transportation is therefore promoted to enhance public health (Froehlich et al. 2009).

### 2.3.3.1 Increase the Utility of Public Transportation

The popularization of public transportation is improved, and private transportation is reduced.

This strategy prioritizes public transportation. Private transportation is reduced once the metro line network and public transportation connection system is perfect, which the return of people-oriented environments and mitigation of air pollution. The ways to achieve this are as follows: (1) plan the extension of bus lanes; (2) promote civic minibuses and provide a "last one mile" service to the community; (3) use buses with a low chassis to conserve energy and make them convenient for senior citizens and vulnerable groups; (4) support the mass rapid transport (MRT) system with buses, increase feeder cars, and shorten the original bus routes to reduce the need for private transportation; (5) develop a whole set of integrated land use and transportation strategies to reduce commuter driving rates and increase the mixture of land use, and provide space to meet people's basic living needs within a certain distance; (6) it is recommended that automobile users pay for the environmental costs caused by that use—air pollution and traffic jams—through tax or parking fees; and (7) evaluate and charge congestion fees in central areas and assess possibilities and management practices.

### 2.3.3.2 Improve Sidewalks and Bike Paths

Sidewalk quantity and quality are improved, and public consciousness of bike transportation is strengthened.

Within neighborhoods, transportation pays attention to safety and comfort, as well as the environment, for biking. After completing the MRT system, public transportation is improved and motorways are reduced, enlarging the areas of sidewalks, sidewalk trees, and biking. The streets are returned to the people, allowing provision of shade and fresh air.

Business activities along the streets are stimulated, and consciousness of walking and biking are increased as steps toward zero-pollution transportation. The ways to achieve this are as follows: (1) sidewalks are renewed and permeable paving surfaces are applied; (2) traffic calming is planned for roads in communities, and the speed limit for automobiles is reduced to encourage people-oriented transportation; (3) prohibit against motor vehicle parking on the arcades and sidewalks to ensure safety and leisure space for walking; (4) road morphology is reviewed and road areas are reduced, with widened sidewalks and increased bike lanes; (5) lines for pedestrian priority are marked in streets surrounding schools and markets with a width of less than 8 meters; and (6) facilities are set up for bike parking, with rental services at MRT stations, schools, and large public facilities for the convenient links with other modes of public transport, and an off-site system of bike returns is planned for all city markets, office areas, stations, schools, etc.

## 2.3.4 Green Blocks

Green blocks are intended to achieve a live model of an ecological city vision. Simply speaking, in this kind of settlement, people have access to basic life necessities within a 15-min walking distance, and public transport is available. Commuting and transport are reduced while mitigating the influence of high prices or shortages of fossil fuels. In addition, the people' attitude changes and environmental consciousness also plays an important role in the construction of this ecological life model. Existing life habits can be transformed into environmental conservation and natural symbiosis, gradually strengthening the green life idea and constituting an integrated ecological city from point to full scale (Team Oregon LLC 2009).

### 2.3.4.1 Develop Lifestyle Settlements

Fifteen-minute walking neighborhoods are constructed, and ecological blocks are promoted.

This model aims to reduce transport, foster an environment of carbon neutralization and zero waste, and satisfy people's basic needs within a 15-min walking distance (Figs. 2.5 and 2.6). Walking can fulfill people's basic needs when fossil fuels are expensive or when there are energy shortages. Investigation and analysis are done to identify areas, which are remedied by policies and incentives. For some sites, ecological blocks are applied as comprehensive plans for energy conservation in buildings and facilities, water resources, and to further set standards and demonstrate viability cities are gradually transformed into symbiotic environments of life, production, and ecology through comprehensive review and adjustment of individual sites. (1) Construct a life model of a settlement with a 15-min walking distance.

(2) Construct the standard of a 15-min-walking-distance neighborhood. Incentives or feedback programs are encouraged to remedy deficiencies in basic living needs. (3) Energy conservation and renewable energy strengthen the design of energy conservation in the residential community and create sustainable renewable energy. (4) Ensure there is 100% recycling of waste in the community, and reduce the demand for domestic water by recycling systems for wastewater treatment and rainwater recovery systems. Recycling water is used for irrigation, and equipment with low water consumption is used. (5) Create a people-oriented transport environment with mass-transport-oriented development, connected with sidewalks and biking paths in series. (6) Ensure that 40–60% of the residential space is green land, and encourage green roofs, green facades, and growth of vegetation in open spaces (Origin Ltd., Team Orgon 2009).

#### 2.3.4.2 Promote Environmentally Friendly Education

Incorporate ecological courses into campus education and education in residential communities.

With the goal of environmentally friendly promotion, educational proposals are deployed for both schools and communities to change people's original living styles and attitudes to pay attention to environmental issues and practice the concept of protecting the earth, forming a bottom-up national movement (Figs. 2.5 and 2.6).

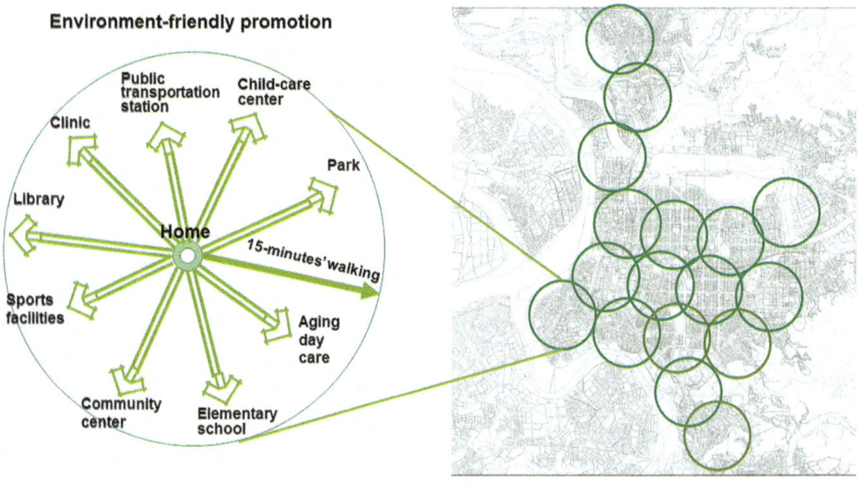


**Fig. 2.5** Concept for a 15-min walking neighborhood for a settlement community

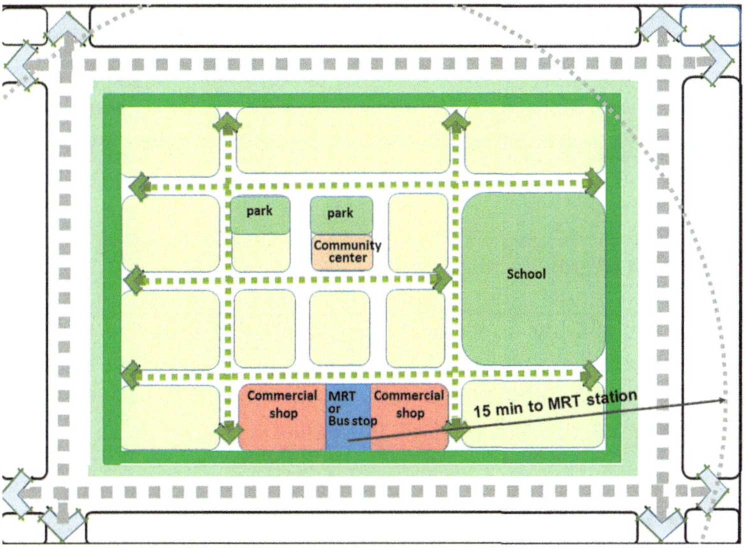


**Fig. 2.6** Green block neighborhood

## *2.3.5 Green Energy*

The global greenhouse effect is mainly caused by the significant increase in greenhouse gas emissions, which stems from city activities and industrial production. From the perspective of city development, the greenhouse effect can be mitigated through energy saving, pollution reduction, etc., with obvious benefits.

It has been pointed out that office buildings consume the most energy in cities. Possible ways to reduce the emission of greenhouse gases include the promotion of green buildings, designs for regional energy production, and use of renewable energy. The most effective way starts with a reduction of unnecessary energy consumption in daily life. The energy conservation measures discussed in this research are relevant to the public and are easily implemented to reduce the emission of greenhouse gases (Broek et al. 2002).

### 2.3.5.1 Conserve Energy

High-efficiency lighting fixtures are applied. Electrical appliances with energy-saving badges are chosen. Indoor air circulation is promoted by opening windows or using fans. Switches are turned off when not in use. Walking, biking, and public transport are encouraged. A few efforts (Chase et al. 2001; Heath et al. 2003;

Pinheiro et al. 2001) have been made to conserve energy in server clusters by tackling the high base power of traditional servers, i.e., the power consumption when the server is on but idle. Other efforts (Bohrer et al. 2002; Elnozahy et al. 2002, 2003) have tackled energy consumed by servers' microprocessors. Finally, the energy consumption of disk array-based servers has received some attention as well (Zhu et al. 2004).

#### 2.3.5.2 Set Up Regional Energy

A demonstration area for a regional energy system is developed. Central energy supplies the energy for buildings and equipment, including steam, hot water, cold water, electricity, etc. Supporting incentives and investment benefits are created. Regional energy-supplying systems are promoted.

#### 2.3.5.3 Use Renewable Energy Effectively

Power generation systems using solar power are set up in parks, green land, squares, and new and old buildings. Waste is planned to be converted into energy using direct combustion, physical conversion, heat transfer, biotransformation, chemical conversion, etc. The benefits of existing renewable energy are refined, including improvements in incineration plants, generation of biogas, and power generation using reservoirs. Industrial parks for renewable energy are planned, including solar/silicon industry, biomass energy, and geothermal research centers. Renewable energy is considered the key strategy for economic development. Environmentally friendly power generation systems are encouraged, with incentives for renewable energy, facilitating growth of the renewable energy industry.

There is a current prototype for a green energy greenhouse system, which can be installed in suspension outside a window in order to automatically supply electric power for its internal operations by means of a solar power generation device configured therein, such that a plant-based ecological environmental system, similar to a greenhouse, can be maintained within a window box (Hung and Peng 2016).

#### 2.3.5.4 Facilitate Green Transport and Fuel

Promote the use of mass transit vehicles, reduce energy consumption and pollution by transport, and introduce new transport modalities with low energy consumption and low pollution.

### *2.3.6 Green Information and Communications Technology*

Digital applications are enlarged. Based on the axis of an intelligent city and a ubiquitous integrated service, functions are performed through electronic government, network communities, and digital life. Wireless and fiber broadband networks are constructed to provide services including travel cards, an intelligent transport system (ITS), a single municipal web-based portal, location-based services (LBS), and health care for citizens, etc. The one most closely related to the ecological city is ITS, which aims to improve the convenience of mass transport. Intelligent information technology is also applied to prevent disasters and reduce commuter traffic flows during traffic peaks.

The application of intelligent technology in disaster prevention includes geographic information system (GIS), global positioning system (GPS), remote sensing (RS), management information system (MIS), etc. Electronic commuting allows more flexible selection geographically for individuals or enterprises. Real-time communication and video conferencing contribute to reducing traffic (energy conservation) and travel times within cities, further mitigating air pollution. ITS has improved the use of mass transport systems and decreased the oil consumption involved in searching for parking. Development of digital technology can be applied to assist in achieving a green city (Shim et al. 2009).

#### 2.3.6.1 Accelerate Construction of Wireless and Fiber-Optic Cable Networks

(1) Pursue continuous construction of a 5G city wireless broadband network. (2) Finish cable laying off main roads for supply of fiber-optic cabling to homes.

#### 2.3.6.2 Apply Intelligent Technology in Disaster Prevention and Relief

(1) To continue database construction and management of disaster prevention and relief: construction and updating are continued for the disaster database, disaster mode database, real-time data monitoring, data on disaster prevention and relief resources, data on areas with fire rescue difficulties, and the database on disaster relief energy. (2) To continue the construction of a real-time monitoring system: a monitoring system for potential landslides and dangerous stream flooding is constructed, and flood monitoring and warning facilities and installations are set up in suitable locations in easily flooded and low-lying areas. (3) Personnel training for application of procedures and technology for disaster prevention and relief: for effective application of database information, the personnel should be trained in the "four S's": GIS, GPS, RS, and MIS.

#### 2.3.6.3 Facilitate Electronic Commuting

Companies with electronic commuting are used as a demonstration. Consulting is provided for companies contemplating introduction of electronic commuting.

#### 2.3.6.4 Develop an Intelligent Transport System

Intelligent transportation systems (ITS) are state-of-the-art approaches based on information, communication, and satellite technologies in mitigating traffic congestion, enhancing safety, and improving the quality of the environment (Shah and Lee 2007; Ambak et al. 2009). The term "ITS" refers to integrated applications employing combinations of information, communications, computing, and sensor and control technologies, which aim to improve transport safety and mobility and reduce vehicle emissions. Many technologies have been developed to enhance vehicle safety by preventing crashes, reducing trauma during crashes, or reducing trauma following crashes. ITS have significant potential to enhance traffic safety (Ambak et al. 2009). Some of the main requirements are as follows: (1) build the equipment needed for a dynamic system of buses and intelligent stop signs; (2) integrate transfer, payment, and dynamic information systems for buses; (3) establish guidance systems for parking information; and (4) establish a system of mass transport and transport information inquiry by a person.

## 2.4 Conclusions and Suggestions

The green city is a concept that calls for joint effort and action.

In a city, people tend to have rather simple demands regarding their living environment and often ignore environmental issues that have not recently had a serious influence on their living environment. However, as greenhouse effects and heat island effects become more prominent, the close relationships between ecology, the living environment, and life are gradually better understood. The issue of creating a green city requires mostly recognition and action, but also entails apprehension and needs to be valued and subsequently maintained.

Environmental quality and consciousness could be improved by green volunteers if citizens attach importance to and care about the living environment. Environmental consciousness calls for substantial changes to promotion and publicity. The existing environmental ecology can be greatly benefited by small changes in the living environment, which should be demonstrated to the public. The optimal way to promote environmental consciousness is to let people be accustomed to, deploy and see the green landscape of the environment.

## References

Ambak K, Atiq R, Ismail R (2009) Intelligent transport system for motorcycle safety and issues. Eur J Sci Res 28(4):600–611
Asian Development Bank (2015) Green city development tool kit, Mandaluyong City. Asian Development Bank, Metro Manila
Benedict and McMahon (2001) Green infrastructure: smart conservation for the 21st century. Sprawl Watch Clearinghouse, Washington, DC
Bohrer PE Elnozahy T, Keller M, Kistler C, Lefurgy CM, Rajamony R (2002) The case for power management in web servers. In: Graybill R, Melhem R (eds) Power-aware computing. Kluwer Academic Publishers, Boston
Booth DB, Jackson R (1997) Urbanization of aquatic systems: degradation thresholds, stormwater detection and the limits of mitigation. J Am Water Resour Assoc 33(5):1077–1089
Broek R v d, Treffers D-J, Meeusen M, Ad v W, Nieuwlaar E, Turkenburg W (2002) Green energy or organic food? A life-cycle assessment comparing two uses of set-aside land. J Ind Ecol 5(3):65–87
Chang H-T (2010) Green infrastructure master plan of Taipei metropolitan. Department of Urban Development, Taipei City Government
Chase J, Anderson D, Thackar P, Vahdat A, Boyle R (2001) Managing energy and server resources in hosting centers. In: Proceedings of the 18th symposium on Operating Systems Principles, October
Chris Blandford Associates (2007) Green infrastructure strategy—a proposed vision for connecting people, places and nature. Greater Norwich Development Partnership
Dietz ME (2007) Low impact development practices: a review of current research and recommendations for future directions. Water Air Soil Pollut 186:351–363
Elnozahy EN, Kistler M, Rajamony R (2002) Energy-efficient server clusters. In: Proceedings of the 2nd workshop on Power-Aware Computing Systems, February
Elnozahy M, Kistler M, Rajamony R (2003) Energy conservation policies for web servers. In: Proceedings of the 4th USENIX symposium on Internet Technologies and Systems, March
Ewing B, Moore D, Goldfinger S, Oursler A, Reed A, Wackernagel M (2010) The ecological footprint atlas 2010. Global Footprint Network, Oakland
Froehlich J, Dillahunt T, Klasnja P, Mankoff J, Consolvo S, Harrison B, Landay JA (2009) UBI green: investigating a mobile tool for tracking and supporting green transportation habits. Human-Computer Interaction Institute. Paper 119
Heath T, Diniz B, Carrera EV, Meira Jr W, Bianchini R (2003) Self-configuring heterogeneous server clusters. In: Proceedings of the workshop on Compilers and Operating Systems for Low Power. September 2003
Hung P, Peng KP (2016) Green-energy, water-autonomous greenhouse system: an alternative-technology approach towards sustainable smart-green vertical greening in smart cities. Int Rev Spat Plann Sustain Dev 5(1):55–70
Joss S (2011) Eco-cities: the mainstreaming of urban sustainability: key characteristics and driving factors. Int J Sustain Dev Plan 6(3):268–269
Joss S (ed) (2012) Tomorrow's city today, eco-city indicators, standards and frameworks. Bellagio conference report. University of Westminster, London
Ma M, Jin Y (2014) Understanding Beijing's moving urban fringe through a spatial equilibrium model. Int Rev Spat Plann Sustain Dev 2(2):14–38. ISSN: 2187–3666 (online). SPSD Press from 2010, SPSD Press, Kanazawa
Meadows D (1999) Indicators and information systems for sustainable development. In: Satterthwaite D (ed) The Earthscan reader in sustainable cities. Earthscan, London
Morley SA, Karr JR (2002) Assessing and restoring the health of urban streams in the Peuget Sound Basin. Conserv Biol 16:1489–1509
OECD and China Development Research Foundation (2010) Trends in urbanisation and urban policies in OECD countries: what lessons for China? OECD Publishing, Paris

Origin Ltd, Team Orgon (2009) Urban development vision and master plans—eco-city development strategy. In: Department of Urban Development. Government, Taipei
Paul MJ, Meyer JL (2001) Streams in the urban landscape. Annu Rev Ecol Syst 32:333–365
Pinheiro E, Bianchini R, Carrera E, Heath T (2001) Dynamic cluster reconfiguration for power and performance. In: Benini L, Kandemir M, Ramanujam J (eds) Compilers and operating systems for low power. Kluwer Academic Publishers, September 2003. Earlier version published as "Load balancing and unbalancing for power and performance in cluster-based systems" in the Proceedings of the workshop on Compilers and Operating Systems for Low Power, September
Reese AJ (2009) Volume-based hydrology. Examining the shift in focus from peak flows and pollution treatment to mimicking predevelopment volumes. http://www.stormh2o.com/september-2009/volume-based-hydrology-1.aspx. Accessed 12 Dec 2010
Register R (1987) Ecocity Berkeley:building cities for a healthy future. North Atlantic Books, Berkeley, ISBN 1-55643-009-4
Shah AA, Lee JD (2007) Intelligent transportation systems in transitional and developing countries. IEEE A&E Syst Mag:27–33
Shim YH, Kim KY, Cho JY, Park JK, Lee BG (2009) Strategic priority of green ICT policy in Korea: applying analytic hierarchy process. World Acad Sci Eng Technol 58:16–20
United Nations Environment Programme (2011) Cities, investing in energy and resource efficiency: towards a green economy. United Nations Publishing
Walsh CJ (2000) Urban impacts on the ecology of receiving waters: a framework for assessment, conservation and restoration. Hydrobiologia 431:107–114
Wikantiyoso R, Tutuko P (2013) Planning review: green city design approach for global warming anticipatory: Surabaya's development plan. Int Rev Spat Plann Sustain Dev 1(3):4–18. ISSN: 2187–3666 (online). SPSD Press from 2010. SPSD Press, Kanazawa
Zhu Q, Shankar A, Zhou Y (2004) PB-LRU: a self-tuning power aware storage cache replacement algorithm for conserving disk energy. In: Proceedings of the 18th international conference on supercomputing, June
Zou X, Li Y (2014) How eco are China's eco-cities? An international perspective. Int Rev Spat Plann Sustain Dev 2(3):18–30

# Chapter 3
# *How Green Is Your Smart House:* Looking Back to the Original Concept of the Smart House

**Puteri Fitriaty, Zhenjiang Shen, and Kenichi Sugihara**

**Abstract** The smart house has been a growing field of interest among researchers and entrepreneurs in recent years. The smart house is designed to create a high quality of life based on technologies of functional automation within residential buildings. It provides a platform to monitor and control certain appliances in the house environment. The concept of the smart house is discussed in this chapter.

An in-depth literature study of the smart house was conducted to establish smart house principles and parameters, then these were compared with the principles and parameters of the green house to identified the similarities and the differences between those two concepts. Concept and parameter implementation of the smart house in practice was investigated through a field survey and online materials. Japan, as one of the leading countries in smart house technology, was chosen as a case study. Six smart houses were selected to evaluate the concept implementation and compare smart house concepts with those of the green house.

The results showed that the smart house concept and the green house concept have an intersection of sets relationship. The intersection set parameters are the parameters included in the principles of respect for users, conservation of energy and water, and working with the climate.

**Keywords** Energy efficient · Green concept · Smart house parameters · Quality of life · ICT · HEMS

---

P. Fitriaty
Architectural Department, Engineering Faculty, Tadulako University, Palu, Indonesia

Urban Planning Laboratory, Environmental Design Division, Graduate School of Natural Science and Technology, Kanazawa University, Kanazawa, Japan

Z. Shen (✉)
School of Natural Science and Technology, Environmental Design Division, Kanazawa University, Kanazawa City, Japan
e-mail: shenzhe@se.kanazawa-u.ac.jp

K. Sugihara
Faculty of Information Media, Gifu Keizai University, Ogaki, Japan



Z. Shen et al. (eds.), *Green City Planning and Practices in Asian Cities*, Strategies for Sustainability, https://doi.org/10.1007/978-3-319-70025-0_3

## 3.1 Introduction

A smart house is expected to provide a high quality of life (GhaffarianHoseini et al. 2013) by employing intelligent energy management interfaces in creating a comfortable living environment (Lee et al. 2016). This intelligent energy management has since become known as the home energy management system (HEMS). The device is believed to be a useful tool to help occupants manage and balance their energy consumption without hindering their own comfort. Unfortunately, the smart house is an emerging field full of promises and aspirations accompanied by very little empirical, social, or cultural research (Wilson et al. 2015). There are many varieties of smart house in the global market and features are continually being developed, but the boundary of the smart house concept has remained unclear (Chan et al. 2009; Strengers and Nicholls 2017). This has encouraged the authors to investigate the original concept of the smart house through its history and research development.

In its development, the smart house has been seen from two points of view. The first view is the researcher's view and the latter is the industry stakeholder's view. The researcher's point of view puts forward ideal thinking related to energy efficiency as the driver (Bhati et al. 2017; Fotouhi Ghazvini et al. 2017; Fabi et al. 2017; Sintov and Schultz 2017), while the industry stakeholders promise user convenience (Strengers and Nicholls 2017). But both views agree with the idea of the smart house employing information and communications technology (ICT) as the key to house management. They also concur that the system does not rely only on efficient use of energy but also on clean energy sources, where photovoltaic energy is the most favorite type used in the smart house.

Another issue is the relationship between the smart house concept and the green house concept. It has been mentioned in some literature that the smart house aims at energy efficiency in the house, which is also included in the green house concept (Kofler et al. 2012; Fensel et al. 2013; Solaimani et al. 2015; GhaffarianHoseini et al. 2013). However, employing many devices for occupant convenience leads to more energy consumption, because the pursuit of convenience has often failed to save energy (Strengers and Nicholls 2017). This fact actually is contrary to the green house concept. In comparing the concepts of the smart house and the green house, an in-depth literature study was needed to identify the principles and parameters of the smart house and to evaluate its implementation in the housing market.

Several smart houses in Japan were used as a study case to evaluate the concept implementation. Japan was chosen for the case study because it is one of the leading countries in smart house technology. The case study was performed by a field survey and study of online materials. To assess the implementation of the concept in an international context, it is necessary to rely on a benchmark. However, such a universally used ranking does not exist yet. Therefore, the contribution of the study is expected to provide parameters of assessment for measuring the degree of smartness. In addition, this study contributes to the literature by comparing the concepts of the smart house and the green house together with the smart house concept implemented in the housing market.

## 3.2 Method and Materials

### *3.2.1 Literature Review Method*

By adopting the method employed by Strengers and Nicholls (2017), Wilson et al. (2015), and Solaimani et al. (2015), the smart house concept was investigated through an international content analysis. The content included magazine articles, online articles, and scientific papers written about common similarities and differences between smart house practices and research. Three online databases were used—ScienceDirect, Scopus, and Google—to undertake the search by using common search keywords, namely "home automation," "smart house," "smart home," and "home energy management system." The terms "autonomous house," "intelligent house," and "domotics" were allowed. However, there was a large overlap between these three online databases' results. Only 78 research publications, 15 magazine and online articles, and 3 books, including book chapters, were chosen based on the research purpose. The selected data were then analyzed by a qualitative inductive method.

### *3.2.2 Analysis of the Smart House Concept*

The qualitative inductive method involved several steps (Fig. 3.1). The original concept was gained by a comprehensive analysis of the history, development, definition, and service functions of the smart house. The history and development could lead to the original purpose of the smart house concept. The definition could be a guide to the scope of the smart house, and the functions of the smart house could be deduced from its service applications. Subsequently, the principles of the smart house could be derived by analyzing the original purpose, scope, and functions of the smart house.

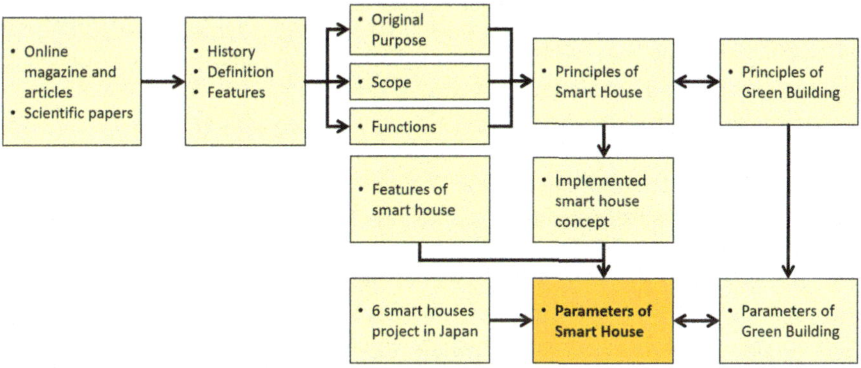


**Fig. 3.1** Research design of the smart house concept evaluation

The key to tracing the original concept of the smart house was time; thus, the timeline was treated as an important indicator for evaluation. For achieving the objective of the study, data were sorted in chronological order. Hence, the transformation of the smart house concept could be easily analyzed. Afterward the principles of the smart house extracted from the first step were compared with the principles of the green house, which acted as control principles.

The implemented principles in the smart house project were analyzed together with their general features to establish smart house parameters. These parameters were then compared with green house parameters to determine the smart house concept standpoint against the green concept. The Comprehensive Assessment System for Built Environment Efficiency (CASBEE) was chosen as the comparative tool in the smart house parameter evaluation.

## 3.3 State of the Art of the Smart House

### *3.3.1 Definition*

The smart house is also known as the "smart home," "autonomous house," or "intelligent house." The original term is "smart house," which was officially introduced for the first time by the National Association of Home Builders in 1984. The term was commonly used to refer to any living or working environment that had been constructed to assist people in carrying out required activities (Allen 1996). The concept gradually transformed into a more holistic perspective, where entire house spaces were managed by a centralized controller, which was designed to interpret user needs in an efficient and well-defined way.

One type of centralized controller later became known as a "home energy management system" or HEMS device (Fabi et al. 2017). The main objective is to reduce electricity costs without sacrificing the occupants' comfort (Shakeri et al. 2017). Designed to optimize consumer costs, the HEMS is able to manage energy demands, communicate with utilities, calculate the use of local generation (Roda et al. 2014), and allow connecting appliances to be remotely managed (Vega et al. 2015). This is an intelligent system that performs planning, monitoring, and controlling functions for energy utilization within the premises (Khan et al. 2015) and includes all necessary elements to achieve a reduction in electricity consumption.

Several definitions of the smart house have been proposed by scholars, as presented in Table 3.1. From the definitions, it can be concluded that the smart house scope is inseparable from the use of technology. Technology is employed to perform various functions related to assisted living, comfort, convenience, health, security, and energy efficiency. An interesting part of the definition timeline is the fact that the keyword term "energy efficiency" did not appear until 2011. Before that, all definitions referred to technology, assisted living, comfort, convenience, security, and entertainment.

**Table 3.1** Definitions of the smart house

| No. | Authors | Definition | Scope |
|---|---|---|---|
| 1 | Allen (1996) | Any living or working environment that has been constructed to assist people in carrying out required activities | Assisted living |
| 2 | Van Berlo (1999) | Home or working environment that includes the technology to allow for devices and systems to be controlled automatically | Technology |
| 3 | Pragnell et al. (2000) | Use of electronic networking technology to integrate the various devices and appliances found in homes, so that an entire home can be controlled centrally or remotely as a single machine | Technology |
| 4 | Aldrich (2003) | Residence equipped with computing and information technology, which anticipates and responds to the needs of the occupants, working to promote their comfort, convenience, security, and entertainment through the management of technology within the home and connections to the world beyond | Technology<br>Comfort<br>Convenience<br>Security<br>Entertainment |
| 5 | Spigel (2005) | Networked house where appliances interact with each other, adapt to dwellers, and allow residents, via the internet, to communicate with the outside world and to speak to the home while away at work or traveling | Technology<br>Assisted living |
| 6 | Chan et al. (2009) | Residence equipped with technology that allows monitoring of its inhabitants and/or encourages independence and the maintenance of good health | Health<br>Technology<br>Assisted living |
| 7 | Ding et al. (2011) | Residence augmented with sensors to observe the environment and devices/actuators to provide proactive services with the goal to improve the occupant's convenience, improving security and saving energy | Convenience<br>Security<br>Energy efficiency |
| 8 | De Silva et al. (2012) | A smart home is a home-like environment that possesses ambient intelligence and automatic control, which allow it to respond to the behavior of residents and provide them with various facilities | Convenience<br>Technology |
| 9 | Cook (2012) | Computer software playing the role of an intelligent agent perceives the state of the physical environment and residents using sensors and then takes actions to achieve specified goals, such as maximizing the comfort of the residents, minimizing the consumption of resources, and maintaining the health and safety of the home and residents | Technology<br>Comfort<br>Energy efficiency<br>Resource efficiency<br>Health<br>Security |
| 10 | Ozkan (2015) | Residence equipped with smart electrical appliances, sensors, and distributed power units containing renewable energy sources | Technology<br>Energy efficiency |
| 11 | Luor et al. (2015) | Residence equipped with technology, which allows the monitoring of its residents and/or encourages independence and the maintenance of normal healthy conditions | Technology<br>Assisted living<br>Health |
| 12 | Fabi et al. (2017) | Connection of sensors, appliances, and devices through a communications network, in order to remotely monitor, access, or control the residential environment and to provide services that respond to the users' needs | Technology<br>Assisted living |

### 3.3.2 History of the Smart House

The smart house concept has existed for many years. It appeared independently at the beginning of the twentieth century, long before the rise of ICT (Badica et al. 2013). The key to the smart house in practice emerged from home automation by labor-saving machines in the early 1900s with the introduction of electric or gas-powered home appliances (Association Franco-Chinoise du Developpement Urbain Durable n.d.; Rothfeld 2015; King 2015; Bounegru 2009). In 1966 the first home automation machine was built to compute a shopping list and to control temperature and home appliances (Association Franco-Chinoise du Developpement Urbain Durable n.d.; Rothfeld 2015).

In 1975, the first automation network technology, X10, was developed. It is a communications protocol for electronic devices, using electric power transmission wiring for control and signaling. The signal is in the form of a brief radiofrequency burst of digital data (Wikipedia 2017; Bounegru 2009; Rothfeld 2015). In the same year, Brenda and Robert Vale published a book titled *The Autonomous House*, which suggested the idea of low-pollution homes, and turned their ideas into reality in the early 1990s. It is an environmentally friendly four-bedroom house with solar panel energy generation and rainwater harvesting, located in the small town of Southwell in the British Midlands (Fig. 3.2).

In 1984, interest in home automation and control systems evolved and led to the initiation of the Smart House Project by the National Association of Home Builders (Bounegru 2009; Rothfeld 2015; King 2015; El-Sebaii et al. 2010). This year was noted in many literature publications as the turning point of the smart house concept, when the term "smart house" was introduced for the very first time. Appearing from another path, gerontechnology was developed to assist elderly people in their daily life in 1991 (Hendricks 2014; Cassie 2015). In 1994, a smart room that acted like an invisible butler was developed by the MIT Media Lab. Several devices were installed to interpret user behavior and help them in advance. The devices included cameras, microphones, and other sensors (MIT Media Lab 1995).

From 1998 to the early 2000s, smart houses began to increase in popularity (Hendricks 2014; Cassie 2015). In 1999, construction of the Aware Home, located in Georgia, USA—one of the pioneering smart houses—was finished. The target user of this smart house was elderly people. In 2003 the University of Florida launched the Mathilda Smart House, followed by the Gator Tech Smart House in 2005. Both houses were designed for the care of elderly and disabled people. In 2006, Duke University also finished their smart house construction in Durham, NC, USA. Unlike the previous houses, this smart house was dedicated to energy efficiency.

In 2010, Nest, a business behind mobile phone–controlled heating, was founded. At about the same time, smart meters from various companies were marketed (King 2015; Rothfeld 2015). In the same year, Japan initiated a smart city project, and a year later the first smart house project was launched. By 2012, the SmartThing application was released with expectations of link-connected gadgets in the house.

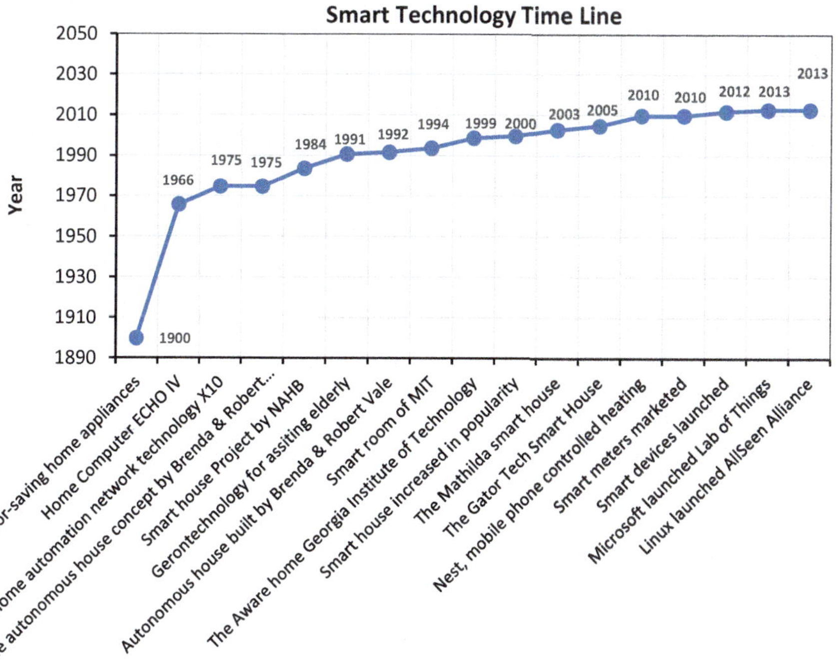


**Fig. 3.2** Timeline of smart house history

Microsoft introduced the Lab of Things (Internet of Things [IoT]) in 2013, which provided a virtual dashboard for monitoring and controlling different appliances and offered standards for home-related apps. The purpose was to encourage researchers to explore myriad studies of home automation. In the same year, the Linux Foundation initiated the AllSeen Alliance, a vendor-neutral approach to creating open-source software for connected technology development. Hence, all products in the smart house can work together (Rothfeld 2015). Over the following years, IoT has developed rapidly.

### 3.3.3 *Developments in Smart House Research*

Beside the history, the development of the smart house concept was also influenced by research from many disciplines including ICT, architecture, building science, interior design, health, energy conservation, and social–behavioral science. Therefore, research publications that directly or indirectly led to the smart house concept were included in the analysis of the smart house concept timeline (Table 3.2). These publications were analyzed by their service functions, which had already been identified by the scope of the smart house in the definition analysis. The service functions included energy efficiency, health, comfort, assisted living, safety and security, and finally convenience.

The 78 scientific publications listed in Table 3.2 were classified into three time periods, representing the stage development of the smart house concept. This division of the periods of research was very helpful in identifying the evolution of the smart house concept within its development (Fig. 3.3).

The first stage was basic implementation within the time period of 1991–2000. In this period, research was mainly focused on the safety and security of elderly and disabled people, their health conditions, and assistive devices for independent living. This was confirmed with the history of smart technology, where gerontechnology for assisting elderly people was developed in 1991. Additionally, construction of the pioneering smart house, the Aware House, which was dedicated to elderly users, was finished in this period and occupied in 2000.

The second period constituted the intermediate stage of smart house realization, which took place in 2001–2010. Health functions dominated smart house services, following by assisted living. Even though most of the services focused on elderly and disabled people, the assisted living function and health monitoring were shifted so that people other than elderly or disabled persons could also be the subject. In this period, two pioneers of the smart house were built: the Mathilda Smart House in 2003 and the Gator Tech Smart House in 2005, which still focused on elderly and disabled people. In the second half of this period, another type of smart house was built, which was focused on people of all ages and had an emphasis on energy efficiency. From this second half of the period, the concept of the smart house shifted to energy efficiency and met the concept of the green house, which arrived from another direction.

The last, but not least, period occurred from 2011 to 2017. This period is regarded as the advanced period, where smart technology leaped to virtual reality and mixed reality, involving IoT. The research publications in this period were already dominated by the energy efficiency functions of the smart house. The catalyst for this was the development of smart meters and HEMS in the market. Thus, the concept of the smart house was shifted from focusing only on assisted living and health to also include energy efficiency for sustainability of the living environment. With this, the definition of the smart house was also transformed by adding energy efficiency within the boundary of the smart concept.

The analysis results showed that the original concept of the smart house was assisted living for daily activities, and safety for elderly and disabled people's health care. The concept then shifted from focusing on the user in need to include users without such needs. The concept was also modified to a greener concept where energy efficiency was also included in smart house considerations.

It is interesting to see that the concept of the smart house was indeed evolving through changes in technology, time, and users. This evolution not only included functions but also involved their components, where the first developments were technology centric then spread to include user-centric (Wilson et al. 2015) and environment-centric developments (GhaffarianHoseini et al. 2013). Judging by recent technology developments, this concept may still evolve in the future, leading to greater convenience.

**Table 3.2** List of publications on the smart house between 1991 and 2017

| | | Service functions of the smart house | | | | | |
|---|---|---|---|---|---|---|---|
| No. | Publication | Energy efficiency | Health | Comfort | Assisted living | Safety and security | Convenience |
| 1 | Stauffer (1991) | √ | | √ | | √ | |
| 2 | Lutolf (1992) | √ | | | | | |
| 3 | Chan et al. (1995) | | √ | | | √ | |
| 4 | Allen (1996) | | √ | | √ | √ | |
| 5 | Mozer (1998) | √ | | √ | | | |
| 6 | Warren and Craft (1999) | | √ | | | | |
| 7 | Tang and Venables (2000) | | √ | | √ | | |
| 8 | Al-Muhtadi et al. (2000) | | | | | √ | |
| 9 | Edge et al. (2000) | | √ | | √ | √ | |
| 10 | Covington et al. (2000) | | | | √ | √ | √ |
| 11 | Kuusik (2001) | | √ | | | √ | |
| 12 | Demongeot et al. (2002) | | √ | | | √ | |
| 13 | Friedewald et al. (2005) | | | | √ | √ | √ |
| 14 | Wood and Newborough (2007) | √ | | | | | |
| 15 | Arcelus et al. (2007) | | √ | | | √ | |
| 16 | Coughlin et al. (2007) | | √ | | | | |
| 17 | Chan et al. (2008) | | √ | √ | | | √ |
| 18 | Park et al. (2008) | | | √ | √ | | √ |
| 19 | Pensas et al. (2009) | √ | | | | | |
| 20 | Chan et al. (2009) | | √ | | √ | | |
| 21 | Hong et al. (2009) | | | | √ | | |
| 22 | Skubic et al. (2009) | | √ | | | √ | |
| 23 | Helal et al. (2009) | | √ | | √ | | |
| 24 | Jahn et al. (2010) | √ | | | | | |
| 25 | Farella et al. (2010) | | √ | | √ | √ | |
| 26 | Tomita et al. (2010) | | √ | | | √ | |
| 27 | Cocco (2011) | | √ | | | | |
| 28 | Han et al. (2011) | √ | | | | | |
| 29 | Ding et al. (2011) | √ | | | | √ | √ |
| 30 | Ziefle et al. (2011) | | √ | | | | |
| 31 | Kofler et al. (2012) | √ | | | | | |
| 32 | Kim et al. (2012) | √ | | | | | |
| 33 | De Silva et al. (2012) | √ | | | √ | | √ |
| 34 | Rokach (2012) | √ | | | | | |

(continued)

Table 3.2 (continued)

| No. | Publication | Service functions of the smart house | | | | | |
|---|---|---|---|---|---|---|---|
| | | Energy efficiency | Health | Comfort | Assisted living | Safety and security | Convenience |
| 35 | Baraka et al. (2013) | √ | | √ | | √ | |
| 36 | Behan and Krejcar (2013) | | | √ | √ | | √ |
| 37 | GhaffarianHoseini et al. (2013) | √ | √ | √ | √ | √ | √ |
| 38 | Balta-Ozkan et al. (2013b) | √ | | √ | √ | √ | |
| 39 | Yoza et al. (2013) | √ | | | | | |
| 40 | Firth et al. (2013) | √ | | | | | |
| 41 | Fensel et al. (2013) | √ | | | | | |
| 42 | Kim et al. (2013) | | √ | | √ | | |
| 43 | Badica et al. (2013) | √ | √ | √ | √ | √ | √ |
| 44 | Morris et al. (2013) | | √ | | √ | √ | |
| 45 | Han et al. (2014) | √ | | | | | |
| 46 | Rollins et al. (2014) | √ | | | | | |
| 47 | Roda et al. (2014) | √ | | | | | |
| 48 | Yoza et al. (2014) | √ | | | | | |
| 49 | Missaoui et al. (2014) | √ | | √ | | | |
| 50 | Zhang et al. (2015) | √ | | | | | |
| 51 | Ozkan (2015) | √ | | | | | |
| 52 | Honold et al. (2015) | √ | | | | | |
| 53 | Sundstrom and Krysander (2015) | √ | | | | | |
| 54 | Louis et al. (2015) | √ | | | | | |
| 55 | Wang et al. (2015) | √ | | | | | |
| 56 | El-Baz and Tzscheutschler (2015) | √ | | | | | |
| 57 | Abushnaf et al. (2015) | √ | | | | | |
| 58 | Iwafune et al. (2015) | √ | | | | | |
| 59 | Beaudin and Zareipour (2015) | √ | | | | | |
| 60 | Vega et al. (2015) | √ | | | | | |
| 61 | Luor et al. (2015) | | | | √ | √ | √ |
| 62 | Gillani Fahad et al. (2015) | | √ | | | | |
| 63 | Alaiad and Al-ayyad (2015) | | √ | | | | |
| 64 | Lobaccaro et al. (2016) | √ | | | | | |
| 65 | Zhou et al. (2016) | √ | | | | | |

(continued)

**Table 3.2** (continued)

| | | Service functions of the smart house | | | | | |
|---|---|---|---|---|---|---|---|
| No. | Publication | Energy efficiency | Health | Comfort | Assisted living | Safety and security | Convenience |
| 66 | Premarathne et al. (2016) | √ | | | | | |
| 67 | Lee et al. (2016) | | | | | | √ |
| 68 | Mano et al. (2016) | | √ | | | | |
| 69 | Gayathri and Easwarakumar (2016) | | √ | | √ | √ | |
| 70 | Risteska Stojkoska and Trivodaliev (2017) | √ | | | | | √ |
| 71 | Jurenoks and Joki (2017) | √ | | | | | |
| 72 | Keskin Arabul et al. (2017) | √ | | | | | |
| 73 | Fabi et al. (2017) | √ | √ | √ | √ | √ | |
| 74 | Bhati et al. (2017) | √ | | | | | |
| 75 | Fotouhi Ghazvini et al. (2017) | √ | | | | | |
| 76 | Strengers and Nicholls (2017) | | | √ | | | √ |
| 77 | Sintov and Schultz (2017) | √ | | | | | |
| 78 | Shakeri et al. (2017) | √ | | | | | |

### 3.3.4 Functions and Features of the Smart House

From the previous results, a conclusion can be drawn that the functions of the smart house consist of assisted living, health care, surveillance for safety and security, energy efficiency, comfort, and convenience. Basically, the features of the smart house are based on its functions. Some features may serve several functions, or the inverse may occur, where one function may need several features.

The assisted living function needs features related to ubiquitous assistance focused on the task of assistance during daily activities, especially for users with disabilities or cognitive limitations (Badica et al. 2013). The features can include cameras for activity monitoring, sensors for detecting unusual activity, signaling of potential accidents or illness, finding lost objects, and providing an alarm for calling other people in the event of an accident. Assisted living can also include artificial intelligence by employing assistive robots for several task, reminder, and alarm functions.

For health care functions, the features of the smart house may include sensors for vital signs and movement detection. They may also include reminder alarms for taking medicine (GhaffarianHoseini et al. 2013). For safety and security, the features

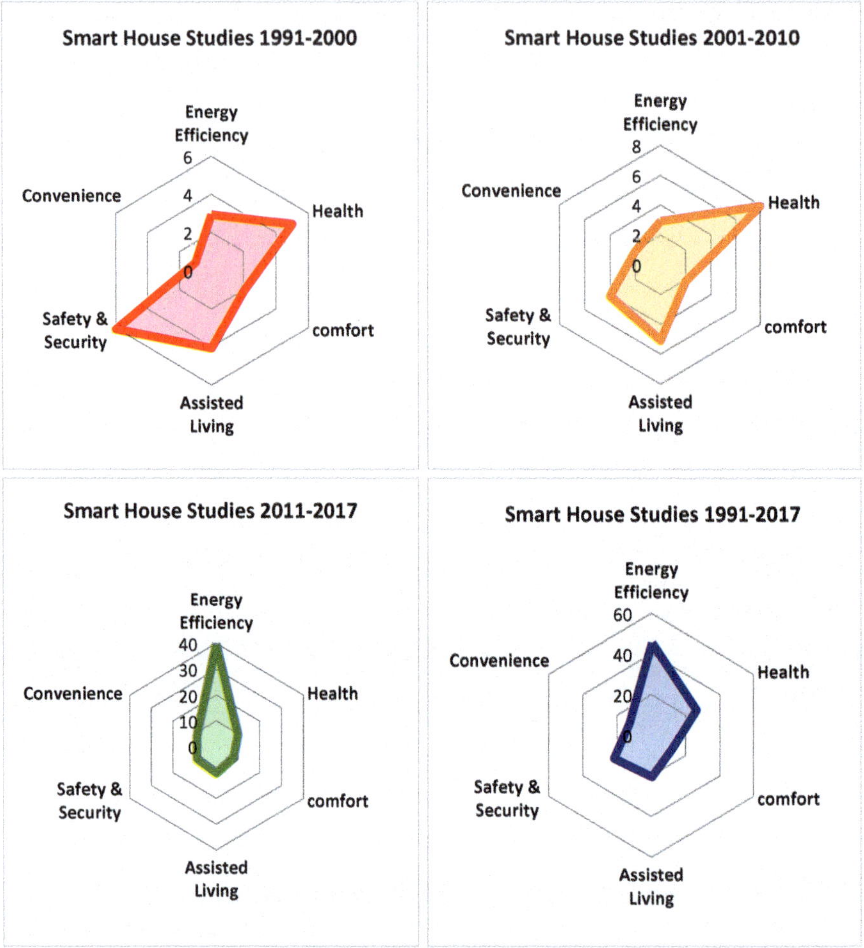


**Fig. 3.3** Evolution of the smart house concept

may include camera surveillance and alarms, remote monitoring, and emergency response for potential intruders, or sensors for potential accidents, illness, or disaster (Badica et al. 2013).

Energy efficiency functions require several features included in the HEMS. The features generally consist of centralized remote controls for monitoring and managing energy usage, data recording, smart connected appliances with low energy consumption, smart meters, renewable energy generation, and energy storage (Balta-Ozkan et al. 2013a).

Comfort and convenience functions can be supported by sensors for lighting, air temperature and humidity, and water supply (which are optimized based on ambient climate conditions and occupant comfort); providing long-distance or mobile-based remote control for home appliances and home environment monitoring; and entertainment and edutainment autoselection channels adjusted from the occupants' favorite selections.

## 3.4 The Concept of the Smart House

For anticipating the possibility of the smart house concept leading to inefficient use of energy and resources, it is important to mark a clear boundary for the smart concept to participate in sustainable development through smart living. For this purpose, the authors proposed a concept boundary for navigation of future smart house developments to optimize user convenience with energy and resource efficiency. Hence, in establishing the boundary, the smart house concept was compared with the green concept, which served as a control concept leading to sustainable smart living.

The concept of the smart house was evaluated by its principles and parameters. The principles of the smart house concept, as mentioned before, consist of assisted living, health care, safety and security, energy efficiency, comfort, and convenience. The health care function is mainly performed by the assisted living function, and the convenience function can be combined into the comfort function. Thus, the service functions of the smart house can be deduced as four functions (Fig. 3.4).

Figure 3.4 shows that two smart house functions dominate the other functions, which makes it very rational to deduce the functions as two principles. Following the term introduced by Brenda and Robert Vale (Vale and Vale 1996), the first principle is "respect for the user" and the second is "conserving energy and water." Assisted living, comfort, convenience, and safety and security are included in the

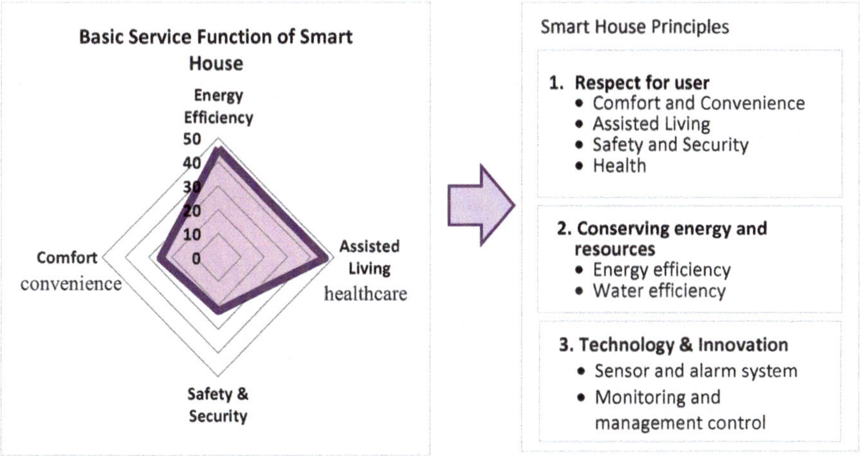


**Fig. 3.4** Principles of the smart house deduced from its functions

**Table 3.3** Principles of the smart house and the green house

| Green building principles | | | |
|---|---|---|---|
| Vale & Vale (1996) | LEED for detached house | CASBEE for detached house (2007) | Smart house principles |
| 1. Conserving energy | 1. Sustainable site | 1. Comfortable, healthy, and safe indoor environment | 1. Respect for users |
| 2. Working with the climate | 2. Water efficiency | 2. Ensuring a long service life | 2. Conserving energy and water |
| 3. Respect for the site | 3. Energy and atmosphere | 3. Creating a richer townscape and ecosystem | 3. Efficient and smart technology |
| 4. Respect for users | 4. Materials and resources | 4. Conserving energy and water | |
| 5. Limiting new resources | 5. Indoor environmental quality | 5. Using resources sparingly and reducing waste | |
| | 6. Innovation | 6. Consideration of global, local, and surrounding environment | |

principle of "respect for the user," while energy efficiency is included in the principle of "conserving energy and water." Technology, as the brain that supports the functions in the smart house, will be one of the principles, namely "efficient and smart technology."

### *3.4.1 Principles of the Smart House*

Three principles of the green house were taken as control concepts for the smart house concept comparison. The selected principles were taken from *Green Architecture* by Brenda and Robert Vale (Vale and Vale 1996), Leadership in Energy and Environmental Design (LEED) for detached houses, and CASBEE for detached houses. The order of smart house principles followed the definition, where the essential function of the smart house was to assist the user. However, assisting the user should be done in a very efficient way, using minimum energy and water through efficient and smart technology (Table 3.3).

*CASBEE* Comprehensive Assessment System for Built Environment Efficiency, *LEED* Leadership in Energy and Environmental Design

### *3.4.2 The Smart House Concept Versus the Green House Concept*

The order of smart house principles is that the user comes first and the environment comes afterward. This is different from Brenda and Robert Vale's green architecture principles and the LEED principles. However, the concept is similar to the CASBEE

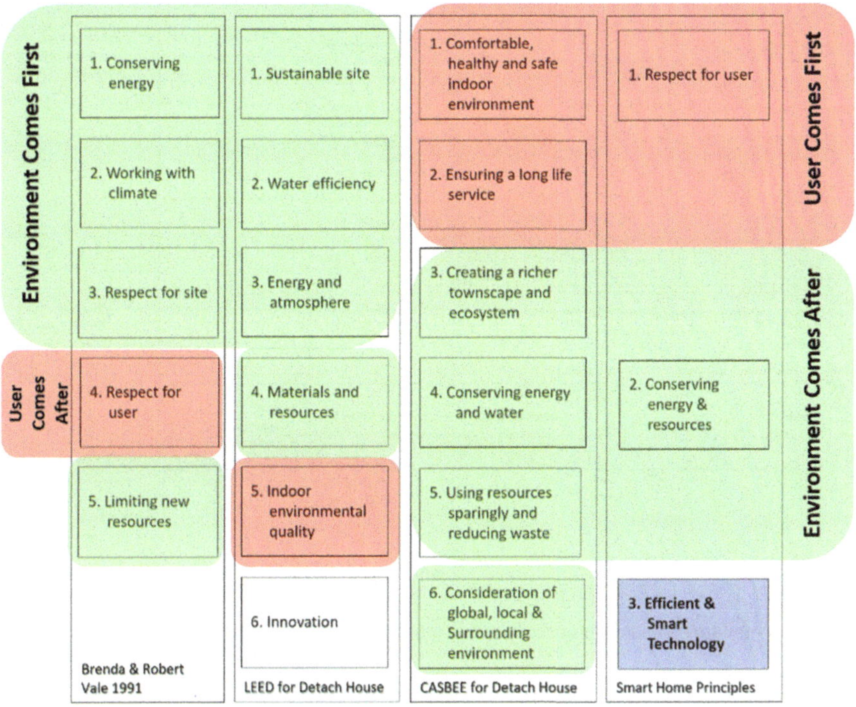


**Fig. 3.5** Order of smart house principles and green concept principles

principles, where the user is the first priority (Fig. 3.5). The principles of the smart house have similarities to and differences from the green concept principles (Fig. 3.6). The similarities are the principles of respect for the user and conservation of energy and water, while the differences are working with the climate, respect for the site, limiting new resources, and technology.

From the scale of the principles, it is obviously seen that the green house concept is bigger than the smart house concept. However, the green house concept does not include the principle of technology, which makes the smart house concept different and special. It is the core of the smart house, distributed throughout the rooms, devices, and systems (heating, ventilation, air conditioning, lighting, sensors, and alarms), conveying information to users and receiving orders from users, or automatically managing the domestic environment (Wilson et al. 2015).

The green concept may use technology to reduce environmental degradation as an optional tool for achieving sustainable living. Without technology, a house can be labeled as a green building. Conversely, technology is the brain of the smart house and without it, a house cannot be labeled as a smart house. The same principles between the two concepts may lead to an opportunity for future development of the smart house to evolve from being user/technology centric to being environment centric for a smart and sustainable living environment.

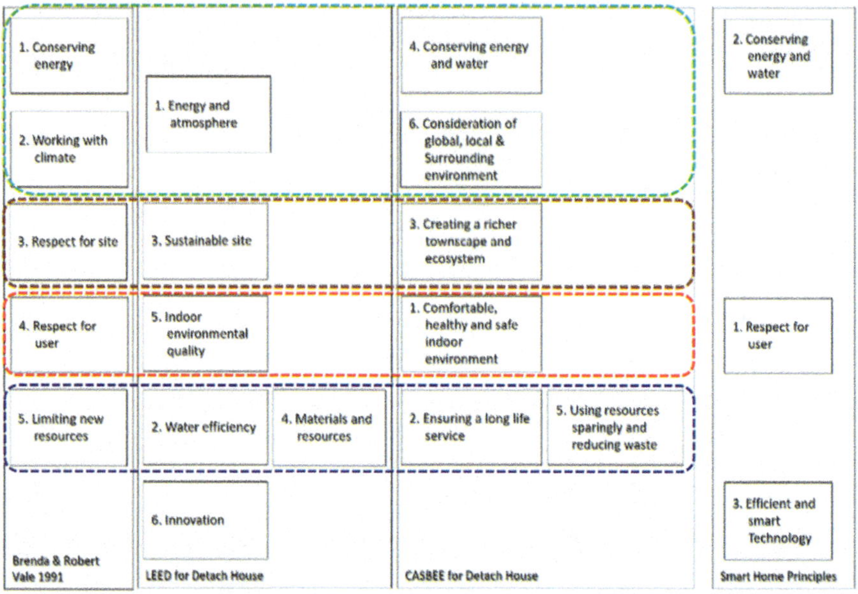


**Fig. 3.6** Similarities and differences between green house concepts and smart house concepts

## 3.5 Implementation of the Smart House Concept in Japan

The literature review underlined that the core of the smart house is characterized by the smart technology employed for functional automation (GhaffarianHoseini et al. 2013). This study evaluated six smart house pilot projects in Japan to achieve the purpose of the study to establish the parameters of the smart house. These parameters were evaluated based on relevant information such as the functions, features, installed technology, and building construction of the smart house pilot project.

### *3.5.1 Smart House Examples in Japan*

The smart houses in Japan were initiated by the private sector with support from the government. To promote energy efficiency in the housing sector, the Ministry of Economy, Trade and Industry (METI) provides subsidies specifically for introducing energy management systems into homes, known as HEMS. This subsidy program is implemented by the Sustainable Open Innovation Initiative (SII) under the budget of METI. Hence, it is likely that the smart house in Japan is focused on energy conservation principles. The features that are mostly found in smart house projects in Japan include clean energy sources, storage batteries, smart metering, HEMS, and electric vehicles (EVs).

#### 3.5.1.1 The Toyota Smart House

The Toyota Smart House has been developed by Toyota Home for a normal single family with two children, aimed at demonstrating a prototype of a smart house in a smart community. It is equipped with numerous devices for saving, creating, and storing energy, including a plug-in hybrid vehicle (PHV) or EV. It also includes a centralized HEMS for controlling devices and facilities in the house, a photovoltaic system, and a storage battery. The HEMS allows spare electricity produced by the photovoltaic system to be stored in batteries and used at a time when more electricity is being consumed. This allows home owners to use their electricity more efficiently. The house also has the possibility of using the storage batteries mounted in the PHV as an emergency power supply for electrical devices. The data obtained from the Toyota Smart House is used for optimizing energy use within the home, building a low-carbon traffic system, optimizing energy use outside the home, and optimizing behavior across the entire domain of everyday living.

#### 3.5.1.2 The COMMA House in Komaba, Meguro, Tokyo

The COMMA House was designed and developed at the University of Tokyo. It is a 93.31 $m^2$ experimental smart house that represents a typical urban detached house in Japan. The construction was completed in August 2011 with improved air tightness and thermal insulation, with a Q value adjustable between 1.6 and 2.4 $W/m^2$ K. The aim of the project is to establish concepts and technologies for smart houses that will become popular by 2020 (Institute of Industrial Science Tokyo University 2014).

The programs cover various aspects including architectural design, facilities, appliances, multivendor system integration, and data management. Accomplishments in these areas should lead to the realization of a smart house that achieves comfort and sustainability while attaining efficient use of energy and coordination within a total energy system (Fig. 3.7).

#### 3.5.1.3 The Daiwa Smart House

The Daiwa Smart House is a single-family house offering smart, ecologically friendly living that emphasizes convenience and comfort (Daiwa House Group n.d.). The construction of this smart house includes heat insulation at the same level as cold climate specifications (Q value = 1.9 $W/m^2$ K) (Fig. 3.8).

The core features are a 4.5 kW solar power system and a HEMS. The energy generated by the photovoltaic system is visualized using the HEMS to support everyday living. The HEMS controls the energy throughout the home, linked with air conditioning, batteries, the front door, and other systems. A tablet device is used to offer intuitive control and can even be used to watch or record television programs. A combination of a 6.2 kWh lithium ion storage battery for home use, the

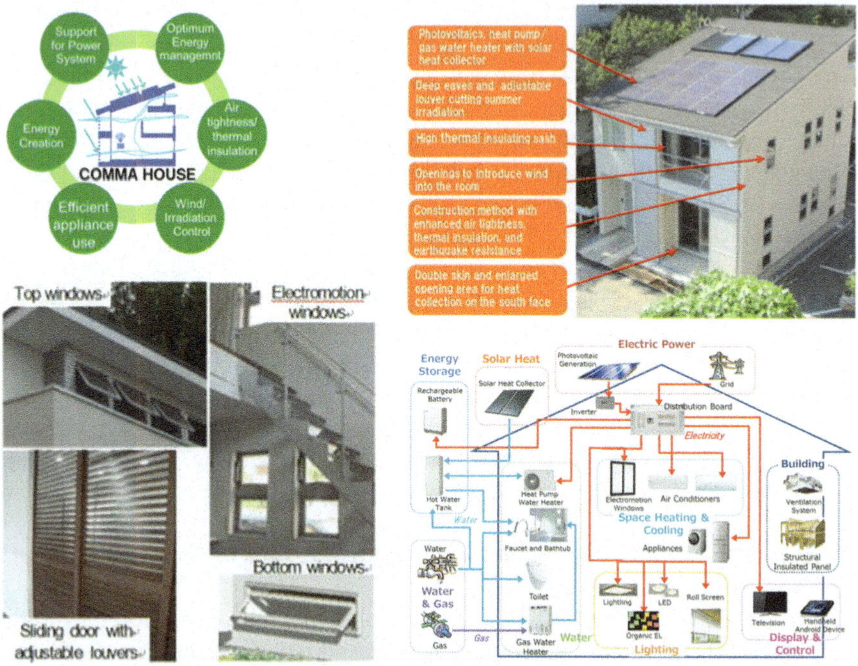


Fig. 3.7 Features of the COMMA Smart House (Institute of Industrial Science Tokyo University 2014)

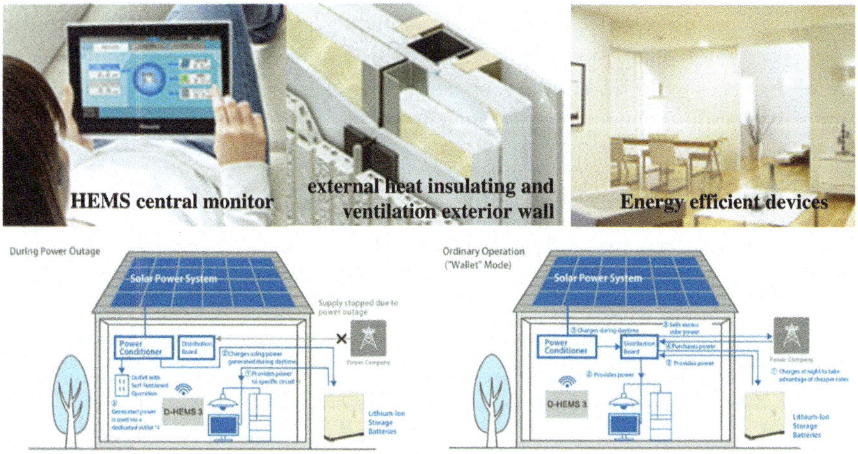


Fig. 3.8 Features of the Daiwa Smart House

HEMS, and a solar power system provides the home with the emergency power necessary for life during an outage caused by an earthquake, lightning strike, or other problems.

#### 3.5.1.4 The PanaHome Smart House

The PanaHome Smart House was developed in 2010 to pursue energy technologies for the entire home, creating a comfortable living space in a zero-carbon-footprint house. The house is designed to harness sunlight, natural breezes, and heat to create a comfortable lifestyle with virtually zero $CO_2$ emissions (Panasonic 2012). Energy efficiency is improved by fully insulated construction (ceilings, exterior walls, and foundations), low-emissivity double glazing, and combined geothermal and integrated hybrid ventilation systems.

Energy efficiency is also gained by adopting an energy creation–storage linked system. The system integrates a generation system, storage battery, and smart appliances with a central HEMS controller for energy management of the entire house. Consequently, greater energy independence during power outages can be achieved. The power generation system combines photovoltaic power generation and ENE-FARM fuel cells. Occupant comfort and eco-friendly living are obtained not only by the energy management system but also through passive technology and building strength (Fig. 3.9).

#### 3.5.1.5 The $U^2$ Home in Noda, Chiba

The $U^2$ Home is a two-story experimental smart house developed by LIXIL. The house was originally constructed in 1996 and renovated in 2015, with a total building area of 200.47 $m^2$. The $U^2$ Home addresses the IoT era and demonstrates experiments on various ideas of future houses. More than 200 sensors are installed indoors and outdoors to collect daily living–related data, such as the occupancy status. They are installed in certain places such as the gate, outer walls, windows, ceilings, walls, doors, kitchen storage, faucets, bathtub, toilet, etc. The temperature and humidity of each room, with opening and closing of doors, are networked. Additionally, information that can be obtained from the outdoor sensors are the weather conditions, wind speed and direction, fine particulate matter content in the air ($PM_{2.5}$), and ultraviolet ray levels.

The utilization of the $U^2$ Home life data is managed at three levels. The first level is "transmission to people," the second is "control of living environments," and the last is "advanced use." On the second level, experiments are conducted aimed at total living value improvements such as security, safety, health, and energy saving. On the last level, research is conducted aimed at more advanced utilization of advanced technologies, such as big data and artificial intelligence (AI), which are expected to be applied in many fields such as nursing care, medical care, crime prevention, energy, life services, education, and the media.

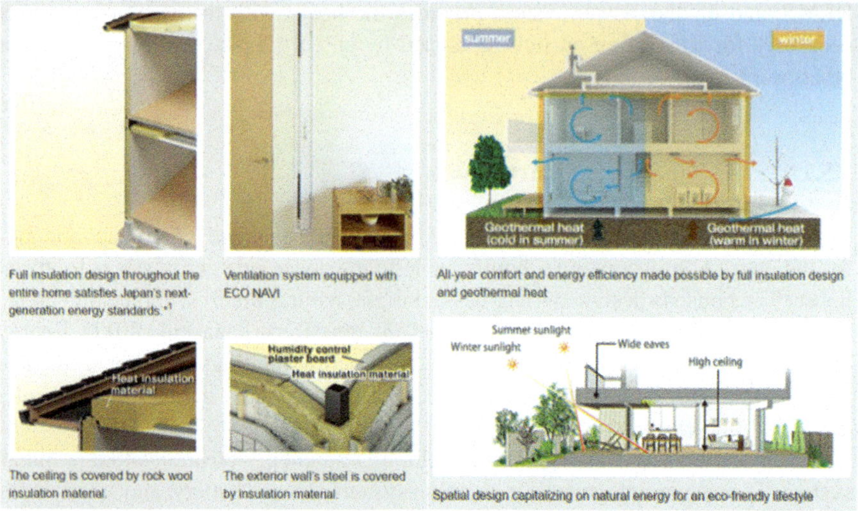


**Fig. 3.9** Design specifications of the PanaHome Smart House

A summary of smart house example specifications and features can be seen in Table 3.4. From the smart house specifications, it can be seen that there is an additional principle of the smart house that is not explicitly described in the literature. This principle is "climate responsiveness" or "working with the climate." Accordingly, the concept of the smart house has four principles. The author proposes equal values for these principles in the following order: respect for the user, responsiveness to the climate, energy and water conservation, and efficient and smart technology.

### *3.5.2 Parameters of the Smart House*

The parameters of the smart house are deduced from smart house functions, features, and specifications directly or indirectly discussed in the selected literature, combined with information from smart house examples. For establishing these parameters, the CASBEE assessment tool for detached houses was used as a control parameter. The resulting parameters of the smart house are shown in Table 3.5. and Fig. 3.10.

**Table 3.4** Specifications of smart house examples

| Example house | Technology | Services | Features | Building specifications |
|---|---|---|---|---|
| Toyota Smart House<br><br>Location:<br>35°04′53.0″N<br>137°09′28.9″E<br>Toyota-Shi, Aichi-Ken<br><br> | 7-inch control unit and touch panel monitor<br>Real-time power consumption (cost and $CO_2$ emission)<br>Historical energy data backup<br>Circuit breaker control (preventing tripping/overload)<br>Smart grid and EDMS-connected system<br>5-minute battery charger can make EV travel up to 20 km<br>Mobile energy and security control and monitoring | Energy efficiency<br>Comfort<br>Security | HEMS monitoring and control unit<br>Photovoltaic generation<br>8.4 kWh storage battery capacity<br>Smart metering<br>PHV or EV<br>Smart sensors for security | Building envelope insulation<br>Earthquake resistance grade 3 Daylighting<br>Natural ventilation (side and stack effect ventilation)<br>Rainwater harvesting<br>Blind shutter with security function |

(continued)

**Table 3.4** (continued)

| Example house | Technology | Services | Features | Building specifications |
|---|---|---|---|---|
| COMMA house (August 2011)<br><br>Location:<br>35°39′37.4″N<br>139°40′40.3″E<br>Meguro-ku, Tokyo-To<br>Building area: 93.31 $m^2$<br><br>(Institute of Industrial Science, Tokyo University, 2014) | The servers have interfaces to retrieve power demand forecasts, weather conditions/forecasts, solar power/heat forecasts, internal temperature/humidity, and water usage, which are utilized to control the system<br>Wall-mounted temperature sensor<br>Cloud-based energy management system<br>Combination of gas, heat pump, and solar heat used for heating water | Energy efficiency<br>Comfort | HEMS monitoring and control unit<br>Photovoltaic generation<br>Solar heat collector<br>Battery<br>Distribution board<br>Electro-motion windows<br>EV/PHV plug-in charger | Air tightness and thermally insulated envelope<br>Earthquake-resistant construction<br>Natural ventilation (side and stack effect ventilation, sliding door)<br>Deep eaves<br>Adjustable louvers<br>Double-skin south facade |

| | | | | |
|---|---|---|---|---|
| Daiwa Smart House<br><br>Location:<br>35°04′54.2″N<br>137°09′32.2″E<br>Toyota-Shi, Aichi-Ken<br><br>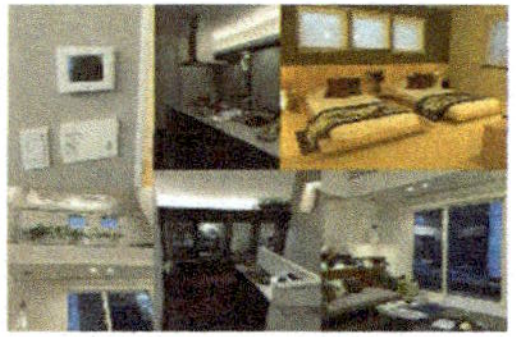 | Centralized house control (air conditioning, entrance door)<br>Mobile energy monitoring control (power generation and sale, amount of electricity used)<br>Enabling photovoltaic power generation and sale<br>Energy-efficient air conditioning, water heating, and electric lighting<br>Terrestrial digital broadcast recording (320 GB HDD for 435 hours) | Energy efficiency<br>Comfort<br>Convenience<br>Security | HEMS monitoring and control unit<br>6.2 kWh lithium ion storage battery<br>4.5 kW photovoltaic generation<br>Crime prevention support<br>Wall-mounted weather forecast display | High thermal insulation of envelope<br>Efficient natural ventilation system<br>Triple glass vacuum layer, low-emissivity glass<br>Daylighting |

(continued)

**Table 3.4** (continued)

| Example house | Technology | Services | Features | Building specifications |
|---|---|---|---|---|
| PanaHome Smart House<br><br>Location:<br>35°20′09.6″N<br>139°27′54.2″E<br>Fujisawa-Shi, Kanagawa-Ken<br>(Panasonic 2012)<br><br> | Autoselect ventilation mode by temperature sensor<br>0.3–1 kW of electricity by ENE-FARM[a]<br>Mobile energy monitoring (wireless connection through TV or smartphone)<br>AiSEG energy measurement unit plus overcurrent detection with current transformer | Energy efficiency<br>Comfort | Photovoltaics<br>Battery storage<br>HEMS<br>Low-consumption appliances<br>Home use fuel cell (ENE-FARM)[a]<br>Amounts of electricity purchased, generated, sold, and consumed, and gas and water consumption | Daylighting<br>Natural ventilation<br>Full thermal insulation design (ceilings, exterior walls, and foundations)<br>Low-emissivity double glazing (air between two glass panes)<br>Geothermal heat<br>Hybrid ventilation system |

| | | | | |
|---|---|---|---|---|
| NICE house<br><br>Location:<br>35°04′54.9″N<br>137°09′32.5″E<br>Toyota-Shi, Aichi-Ken<br><br><br>(Nice Home Co 2017; Toyota Ecoful Town 2017) | 10 kW of photovoltaic power generation<br>5 kWh battery storage<br>In power outage conditions can provide electricity of up to 265 W for 15 hours<br>Security camera | Energy efficiency<br>Comfort | Photovoltaics<br>Battery storage<br>HEMS<br>Security camera<br>Trash management | Daylighting<br>Natural ventilation<br>Earthquake resistance<br>Double-skin facade on west wall with adjustable shading<br>Thermally insulated building envelope<br>Use of wood fiber material ("cellulose fiber") with high thermal insulation and sound insulation<br>80% of the raw materials are made of newspaper<br>80% domestic timber is used |

(continued)

**Table 3.4** (continued)

| Example house | Technology | Services | Features | Building specifications |
|---|---|---|---|---|
| U² Home<br>Location:<br>Noda, Chiba<br>LIXIL Corporation<br>Building area: 137.2 $m^2$<br><br>(LIXIL 2015) | 200 sensors installed indoors and outdoors to collect data (outdoors: temperature/humidity, brightness, wind direction/wind speed, ultraviolet rays, pollen/dust, $PM_{2.5}$; indoors: temperature/humidity, brightness, human movement, building material movement [i.e., doors, windows])<br>Recording: equipment operation, electricity/gas/water/heat energy data<br>Mobile control and monitoring connected to the cloud | Energy efficiency<br>Comfort<br>Convenience<br>Aid<br>Security<br>Health | Photovoltaics<br>HEMS<br>Battery storage<br>Data communication (panel displays and interactive robots)<br>Voice and warning sounds emitted by the robot, flickering and displaying colors, video transmitted in various ways<br>Emergency call button<br>Surveillance camera for crime prevention and for monitoring children and pets | Natural ventilation with automatic open–close mode based on outdoor weather data<br>Electric shutters and sunscreen automatically controlled via the network |

*EDMS* energy data management system, *HEMS* home energy management system, *HDD* hard disk drive, *EV* electric vehicle, *PHV* plug-in hybrid vehicle, $PM_{2.5}$ fine particulate matter content in the air

[a]ENE-FARM" is the nickname given to the world's first fuel cell for practical home use, which has made its debut in Japan. This fuel cell is a new energy system, which extracts hydrogen from liquified petroleum gas and combines it with ambient oxygen to generate electrical power, while simultaneously capturing residual heat, which is used to heat water. In comparison with conventional electrical supply grid systems, it has the capability for a very high efficiency ratio and a significant reduction in $CO_2$ gas emissions (Japan LP Gas Association 2015)

**Table 3.5** Parameters of the smart house compared with Comprehensive Assessment System for Built Environment Efficiency (CASBEE) parameters for detached houses

| **Smart house parameter** | | | | | Vs | **CASBEE parameter for detached houses** | | | |
|---|---|---|---|---|---|---|---|---|---|
| **I** | **Respect for the user** | | | | | **$Q_H$ Environmental quality of the building** | | | |
| **1** | **Assisted living** | | | | | **$Q_H1$** | **Comfortable, healthy and safe indoor environment** | | |
| | **1.1** | Automatic labor-saving devices | | | | **1** | **Heating and cooling** | | |
| | | 1 | Domestic task assistance | Optional | | | **1.1** | Basic performance | |
| | | 2 | Child, elderly, and disabled monitoring assistance | Need-based | | | | 1 | Ensuring thermal insulation and airtightness performance |
| | **1.2** | Artificial intelligence (robot assistant) | | | | | | 2 | Sunlight adjustment capability |
| | | 1 | Providing basic care for elderly and disabled | Need-based | | | **1.2** | Preventing summer heat | |
| | | 2 | Helping with mobility | Need-based | | | | 1 | Allowing breezes in and heat out |
| | | 3 | Providing household maintenance | Need-based | | | | 2 | Proper planning for cooling |
| **2** | **Health care** | | | | | | **1.3** | Preventing winter cold | |
| | **2.1** | Free of chemical contaminants | | Required | | | | 1 | Proper planning for heating |
| | **2.2** | Fresh air circulation | | Required | | **2** | **Health, safety, and security** | | |
| | **2.3** | Vital sign detection and sleep patterns | | Need-based | | | **2.1** | Countermeasures against chemical contaminants | |
| | **2.4** | Wearable health care assistant that adapts to user need | | Need-based | | | **2.2** | Proper planning for ventilation | |
| **3** | **Safety and security** | | | | | | **2.3** | Precautions against crime | |
| | **3.1** | Crime prevention | | Required | | **3** | **Brightness** | | |
| | **3.2** | Accident prevention | | Optional | | | **3.1** | Use of daylight | |
| | **3.3** | Resistance against fire and natural disasters | | Required | | **4** | **Quietness** | | |
| **4** | **Comfort and convenience** | | | | | **$Q_H2$** | **Ensuring a long service life** | | |
| | **4.1** | Thermal comfort range | | Required | | **1** | **Basic life performance** | | |
| | **4.2** | Standard illuminance level and glare | | Required | | | **1.1** | Building frames | |
| | **4.3** | Entertainment and edutainment | | Optional | | | **1.2** | Exterior wall materials | |

(continued)

Table 3.5 (continued)

| Smart house parameter | | | | | Vs | CASBEE parameter for detached houses | | | |
|---|---|---|---|---|---|---|---|---|---|
| **II** | **Responsiveness to the climate** | | | | | | 1.3 | Roof materials/flat roof | |
| **1** | **Thermal performance** | | | | | | 1.4 | Resistance against natural disasters | |
| | 1.1 | Thermal insulation and airtightness | | Required | | | 1.5 | Fire preparedness | |
| | 1.2 | Control of thermal load of building | | Required | | | | 1 | Fire-resistant structure (excluding openings) |
| | 1.3 | Climate-responsive building form | | Required | | | | 2 | Early detection of fire |
| | 1.4 | Solar heat gain protection and adjustment of openings | | Required | | **2** | **Maintenance** | | |
| **2** | **Ventilation performance** | | | | | | 2.1 | Ease of maintenance | |
| | 2.1 | Ventilation opening ratio | | Required | | | 2.2 | Maintenance system | |
| | 2.2 | Ventilation rate | | Required | | **3** | **Functionality** | | |
| | 2.3 | Hybrid ventilation system | | Optional | | | 3.1 | Size and layout of rooms | |
| **3** | **Daylighting performance** | | | | | | 3.2 | Barrier-free design | |
| | 3.1 | Accessibility to sky | | Required | | **$Q_H3$** | **Creating a richer townscape and ecosystem** | | |
| | 3.2 | Depth of rooms | | Required | | **1** | **Consideration of the townscape and landscape** | | |
| | 3.3 | Daylight opening ratio | | Required | | **2** | **Creating a biological environment** | | |
| **III** | **Conservation of energy and water** | | | | | | 2.1 | Greening of the premises | |
| **1** | **Energy generation system** | | | | | | 2.2 | Ensuring a biological habitat | |
| | 1.1 | Onsite renewable energy production | | Required | | **3** | **Safety and security of the region** | | |
| | 1.2 | Support for smart grid buy and sell mode | | Optional | | **4** | **Utilizing regional resources and inheriting the regional housing culture** | | |
| | 1.3 | Storage battery | | Required | | **$LR_H$ Environmental load reduction of the building** | | | |
| | 1.4 | Automatic and uninterrupted power switches | | Optional | | **$LR_H1$** | **Conserving energy and water** | | |
| **2** | **Energy saving through equipment** | | | | | **1** | **Energy saving through building innovation** | | |
| | 2.1 | Air conditioning systems | | | | | 1.1 | Control of thermal load of building | |
| | | 1 | Efficient heating system | Required | | | 1.2 | Natural energy use | |
| | | 2 | Efficient cooling system | Required | | **2** | **Energy saving through equipment performance** | | |

| | | | | | | | | | |
|---|---|---|---|---|---|---|---|---|---|
| | **2.2** | Hot water equipment | | | | | **2.1** | Air conditioning systems | |
| | | 1 | Hot water supply equipment | Need-based | | | | 1 | Heating system |
| | | 2 | Heat insulation of bathtub | Need-based | | | | 2 | Cooling system |
| | | 3 | Natural source of heat energy | Optional | | | **2.2** | Hot water equipment | |
| | **2.3** | Energy-efficient lighting fixtures | | Required | | | | 1 | Hot water supply equipment |
| | **2.4** | Daylighting integration system | | Optional | | | | 2 | Heat insulation of bathtub |
| | **2.5** | Energy-efficient electric appliances | | Required | | | | 3 | Hot water plumbing |
| **3** | **Energy saving through management** | | | | | | **2.3** | Lighting fixtures, home electric appliances, and kitchen equipment | |
| | **3.1** | Home energy management and control system | | Required | | | **2.4** | Ventilation system | |
| | **3.2** | Smart metering for electrical energy and gas | | Required | | | **2.5** | Highly energy-efficient equipment | |
| **4** | **Water conservation** | | | | | | | 1 | Home cogeneration system |
| | **3.1** | Water-saving systems | | Required | | | | 2 | Solar power generation system |
| | **3.2** | Rainwater harvesting | | Optional | | **3** | **Water conservation** | | |
| | **3.3** | Efficient plumbing fixtures with adjustable and automatic modes | | Optional | | | **3.1** | Water-saving systems | |
| **IV** | **Efficient and smart technology** | | | | | | **3.2** | Rainwater use | |
| **1** | **Technology for management** | | | | | **4** | **Well-informed maintenance and operation schemes** | | |
| | **1.1** | User monitoring and adjustment | | Required | | | **4.1** | Presentation of lifestyle advice | |
| | **1.2** | System adaptation and automatic execution | | Optional | | | **4.2** | Management and control of energy | |
| | **1.3** | Virtual or mixed-reality management system using the Internet of Things | | Optional | | **$LR_H2$** | **Using resources sparingly and reducing waste** | | |
| | **1.4** | Apps and/or cloud-based management system | | Required | | **1** | **Introduction of materials useful for resource saving and waste prevention** | | |
| | **1.5** | Forecasting-based management | | Optional | | | **1.1** | Building frames | |

(continued)

**Table 3.5** (continued)

| **Smart house parameter** | | | | **Vs** | **CASBEE parameter for detached houses** | | | |
|---|---|---|---|---|---|---|---|---|
| 2 | **Technology for detection and prevention** | | | | | | 1 | Wooden house |
| | **2.1** | Ambient parameter sensors (temperature, humidity, radiation, wind) | Required | | | | 2 | Steel-framed house |
| | **2.2** | Illumination sensors | Required | | | | 3 | Concrete house |
| | **2.3** | Movement sensors | Required | | | **1.2** | Ground-reinforcing materials, foundation work, and foundations | |
| | **2.4** | Sound and face recognition | Optional | | | **1.3** | Exterior materials | |
| | **2.5** | Heat, flame, and smoke sensors | Required | | | **1.4** | Interior materials | |
| | **2.6** | Vital signs sensors | Need-based | | | **1.5** | Materials for the external area | |
| 3 | **Alarm and notification system** | | | | **2** | **Reduction of waste in the production and construction stages** | | |
| | **3.1** | Collision alarm | Need-based | | | **2.1** | Production stage (members for building frames) | |
| | **3.2** | Theft and burglary alarm | Required | | | **2.2** | Production stage (members other than those for building frames) | |
| | **3.3** | Fire and gas alarm | Required | | | **2.3** | Construction stage | |
| | **3.4** | Scheduled activity notification | Need-based | | **3** | **Promotion of recycling** | | |
| | **3.5** | Microphone and speaker system for requesting help | Need-based | | | **3.1** | Provision of information on materials used | |
| | **3.6** | Emergency button | Required | | **$LR_H3$** | **Consideration of the global, local, and surrounding environment** | | |
| | | | | | **1** | **Consideration of global warming** | | |
| | | | | | **2** | **Consideration of the local environment** | | |
| | | | | | | **2.1** | Control of the burden on the local infrastructure | |
| | | | | | | **2.2** | Preservation of the existing natural environment | |

| | | | | |
|---|---|---|---|---|
| | **3** | **Consideration of the surrounding environment** | | |
| | | **3.1** | Reduction of noise, vibration, exhaust, and exhaust heat | |
| | | **3.2** | Improvement of the thermal environment of the surrounding area | |
| | **$nLR_H1$ Scores after correction for solar power generation** | | | |
| | **2** | **Energy saving through equipment performance** | | |
| | | **2.1** | Air conditioning systems | |
| | | | 1 | Heating system |
| | | | 2 | Cooling system |
| | | **2.2** | Hot water equipment | |
| | | | 1 | Hot water supply equipment |
| | | | 2 | Heat insulation of bathtub |
| | | | 3 | Hot water plumbing |
| | | **2.3** | Lighting fixtures, home electric appliances, and kitchen equipment | |
| | | **2.4** | Ventilation system | |
| | | **2.5** | Highly energy-efficient equipment | |
| | | | 1 | Home cogeneration system |
| | | | 2 | Solar power generation system |

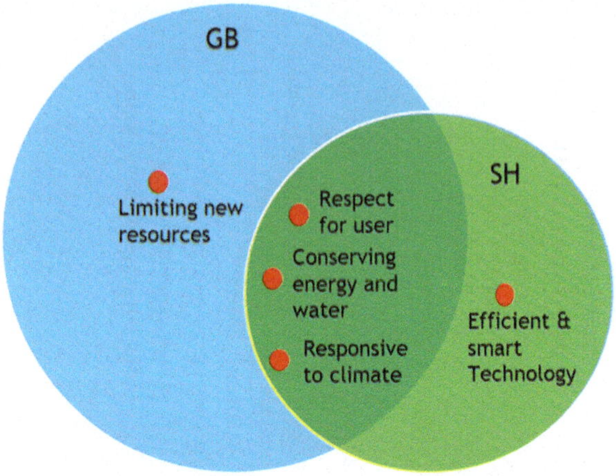


**Fig. 3.10** Smart house standpoint against the green house concept

## 3.6 Conclusion

This study aimed to map the smart house standpoint against the green concept. To achieve the goal, the principles and parameters of the smart house were established according to the literature and smart house practices. An interesting finding in this study is that the concept of the smart house has evolved from the original concept—through changes in technology, time, and users—into a greener concept where energy efficiency is also included in smart house consideration.

This study underlines the fact that technology alone may not guarantee energy efficiency. Therefore, a proper design adapted to climate conditions should be added as a principle of the smart house even though it is not explicitly discussed in the literature. The principles of the smart house proposed in this study consist of respect for users, responsiveness to the climate, conservation of energy and water, and efficient and smart technology.

The evaluation results reveal that the smart house concept and green building concept have an intersection of sets relationship. The intersection set parameters are the parameters included in the principles of respect for users, responsiveness to the climate, and conservation of energy and water.

**Acknowledgments** The authors are grateful to JSPS (C) for providing funds through the Urban Planning Laboratory (project no. 15 K06354), Kanazawa University.

## References


Abushnaf J, Rassau A, Gornisiewicz W (2015) Impact of dynamic energy pricing schemes on a novel multi-user home energy management system. Electr Power Syst Res 125:124–132. https://doi.org/10.1016/j.epsr.2015.04.003

Alaiad A, Al-ayyad M (2015) Smart home healthcare settings: a qualitative study of the domain boundary boundary. Transactions of the International Conference on Health Information Technology Advancement, 3(1), pp 108–121. Available at: http://scholarworks.wmich.edu/ichita_transactions/50/

Aldrich FK (2003) Smart homes past, present and future. In: Harper R (ed) Inside the smart home. Springer, London, pp 17–39

Allen B (1996) An integrated approach to smart house technology for people with disabilities. Med Eng Phys 18(3):203–206

Al-Muhtadi J et al (2000) Secure smart homes using Jini and UIUC SESAME. In: Proceedings – annual computer security applications conference, ACSAC, (January), pp 77–85

Arcelus A et al. (2007) Integration of smart home technologies in a health monitoring system for the elderly. In: BT – 21st international conference on advanced information networking and applications workshops. IEEE Computer Society, pp 820–825

Association Franco-Chinoise du Developpement Urbain Durable (n.d.) The history of smart homes. Available at: http://www.afcdud.com/fr/smart-city/422-how-the-history-of-smart-homes.html. Accessed 31 May 2017

Badica C, Brezovan M, Badica A (2013) An overview of smart home environments: architectures, technologies and applications. CEUR Workshop Proceedings 1036(i):78–85. Available at: http://ceur-ws.org/Vol-1036/p78-Badica.pdf

Balta-Ozkan N et al (2013a) Social barriers to the adoption of smart homes. Energy Policy 63:363–374. https://doi.org/10.1016/j.enpol.2013.08.043

Balta-Ozkan N et al (2013b) The development of smart homes market in the UK. Energy 60:361–372. https://doi.org/10.1016/j.energy.2013.08.004

Baraka K, et al (2013) Low cost Arduino/Android-based energy-efficient home automation system with smart task scheduling. In: 2013 Fifth international conference on computational intelligence, communication systems and networks, pp 296–301

Beaudin M, Zareipour H (2015) Home energy management systems: a review of modelling and complexity. Renew Sust Energ Rev 45:318–335. https://doi.org/10.1016/j.rser.2015.01.046

Behan M, Krejcar O (2013) Vision of smart home point solution as sustainable intelligent house concept, IFAC. Available at: https://doi.org/10.3182/20130925-3-CZ-3023.00057

Bhati A, Hansen M, Chan CM (2017) Energy conservation through smart homes in a smart city: a lesson for Singapore households. Energy Policy 104(February):230–239. https://doi.org/10.1016/j.enpol.2017.01.032

Bounegru L (2009) Smart houses: from managing the house at a distance to the management of life itself. Available at: http://lilianabounegru.org/wp-content/uploads/2009/11/thesis_smart_houses_liliana_bounegru.pdf

Cassie (2015) The history of home automation. p 2. Available at: https://myalarmcenter.com/blog/the-history-of-home-automation/. Accessed 28 May 2017

Chan M et al (1995) Smart house automation system for the elderly and the disabled. In: 1995 IEEE international conference on systems, man and cybernetics intelligent systems for the 21st century, 2, pp 1586–1589

Chan M et al (2008) A review of smart homes—present state and future challenges. Comput Methods Prog Biomed 91(1):55–81

Chan M et al (2009) Smart homes—current features and future perspectives. Maturitas 64(2):90–97

Cocco J (2011) Smart home technology for the elderly and the need for regulation. J Environ Public Health Law 6(1):85–108

Cook D (2012) How smart is your home? Science, 335(6076), pp 1579–1581. Available at: http://www.sciencemag.org.proxy.library.cornell.edu/content/335/6076/1579.full%5Cnfile:///Users/raj/Dropbox/Papers_dropbox/Papers/Cook/Cook_How Smart Is Your Home?_Science_2012.pdf%5Cnpapers://5860649b-6292-421d-b3aa-1b17a5231ec5/Paper/p120433

Coughlin JF et al (2007) Older adult perceptions of smart home technologies: implications for research, policy & market innovations in healthcare. In: Annual international conference of the IEEE engineering in medicine and biology – proceedings, (August), pp 1810–1815

Covington M, Moyer M, Ahamad M (2000) Generalized role-based access control for securing future applications. In: 23rd national information systems security conference, (NISSC 2000). pp 1–10. Available at: http://smartech.gatech.edu/handle/1853/6580

Daiwa House Group (n.d.) Daiwa Smart Houses "SMAxECO." Available at: http://www.daiwahouse.com/English/sustainable/eco/products/smarthouse.html. Accessed 28 May 2017

De Silva LC, Morikawa C, Petra IM (2012) State of the art of smart homes. Eng Appl Artif Intell 25(7):1313–1321. https://doi.org/10.1016/j.engappai.2012.05.002

Demongeot J et al (2002) Multi-sensors acquisition, data fusion, knowledge mining and alarm triggering in health smart homes for elderly people. C R Bio 325(6):673–682

Ding D et al (2011) Sensor technology for smart homes. Maturitas 69(2):131–136. https://doi.org/10.1016/j.maturitas.2011.03.016

Edge M, Taylor B, Dewsbury G, Gordon TR, Groves MA, Columbia B et al (2000) The potential for "Smart Home" systems in meeting the care needs of older persons and people with disabilities. Senior Housing Update 10(1):6–7

El-Baz W, Tzscheutschler P (2015) Short-term smart learning electrical load prediction algorithm for home energy management systems. Appl Energy 147:10–19. https://doi.org/10.1016/j.apenergy.2015.01.122

El-Sebaii AA et al (2010) Global, direct and diffuse solar radiation on horizontal and tilted surfaces in Jeddah, Saudi Arabia. Appl Energy 87(2):568–576. Available at: http://linkinghub.elsevier.com/retrieve/pii/S0306261909002700

Fabi V, Spigliantini G, Corgnati SP (2017) Insights on smart home concept and occupants' interaction with building controls. Energy Procedia 111(September 2016):759–769. Available at: http://www.sciencedirect.com.sdl.idm.oclc.org/science/article/pii/S1876610217302680

Farella E, Falavigna M, Ricc B (2010) Aware and smart environments: the Casattenta project. Microelectron J 41(11):697–702

Fensel A, Tomic S, Kumar V (2013) Sesame-s: semantic smart home system for energy efficiency. Informatik-Spektrum 36(1):46–57. Available at: http://link.springer.com/article/10.1007/s00287-012-0665-9

Firth SK et al (2013) Decision support systems for domestic retrofit provision using smart home data streams. In: CIB W78 30th international conference on applications of IT in the AEC industry. Move towards smart buildings: infrastructures and cities. Beijing: CIB, p 10 pages. Available at: https://dspace.lboro.ac.uk/dspace-jspui/handle/2134/15904

Fotouhi Ghazvini MA et al (2017) Demand response implementation in smart households. Energ Buildings 143:129–148. Available at: http://linkinghub.elsevier.com/retrieve/pii/S037877881730823X

Friedewald M et al (2005) Perspectives of ambient intelligence in the home environment. Telematics Inform 22(3):221–238

Gayathri KS, Easwarakumar KS (2016) Intelligent decision support system for dementia care through smart home. Procedia Comput Sci, 93(September), pp 947–955. Available at: https://doi.org/10.1016/j.procs.2016.07.281

GhaffarianHoseini AH et al (2013) The essence of future smart houses: from embedding ICT to adapting to sustainability principles. Renew Sust Energ Rev 24:593–607. https://doi.org/10.1016/j.rser.2013.02.032

Gillani Fahad L, Khan A, Rajarajan M (2015) Activity recognition in smart homes with self verification of assignments. Neurocomputing, 149(PC), pp 1286–1298. Available at: https://doi.org/10.1016/j.neucom.2014.08.069

Han J, Choi C s, Lee I (2011) More efficient home energy management system based on ZigBee communication and infrared remote controls. IEEE Trans Consum Electron 57(1):85–89

Han J et al (2014) Smart home energy management system including renewable energy based on ZigBee and PLC. IEEE Trans Consum Electron 60(2):198–202

Helal A, Cook DJ, Schmalz M (2009) Smart home-based health platform for behavioral monitoring and alteration of diabetes patients. J Diabetes Sci Technol (Online) 3(1):141–148

Hendricks D (2014) The history of smart homes. (April 22th). Available at: http://www.iotevolutionworld.com/m2m/articles/376816-history-smart-homes.htm. Accessed 9 May 2017

Hong X et al (2009) Evidential fusion of sensor data for activity recognition in smart homes. Pervasive Mob Comput 5(3):236–252. https://doi.org/10.1016/j.pmcj.2008.05.002

Honold J, Wimmer P, Kandler C (2015) Potential of energy management systems in residential buildings. Energy Procedia 78:2094–2099. Available at: http://www.sciencedirect.com/science/article/pii/S1876610215019773

Institute of Industrial Science Tokyo University (2014) COMMA House., (February). Available at: http://www.commahouse.iis.u-tokyo.ac.jp/sozai/publication/release_20140204.pdf. Accessed 29 May 2017

Iwafune Y et al (2015) Cooperative home energy management using batteries for a photovoltaic system considering the diversity of households. Energy Convers Manag 96:322–329. https://doi.org/10.1016/j.enconman.2015.02.083

Jahn M et al (2010) The energy aware smart home. In: 2010 5th international conference on future information technology, pp 1–8

Japan LP Gas Association (2015) Home-use fuel cell (ENE-FARM). Available at: http://www.j-lpgas.gr.jp/en/appliances/. Accessed 29 May 2017

Jurenoks A, Joki D (2017) Sensor network information flow control method with static coordinator within internet of things in smart house environment. Procedia Comput Sci 104:385–392

Keskin Arabul F et al (2017) Providing energy management of a fuel cell–battery–wind turbine–solar panel hybrid off grid smart home system. Int J Hydrog Energy. Available at: http://www.sciencedirect.com/science/article/pii/S0360319917308303

Khan AA et al (2015) HEMSs and enabled demand response in electricity market: an overview. Renew Sust Energ Rev 42:773–785

Kim H et al (2012) Implementing home energy management system with UPnP and mobile applications. Comput Commun 36(1):51–62. https://doi.org/10.1016/j.comcom.2012.01.007

Kim MJ et al (2013) A critical review of user studies on healthy smart homes. Indoor and Built Environment 22(1):260–270. Available at: http://journals.sagepub.com/doi/10.1177/1420326X12469733

King L (2015) The evolution of the smart home., (April 15th). Available at: https://www.raconteur.net/technology/the-evolution-of-the-smart-home. Accessed 9 May 2017

Kofler MJ, Reinisch C, Kastner W (2012) A semantic representation of energy-related information in future smart homes. Energ Buildings 47:169–179. https://doi.org/10.1016/j.enbuild.2011.11.044

Kuusik A (2001) Engineering approach for realization of small smart home systems. IFAC Proceedings Volumes 34(9):403–407. Available at: http://linkinghub.elsevier.com/retrieve/pii/S1474667017417393

Lee W et al (2016) Automatic agent generation for IoT-based smart house simulator. Neurocomputing 209:14–24. https://doi.org/10.1016/j.neucom.2015.04.130

LIXIL (2015) LIXILの研究施設「U2-Home ユースクウェアホーム」にてIoTを活用した『住生活の未来』に関する実証実験を実施~2015年度には、ビッグデータ　を活用した高度利用の研究に着手~ (LIXIL demonstration experiment on ‘future of housing life’ employing IoT at LIXIL’s research facility “U2-Home” carried out in 2015 by utilizing big data for advanced use). LIXIL News—Link to Good Living. Available at: http://newsrelease.lixil.co.jp/news/2015/070_company_0331_02.html

Lobaccaro G, Carlucci S, Lofstrom E (2016) A review of systems and technologies for smart homes and smart grids. Energies 9(5):1–33

Louis J-N et al (2015) Environmental impacts and benefits of smart home automation: life cycle assessment of home energy management system. IFAC-PapersOnLine 48(1):880–885. Available at: http://www.sciencedirect.com/science/article/pii/S2405896315001597
Luor T et al (2015) Exploring the critical quality attributes and models of smart homes. Maturitas 82(4):377–386. https://doi.org/10.1016/j.maturitas.2015.07.025
Lutolf R (1992) Smart home concept and the integration of energy meters into a home based system. In: Seventh International Conference on Metering Apparatus and Tariffs for Electricity Supply 1992, pp 277–278
Mano LY et al (2016) Exploiting IoT technologies for enhancing health smart homes through patient identification and emotion recognition. Comput Commun 89–90:178–190. https://doi.org/10.1016/j.comcom.2016.03.010
MIT Media Lab (1995) Smart rooms. Available at: http://vismod.media.mit.edu/vismod/demos/smartroom/. Accessed 31 May 2017
Missaoui R et al (2014) Managing energy smart homes according to energy prices: analysis of a building energy management system. Energy Build 71:155–167. https://doi.org/10.1016/j.enbuild.2013.12.018
Morris ME et al (2013) Smart-home technologies to assist older people to live well at home. Aging Sci 1(1):1–9. https://doi.org/10.4172/jasc.1000101
Mozer MC (1998) The neural network house: an environment that adapts to its inhabitants. In: American Association for Artificial Intelligence Spring Symposium on Intelligent Environments, pp 110–114
Nice Home Co (2017) Making houses for nice homes. pp 1–9. Available at: http://www.nice-home.jp/style/. Accessed 31 May 2017
Ozkan HA (2015) A new real time home power management system. Energ Buildings 97:56–64. https://doi.org/10.1016/j.enbuild.2015.03.038
Panasonic C (2012) How can we build next-generation houses in which comfortable living leads to eco friendliness. PanaHome Smart Houses. Available at: https://panasonic.net/es/solution-works/smart_house/. Accessed 29 May 2017
Park KH, Lee HEL, Bien Z (2008) A steward robot to help daily activities in a smart house environment. In: IFAC. https://doi.org/10.3182/20080706-5-KR-1001.01988
Pensas H, Raula H, Vanhala J (2009) Energy efficient sensor network with service discovery for smart home environments. In: 2009 Third International Conference on Sensor Technologies and Applications, pp 399–404
Pragnell M, Spence L, Moore R (2000) The market potential for smart homes. In: York. Available at: http://opensigle.inist.fr/handle/10068/582797
Premarathne US, Khalil I, Atiquzzaman M (2016) Trust based reliable transmissions strategies for smart home energy consumption management in cognitive radio based smart grid. Ad Hoc Netw 41:15–29. Available at: http://www.sciencedirect.com/science/article/pii/S1570870515300032
Risteska Stojkoska BL, Trivodaliev KV (2017) A review of internet of things for smart home: challenges and solutions. J Clean Prod 140:1454–1464. https://doi.org/10.1016/j.jclepro.2016.10.006
Roda C et al (2014) Home energy management system. Schneider Electric Industries SAS, (May). Available at: https://www.researchgate.net/profile/Cesar_Roda/publication/274780353_Home_Energy_Management_System/links/5529c17b0cf2779ab790ca5d.pdf/download?version=va
Rokach JZ (2012) Smart houses in a world of smart grids. Electr J 25(3):94–97
Rollins S et al (2014) A system for collecting activity annotations for home energy management. Pervasive Mob Comput 15:153–165. https://doi.org/10.1016/j.pmcj.2014.05.008
Rothfeld L (2015) The smart home. (January 8th), pp 28–29. Available at: http://mashable.com/2015/01/08/smart-home-tech-ces/#DgN40603Pkqo. Accessed 9 May 2017
Shakeri M et al (2017) An intelligent system architecture in home energy management systems (HEMS) for efficient demand response in smart grid. Energ Buildings 138:154–164. Available at: http://linkinghub.elsevier.com/retrieve/pii/S0378778816318369

Sintov N, Schultz P (2017) Adjustable green defaults can help make smart homes more sustainable. Sustainability 9(4): 622. Available at: http://www.mdpi.com/2071-1050/9/4/622

Skubic M et al (2009) Smart home application for elder care current status and lessons learned. Technol Health Care 17:183–201

Solaimani S, Keijzer-Broers W, Bouwman H (2015) What we do – and don't – know about the smart home: an analysis of the smart home literature. Indoor and Built Environment 24(3):370–383. Available at: http://journals.sagepub.com/doi/10.1177/1420326X13516350

Spigel L (2005) Designing the smart house: posthuman domesticity and conspicuous production. Eur J Cult Stud 8(4):403–426. Available at: http://ecs.sagepub.com/cgi/doi/10.1177/1367549405057826

Stauffer HB (1991) Smart enabling system for home automation. IEEE Trans Consum Electron 37(2):xxix–xxxv

Strengers Y, Nicholls L (2017) Convenience and energy consumption in the smart home of the future: industry visions from Australia and beyond. Energy Res Soc Sci, p 8 pages. Available at: https://doi.org/10.1016/j.erss.2017.02.008

Sundstrom C, Krysander M (2015) Smart energy usage for vehicle charging and house heating. IFAC-PapersOnLine 28(15):224–229. https://doi.org/10.1016/j.ifacol.2015.10.032

Tang P, Venables T (2000) "Smart" homes and telecare for independent living. J Telemed Telecare 6(1):8–14. Available at: http://journals.sagepub.com/doi/10.1258/1357633001933871

Tomita MR et al (2010) Smart home with healthcare technologies for community-dwelling older adults. In: Al-Qutayri MA (ed) Smart Home System. InTech, Rijeka, pp 139–158. Available at: http://www.intechopen.com/source/pdfs/9628/InTech-Smart_home_with_healthcare_technologies_for_community_dwelling_older_adults.pdf

Toyota Ecoful Town (2017) 出展企業紹介ナイスホーム株式会社 (Exhibitor Introduction of Nice Home Co., Ltd.). Available at: http://toyotaecofultown.com/news/. Accessed 31 May 2017

Vale B, Vale R (1996) Green architecture: design for a sustainable future. Thames and Hudson, London

Van Berlo A (1999) Design guidelines on smart homes. In: A COST 219bis Guidebook. European Co-Operation in the Field of Scientific and Technical Research, COST. Available at: https://books.google.co.jp/books?id=k60HGwAACAAJ

Vega AM, Santamaria F, Rivas E (2015) Modeling for home electric energy management: a review. Renew Sust Energ Rev 52(2015):948–959. https://doi.org/10.1016/j.rser.2015.07.023

Wang C et al (2015) Robust optimization for load scheduling of a smart home with photovoltaic system. Energy Convers Manag 102:247–257. https://doi.org/10.1016/j.enconman.2015.01.053

Warren S, Craft RL (1999) Designing smart health care technology into the home of the future. In: Proceedings of the first joint BMES/EMBS conference. 1999 IEEE Engineering in Medicine and Biology 21st annual conference and the 1999 annual fall meeting of the Biomedical Engineering Society (Cat. No.99CH37015), 2(505), p. 677. Available at: http://ieeexplore.ieee.org/document/803832/

Wikipedia (2017) Home automation. Wikipedia. Available at: https://en.wikipedia.org/wiki/Home_automation [Accessed 29 May 2017]

Wilson C, Hargreaves T, Hauxwell-Baldwin R (2015) Smart homes and their users: a systematic analysis and key challenges. Pers Ubiquit Comput 19(2):463–476

Wood G, Newborough M (2007) Energy-use information transfer for intelligent homes: enabling energy conservation with central and local displays. Energ Buildings 39(4):495–503

Yoza A et al (2013) Optimal operation method of smart house by controllable loads based on smart grid topology. Int J Emerg Electr Power Syst 14(5):411–420. https://doi.org/10.1016/j.renene.2011.07.026

Yoza A et al (2014) Optimal capacity and expansion planning methodology of PV and battery in smart house. Renew Energy 69:25–33. https://doi.org/10.1016/j.renene.2014.03.030

Zhang W et al (2015) Experimental and computational study for household equipment system in a smart house with solar collectors. Energy Procedia 78:3428–3433. https://doi.org/10.1016/j.egypro.2015.12.329

Zhou B et al (2016) Smart home energy management systems: concept, configurations, and scheduling strategies. Renew Sust Energ Rev 61:30–40. Available at: http://www.sciencedirect.com/science/article/pii/S1364032116002823

Ziefle M, Rocker C, Holzinger A (2011) Medical technology in smart homes: exploring the user's perspective on privacy, intimacy and trust. In: Proceedings—35th International Computer Software and Applications Conference, pp 410–415

# Chapter 4
# Green Building Development in China

**Yanchen Liu and Borong Lin**

**Abstract** This chapter introduces the Three-Star evaluation standards in China, the energy performance of green office buildings, a comparison of occupant satisfaction between green and common office buildings, and three typical green building cases in Shenzhen (a hot summer–warm winter zone), Shanghai (a hot summer–cold winter zone), and Tianjin (a cold zone), respectively.

**Keywords** Green buildings · Three-star evaluation standards · Building energy performance · Occupant satisfaction

## 4.1 Introduction

The first Chinese comprehensive green building evaluation standard was developed by the Chinese Academy of Building Research in 2006. After that, green buildings experienced a boom in development. Certificated green buildings in China are expected to be awarded a One-Star to Three-Star green building label. Recently, energy conservation and emission reductions have been increasingly valued by the Chinese government as well as Chinese society. The government has raised its support for green buildings, and laws and regulations have been developed to create a good environment to promote rapid development of green buildings. Gradually, the building energy efficiency standards and green building–related technical codes have covered all climate zones, different building types, and the whole process of planning, design, construction, and operation.

In the 2015 United Nations Climate Change Conference, the Chinese government indicated that China's carbon dioxide emissions would reach a peak around 2030, which was hoped to occur earlier. Carbon dioxide emissions per unit of gross

Y. Liu · B. Lin (✉)
Department of Building Science, Tsinghua University, Beijing, China

Key Laboratory of Eco Planning & Green Building, Ministry of Education, Tsinghua University, Beijing, China
e-mail: linbr@tsinghua.edu.cn



Z. Shen et al. (eds.), *Green City Planning and Practices in Asian Cities*,
Strategies for Sustainability, https://doi.org/10.1007/978-3-319-70025-0_4

domestic product (GDP) in 2030 would be cut by 60–65% versus those in 2005, and nonfossil energy would account for about 20% of primary energy consumption (NDRC 2015). These targets reinforce the prominent role that the rapid development of green buildings plays in China's ecological civilization construction and energy-saving and emissions reduction, which means that the Three-Star evaluation standards are implemented at a wide level.

To promote the sustainable development of buildings, various certification systems have been developed around the world, such as Leadership in Energy and Environmental Design (LEED) in the USA, the Building Research Establishment Environmental Assessment Method (BREEAM) in the UK, Deutsche Gesellschaft für Nachhaltiges Bauen/German Sustainable Building Council (DGNB) in Germany, and the Comprehensive Assessment System for Built Environment Efficiency (CASBEE) in Japan.

LEED was created by the US Green Building Council (USGBC) in 1998 for rating new and existing buildings at all phases of development (USGBC 2009, 2012). LEED v4 is the newest version of LEED to date (USGBC 2013). All LEED products contain five common categories: sustainable sites, water efficiency, energy and atmosphere, materials and resources, and indoor environmental quality.

BREEAM, the world's first green building evaluation system, was first launched in 1990 (BRE 2016). BREEAM encompasses ten credit categories, including energy, health and wellbeing, innovation, land use, materials, management, transport, waste, water, and pollution (BRE 2016).

The DGNB was founded in 2007 by 16 initiators from various subject areas within the construction and real estate sectors. The aim was to promote sustainable and economically efficient building even more strongly in the future. Quality is assessed comprehensively over the entire life cycle of the building. Outstanding fulfillment of up to 50 sustainability criteria from the quality sections of ecology, economy, sociocultural aspects, technology, process work flows, and site are certified (DGNB 2016).

CASBEE was developed by a research committee established in 2001 through the collaboration of academia, industry, and national and local governments, which established the Japan Sustainable Building Consortium (JSBC) under the auspices of the Ministry of Land, Infrastructure, Transport and Tourism (MLIT) (IBEC 2015). CASBEE covers the following four assessment fields: (1) energy efficiency; (2) resource efficiency; (3) local environment; and (4) indoor environment. These four fields are divided into two main assessment categories: built environment quality (Q) and built environment load (L), being evaluated separately to calculate built environment efficiency (BEE).

In China the Three-Star evaluation standard is a semi-mandatory, consensus-based, government- and market-driven program that provides certification of green buildings by a third party, contributing to maximizing conservation of resources (including energy, land, water, and materials), protecting the natural environment, and minimizing pollution (MOHURD 2014). The Three-Star evaluation standard is the most popular green building rating tool in China, and its market share is continuing to grow.

In the following sections, this chapter will introduce the Three-Star evaluation standards, the energy performance of green office buildings, a comparison of occupant satisfaction in green buildings ($S_G$) and occupant satisfaction in common office buildings ($S_C$), and three typical green building cases in Shenzhen (a hot summer–warm winter zone), Shanghai (a hot summer–cold winter zone), and Tianjin (a cold zone), respectively.

## 4.2 Three-Star Evaluation Standards

Since the launch of the Assessment System for Green Buildings of the Beijing Olympics (GBCAS) in 2002, and the implementation of the Assessment Standard for Green Building in 2006, i.e., the Three-Star evaluation standard, the Ministry of Housing and Urban–Rural Development of China (MOHURD) has created a series of different tools (hereafter known as the "Three-Star evaluation standard family"). By the end of 2015, the China Society for Urban Studies indicated that 3979 projects had received Three-Star certification across the country, reaching a total building area of 0.46 billion $m^2$, among which 3775 projects obtained green building design labels and 204 obtained green building operation labels (China Society for Urban Studies 2016). In the Three-Star evaluation standard, green buildings are defined as those that "maximize conservation of resources (including energy, land, water, and materials), protect the natural environment, minimize pollution, provide people with healthy, adaptive, and efficient spaces during the life cycle and coexist in harmony with the natural environment." By the end of 2016, 13 standards for green buildings in the Three-Star family were officially issued by MOHURD, covering the design, construction, operation, and retrofit stages: Assessment Standard for Green Buildings (GB/T 50378-2014); Evaluation Standard for Green Industrial Buildings (GB/T 50878-2013); Evaluation Standard for Green Office Buildings (GB/T 50908-2013); Assessment Standard for Green Store Buildings (GB/T 51100-2015); Evaluation Standard for Green Hospital Buildings (GB/T 51153-2015); Assessment Standard for Green Hotel Buildings (GB/T 51165-2016); Assessment Standard for Green Museums and Exhibition Buildings (GB/T 51148-2016); Evaluation Standard for Green Construction of Buildings (GB/T 50640-2010); Assessment Standard for Green Retrofitting of Existing Buildings (GB/T 51141-2015); Code for Green Design of Civil Buildings (JGJ/T 229-2010); Code for Green Construction of Buildings (GB/T 50905-2014); Technical Code for Operation and Maintenance of Green Buildings (JGJ/T 391-2016); and Technical Specification for Green Production and Management of Ready-Mixed Concrete (JGJ/T 328-2014).

All Three-Star rating tools contain five common sections: (1) land saving and outdoor environment; (2) energy saving and energy utilization; (3) water saving and water resource utilization; (4) material saving and material resource utilization; and (5) indoor environment quality.

In the current version of the Assessment Standard for Green Buildings (GB/T 50378-2014), assessment items involved in the indoor environment quality (IEQ) category are either mandatory or voluntary. The seven required IEQ assessment items address minimum requirements for indoor noise, sound insulation performance of the envelope, indoor lighting, indoor temperature, relative humidity (RH) and fresh air volume in air-conditioned buildings, anti-condensation envelope inner surfaces, insulation performance of roof and east and west external walls, and indoor air pollutants, while voluntary assessment items address higher requirements for the indoor acoustic environment, lighting environment and view, thermal environment, and indoor air quality (IAQ) performance, such as the IAQ monitoring system. Certification benchmarks go from One Star (50–59 points), to Two Stars (60–79 points) and Three Stars (80 points and above), with a total of 100 points being attainable.

## 4.3 Energy Performance of Green Office Buildings in China

Thirty-one green office buildings located in Beijing, Tianjin, Tanggu, Jinan, Langfang, Shanghai, Nanjing, Suzhou, Chongqing, Shenzhen, and Guangzhou were selected to compare their actual energy consumption with the energy consumption requirements in the Chinese Standard for Energy Consumption of Buildings (SECB), LEED-certified buildings in the USA, and common office buildings in China (Lin et al. 2016).

Climate zones in China are usually divided into the severe cold zone, cold zone, hot summer–cold winter zone, hot summer–warm winter zone, and mild zone, which are divided by different heating degree days (HDD) or cooling degree days (CDD). The buildings investigated in this research are located in the cold zone (C), hot summer–cold winter zone (HSCW), and hot summer–warm winter zone (HSWW), as shown in Table 4.1.

**Table 4.1** Climate zones of the different cities

| Location | HDD/CDD | Climate zone | Mean temperature of zone (°C) | |
|---|---|---|---|---|
| | | | Summer | Winter |
| Beijing, Tianjin, Tanggu, Jinan, Langfang | $2000 \leq$ HDD18 °C $< 3800$, CDD26 °C $> 90$ | C | 25.6 | −1.1 |
| Shanghai, Nanjing, Suzhou | $1200 \leq$ HDD18 °C $< 2000$ | HSCW | 25.7 | 6.4 |
| Chongqing | $700 \leq$ HDD18 °C $< 1200$ | | | |
| Shenzhen, Guangzhou | HDD18 °C $< 500$ | HSWW | 28.0 | 15.3 |

*C* cold, *CDD* cooling degree day, *HDD* heating degree day, *HSCW* hot summer–cold winter, *HSWW* hot summer–warm winter

Boerstra et al. (2015) developed a new, updated version of the ISSO 74 (ATG) guideline, in which spaces or buildings were distinguished as type α or type β. Type α refers to free-running situations in summer with openable windows and other adaptive opportunities for the occupants (particularly, a nonstrict clothing policy), whereas type β refers to summer situations that primarily rely on centrally controlled cooling. Similarly, in the Chinese SECB (MOHURD 2016), buildings are divided into type A and type B.

Figure 4.1 shows the actual total site energy consumptions of the 31 office buildings located in the cold climate zone, hot summer–cold winter zone, and hot summer–warm winter zone by two different building types. These results are based on the annual monthly electricity consumption from energy metering systems installed in each building. Seven office buildings out of the 15 cases in the cold zone, shown with orange bars, use electric heat pumps for heating in winter, and the remaining eight buildings use district heating, whose heat consumption is converted into electricity consumption, in accordance with the electricity equivalent conversion method (Yi and Xiu 2010), shown by red bars in Fig. 4.1.

The energy consumption requirements for building type A and building type B in the Chinese SECB are shown in Table 4.2, as well as the median, minimum, and maximum values of measured energy consumption. The statistics indicate that in the hot summer–cold winter zone, the average total energy consumption values for both building types are close to the recommended values of 70 and 80 kWh/(m$^2$·a), which means the energy performance of green office buildings in this climate zone is aligned with the standard. However, in cold zones and in hot summer–warm winter zones, the energy consumption values for both type A and type B are higher than

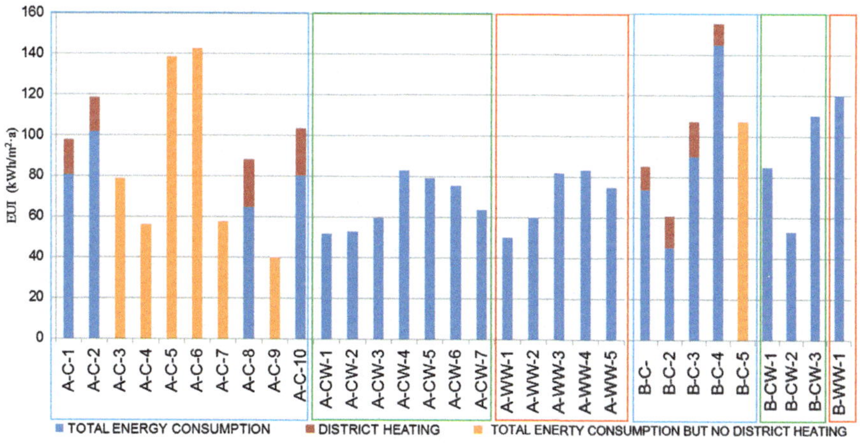


**Fig. 4.1** Total energy consumption of green buildings in China (kWh/(m$^2$·a))
Note:
(a) A/B-C/CW/WW refers to the building type A/B in the cold (C)/hot summer–cold winter (CW)/ hot summer–warm winter (WW) zones
(b) Blue frame refers to cold climate zone, green frame refers to hot summer–cold winter zone, and red frame refers to hot summer–warm winter zone)

**Table 4.2** Comparison of energy consumption of green office buildings according to the Chinese Standard for Energy Consumption of Buildings (SECB)

| | Type A | | | | Type B | | | |
|---|---|---|---|---|---|---|---|---|
| Climate zone | Upper limit (kWh/($m^2$·a)) | Recommended value (kWh/($m^2$·a)) | Green office median (minimum, maximum) | Difference[a] | Upper limit (kWh/($m^2$·a)) | Recommended value (kWh/($m^2$·a)) | Green office median (minimum, maximum) | Difference[a] |
| Cold | 65 | 55 | 79.8 (40.1, 143) | 23% (45%) | 80 | 60 | 90 (45.4, 144.7) | 13% (50%) |
| Hot summer–cold winter | 85 | 70 | 63.7 (52, 83.2) | −25% (−9%) | 110 | 80 | 85 (53, 110) | −23% (6%) |
| Hot summer–warm winter | 80 | 65 | 75.0 (50, 83.5) | −6% (15%) | 100 | 75 | 120 | 20% (60%) |

[a]The difference refers to the percentage difference between the measured median and the upper limit, with the recommended value in parenthesis. A positive difference indicates higher energy consumption in green office buildings

the recommended limit, although the value for type B in the hot summer–warm winter zone may be not representative because of the sample size. It should be noted that the upper limit and recommended value for the cold zone in the SECB do not include district heating energy consumption, and that in Figs. 4.1 and 4.7, the 15 cases in the cold zone shown with orange bars use air conditioning systems for heating and the remaining eight cases use district heating in winter; red bars show the heat consumption converted into electricity consumption. Analysis of the existing data indicates that the heat consumption for district heating of green buildings is about 12–24 kWh/($m^2$·a), significantly lower than the 32–38 kWh/($m^2$·a) value of common buildings (Tsinghua University Building Energy Research Centre 2009).

Figure 4.2 compares energy consumption levels of green office buildings in China and LEED-certified buildings in the USA. No significant correlation is observed between measured energy performances of green office buildings and the certification levels of the buildings in China, which is similar to the case of LEED buildings in the USA, according to Newsham et al. (2009). In general, the total energy consumption of Chinese green office buildings is approximately one third that of LEED-certified buildings in the USA. Figure 4.2 presents the general energy consumption levels of green buildings in China and the USA just for preliminary understanding. Overall comparisons that take into account quality and services of buildings should be considered for further exploration.

Table 4.3 presents a comparison of green and common office buildings in three different climate zones, including 48 green office buildings (among which the additional 17 cases come from a survey by the Chinese Society for Urban Studies) and 481 common office buildings (from a survey by Xiao (2011)). As shown in Table 4.3, the medians for green office buildings are 30% and 19% lower than the figures for common office buildings in the hot summer–cold winter zone and the hot summer–warm winter zone, respectively. However, in the cold zone, the electricity consumption of green office buildings is similar to that of non-green buildings, with only a

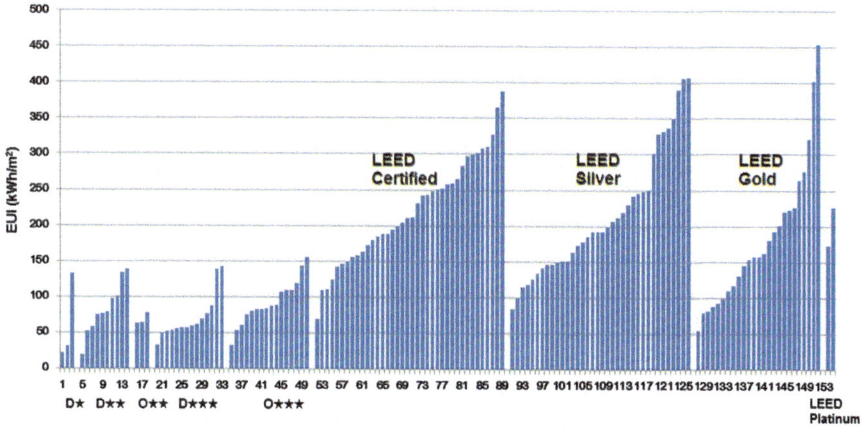


**Fig. 4.2** Energy consumption of green office buildings in China versus the USA. *D* design, *O* operation

**Table 4.3** Electricity consumption intensity comparison in three different climate zones

| | Cold zone | | Hot summer–cold winter zone | | Hot summer–warm winter zone | | Overall | |
|---|---|---|---|---|---|---|---|---|
| | Green | Common | Green | Common | Green | Common | Green | Common |
| *N* | 22 | 121 | 18 | 302 | 8 | 58 | 48 | 481 |
| First quartile (kWh/m²) | 57.7 | 60.1 | 52.8 | 88.1 | 52.6 | 65.7 | 55.1 | 75.8 |
| Median (kWh/m²) | 80.0 | 82.0 | 77.6 | 110.6 | 71.5 | 88.5 | 77.1 | 97.8 |
| Third quartile (kWh/m²) | 100.4 | 111.0 | 110.2 | 136.0 | 80.3 | 123.7 | 101.5 | 128.2 |
| SD | 30.6 | 44.1 | 37.9 | 37.3 | 28.0 | 56.7 | 32.9 | 42.9 |
| Skewness | 0.6 | 1.3 | 0.1 | 0.9 | 0.2 | 1.6 | 0.3 | 1.0 |

*SD* standard deviation

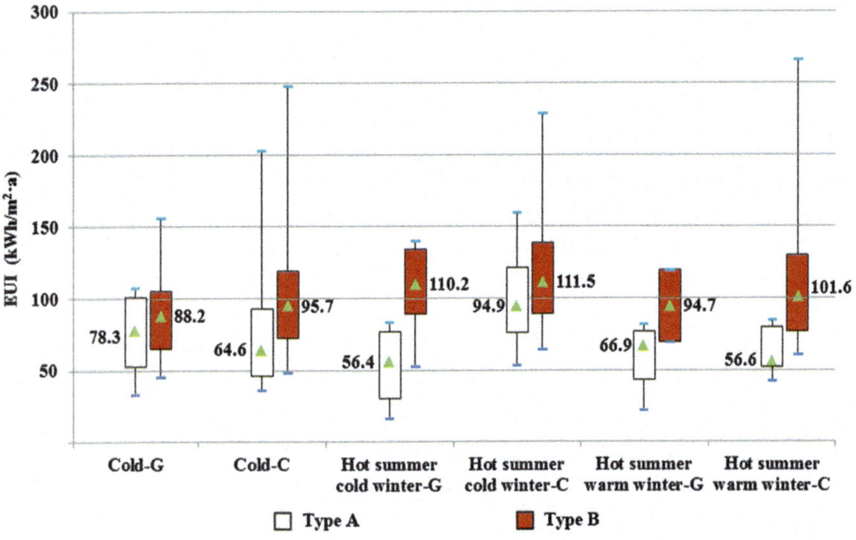


**Fig. 4.3** Differences in actual electricity consumption of green and common office buildings in three different climate zones. (Note: there are only two green cases of building type B in the hot summer–warm winter zone, so the maximum and minimum are the same with the 25th and 75th percentiles, respectively, and the median refers to the mean)

2% difference. For buildings overall, the median for green buildings is 21% lower than the figure for common buildings.

Building type is a key factor affecting a building's actual energy consumption. Figure 4.3 and Table 4.4 present the differences and statistical significance of the median differences between green and common office buildings of type A and

**Table 4.4** Statistical significance of the differences in median energy use intensity (EUI) (ΔEUI: green building minus common building)

| | Type A | | | Type B | | |
|---|---|---|---|---|---|---|
| Climate zone | Cold | Hot summer–cold winter | Hot summer–warm winter | Cold | Hot summer–cold winter | Hot summer–warm winter |
| ΔEUI (green – common) (kWh/(m$^2$·a)) | 13.7 (NS) | −38.5 ($p < 0.001$) | 10.3 (NS) | −7.5 (NS) | −1.3 (NS) | −6.9 (NS) |

*NS* not significant

type B in the three different climate zones. These results indicate that type A buildings have different energy consumption features from type B buildings in terms of the comparison between green and common office buildings. In the comparison of the median energy consumptions in the corresponding climate zones, for building type A, it is only in the hot summer–cold winter zone that the green building value is statistically significantly lower than the common building value. Conversely, for building type B, there is no statistically significant difference in the medians for electricity consumption between green and common office buildings, whatever the climate zone. It is also worth noting that due to improvements in the technology, operation, and management of green office buildings, the maximum energy consumption can be reduced effectively compared with common buildings, whatever the building type (A or B).

## 4.4 Comparison of Occupant Satisfaction Between Green and Common Office Buildings

To evaluate green buildings, the IEQ is as important as energy consumption. In order to analyze the IEQ and occupant satisfaction in green buildings in China, the research group has undertaken a dozen systematic investigations of green office buildings in recent years. Ten of the 31 green office buildings were selected to analyze the IEQ through field investigation and long-term measurement of indoor environment parameters by self-recording instruments, combined with user subjective questionnaire surveys (Pei et al. 2015). The selected green and common buildings for detailed IEQ satisfaction analysis were as similar as possible, e.g., being of the same size and age, having the same number of occupancy hours, being in the same climate zone, and having the same kind of occupants doing similar work. The basic information on the case study office buildings can be seen in Table 4.2. It should be noted that in China, the overall office building stock was built before energy efficiency of buildings was regulated via the Design Standard for Energy Efficiency of Public Buildings (GB 50189-2005). Anonymous questionnaires were distributed by hand to staff in each case study building to collect occupant subjective evaluations.

In order to reflect the objective situation, firstly the questionnaire survey was based on a selection of at least 20% of users in the entire building; meanwhile, we guaranteed that the number of questionnaires collected in each building was more than 20. Occupant satisfaction was scored on a scale from −1 (very dissatisfied) to +1 (very satisfied), as shown in Fig. 4.4 and Table 4.5.

Satisfaction with IEQ and building service performance (BSP) were analyzed through SPSS statistical software, including the thermal environment, IAQ, lighting environment, acoustic environment and overall environment, comfort of facilities, decoration, work/storage space, visual privacy, communication convenience, cleanliness, and operation and maintenance (O&M). The two-sample mean-comparison test we used rejected the null hypothesis at a 5% level that the occupant satisfaction values for the two types of building (green and common) came from different distributions. In each box plot, the bottom and the top of the box represent the 25th and 75th percentiles, respectively, and the triangle near the middle of the box is the median. The ends of the whiskers indicate the minimum and maximum.

Figure 4.5 and Table 4.6 present the differences and the statistical significance of the differences in the medians for satisfaction with the IEQ between respondents in the green and common buildings investigated in this study. It could be found that the median values for all IEQ items in green buildings were statistically significantly higher than those in common buildings. Satisfaction with the IAQ was below neutral for common buildings but above neutral for green buildings. There was little effect of building type on satisfaction with the acoustic environment ($S_C = 0.1$, $S_G = 0.3$); this might have been a logical consequence of open-plan design, which favors daylight penetration but not noise control. In sum, green building users were more satisfied with the IEQ than common building users, with significant differences with regard to the thermal environment and IAQ. For the same type of buildings, satisfaction with light was the highest.

Similarly, the differences and the statistical significance of the differences in satisfaction with BSP are shown in Fig. 4.6 and Table 4.7. It could be found that there was a high level of statistical significance for the differences in all BSP items between green and common buildings. The difference in satisfaction with the comfort of facilities and O&M between building types was significant ($\Delta S = 0.4$). Little difference in work/storage space or visual privacy was found ($\Delta S = 0.1$).

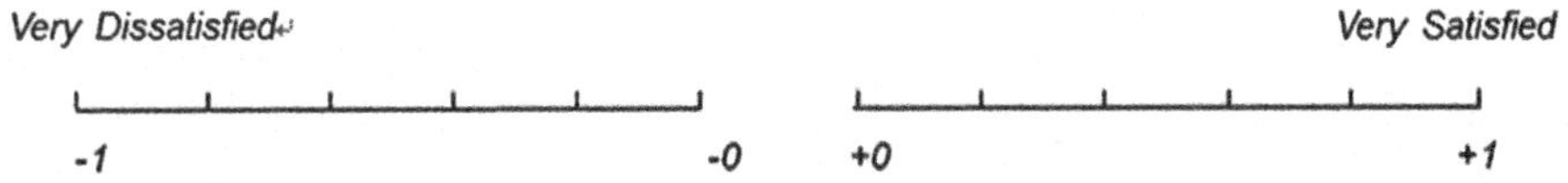


**Fig. 4.4** Scale model of satisfaction

**Table 4.5** Case study building characteristics

| Building type | Data type | Sample size[a] | Climate zone | Occupancy | Years of construction | Main characteristics |
|---|---|---|---|---|---|---|
| Green office building | Energy consumption | 31 | Cold, HSCW, HSWW | 0900–1800 hours on weekdays | 2009–2012 | Many green building techniques, such as a high-performance envelope, renewable energy systems, ventilation/heat recovery, etc. |
| | IEQ questionnaire | 10 (360) | Cold, HSCW | | | |
| Common office building | Energy consumption | 481 | Cold, HSCW, HSWW | 0900–1800 hours on weekdays | 1990–2010 | No green building technology |
| | IEQ questionnaire | 8 (244) | Cold, HSCW | | 2005–2010 | |

*HSCW* hot summer–cold winter, *HSWW* hot summer–warm winter, *IEQ* indoor environment quality

[a]The sample size refers to the number of case study buildings, and the number of valid questionnaires is given in parentheses

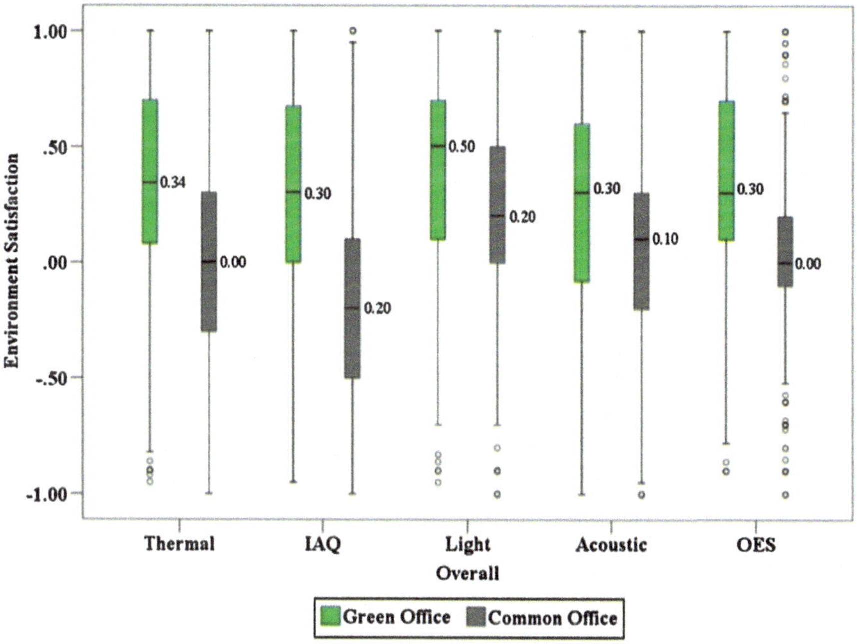


**Fig. 4.5** Satisfaction with indoor environment quality (IEQ). *OES* overall environment satisfaction

**Table 4.6** Statistical significance of the differences in satisfaction with indoor environment quality (IEQ)

| IEQ | Green buildings | | Common buildings | | | ΔS: green – common | ΔS | Dissatisfaction (%) | |
|---|---|---|---|---|---|---|---|---|---|
| | $S_G$ | Variance | $S_C$ | Variance | *p* value | | % | Green buildings | Common buildings |
| Thermal | 0.34 | 0.20 | 0 | 0.20 | 0.0000 | 0.34 | 17 | 20 | 41 |
| IAQ | 0.3 | 0.18 | −0.2 | 0.20 | 0.0000 | 0.5 | 25 | 23 | 64 |
| Light | 0.5 | 0.17 | 0.2 | 0.14 | 0.0000 | 0.3 | 15 | 15 | 19 |
| Acoustic | 0.3 | 0.20 | 0.1 | 0.16 | 0.0000 | 0.2 | 10 | 26 | 39 |
| OES | 0.3 | 0.15 | 0 | 0.13 | 0.0000 | 0.3 | 15 | 15 | 39 |

*ΔS* difference in satisfaction, *IAQ* indoor air quality, *OES* overall environment satisfaction, $S_C$ occupant satisfaction in common buildings, $S_G$ occupant satisfaction in green buildings

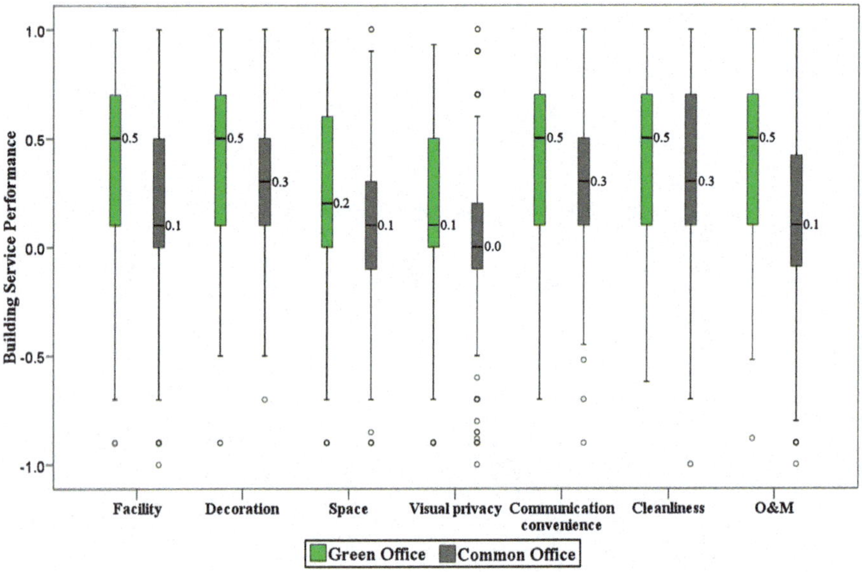


**Fig. 4.6** Satisfaction with building service performance (BSP)

## 4.5 Typical Green Building Cases

### *4.5.1 Headquarters of the Shenzhen Institute of Building Research Co. Ltd*

The headquarter of the Shenzhen Institute of Building Research Co. Ltd. (known as iBR) is located in the Futian District of Shenzhen. It has a total building area of 18,623 $m^2$, with 12 floors aboveground and two floors underground (Tsinghua

**Table 4.7** Statistical significance of the differences in satisfaction with building service performance (BSP)

| | Green buildings | | Common buildings | | | | | Dissatisfaction (%) | |
|---|---|---|---|---|---|---|---|---|---|
| BSP | $S_G$ | Variance | $S_C$ | Variance | *p* value | ΔS (green – common) | ΔS % | Green buildings | Common buildings |
| Facilities | 0.5 | 0.14 | 0.1 | 0.19 | 0.0000 | 0.4 | 0.2 | 9 | 23 |
| Decoration | 0.5 | 0.12 | 0.3 | 0.12 | 0.0000 | 0.2 | 0.1 | 5 | 8 |
| Space | 0.2 | 0.18 | 0.1 | 0.16 | 0.0002 | 0.1 | 0.05 | 20 | 27 |
| Visual privacy | 0.1 | 0.19 | 0.0 | 0.16 | 0.0000 | 0.1 | 0.05 | 25 | 38 |
| Communication convenience | 0.5 | 0.12 | 0.3 | 0.12 | 0.0000 | 0.2 | 0.1 | 6 | 11 |
| Cleanliness | 0.5 | 0.13 | 0.3 | 0.12 | 0.0023 | 0.2 | 0.1 | 7 | 7 |
| O&M | 0.5 | 0.12 | 0.1 | 0.18 | 0.0000 | 0.4 | 0.2 | 5 | 25 |

*ΔS* difference in satisfaction, *O&M* operation and maintenance, $S_C$ occupant satisfaction in common buildings, $S_G$ occupant satisfaction in green buildings

University Building Energy Research Centre 2014). It has been in use since 2009 and has won many awards including a Three-Star green building design assessment in 2009, a Three-Star green building operating assessment in 2011, and the first prize for an innovative green building award from the Ministry of Housing and Urban–Rural. The construction and installation cost was only 4200 yuan/$m^2$ due to use of many local energy-conserving technologies of low cost.

#### 4.5.1.1 Architectural Design Features for Energy Saving

The iBR building has 12 floors aboveground with 10 floors of indoor space, and is about 45 m high and 30 m deep, as shown in Fig. 6.3. The function of the building is mainly offices, with 350–450 occupants. The building's window-to-wall ratio is 0.39, and the global thermal conductivity of the walls is 0.69 W/($m^2$·K). The windows have a thermal conductivity value of 3.5 W/($m^2$·K) and the shading coefficient is 0.34.

The building is composed of two parts: a lower part and a higher part. The lower part has five floors with a main atrium, exhibition areas, labs, meeting rooms, and a lecture hall. In the higher part, the 7th floor to the 10th floor consist of office areas, and the height of the 8th and 10th floors is 7.2 m with mezzanine areas. The 11th and 12th floors contain apartments for visiting experts, common spaces for the occupants' activity, a canteen, and so on. There is an outdoor garden on the 6th floor between the lower and higher parts of the building. In addition, the roof also has a garden, and solar panels and water heaters are used for rooftop shading.

As Shenzhen is located in the south of China with the climate features of a hot summer and mild winter, the iBR building was designed as an open-type building to make use of the features of the local climate. Firstly, the main atrium in the lower part is a large space, which is connected with the outdoor environment by gates (as shown in Fig. 4.8). Secondly, each floor in the higher part is composed of two groups of enclosed spaces linked on each floor by an outside platform (as shown in Fig. 4.9), which can be used as a meeting space (as shown in Fig. 4.10) and supplies a hot water room, the floor's printers, the reception room, and a corridor, elevator lobby, and stairs. There are some enclosed spaces beside the platform, used as washrooms, elevators, and equipment rooms. The windows of each floor are openable and can be controlled by the office occupants. Sunshades have been designed outside the windows (as shown in Fig. 4.11). In addition, there are reflectors inside and outside the windows, which can reflect the sunlight up to the white ceiling to enhance the natural lighting and also avoid dazzling light (Figs. 4.7 and 4.12).

**Fig. 4.7** Architectural appearance of the Institute of Building Research (iBR) building

There is a lecture hall with a high height in the lower part, and it can seat 300 persons. Its remarkable feature is that its walls can be opened to make use of natural ventilation instead of air conditioning when the outdoor temperature is comfortable (as shown in Fig. 4.13). In addition, its stairs are outside (as shown in Fig. 4.14).

#### 4.5.1.2 Air Conditioning System

Several kinds of air conditioning methods have been adopted based on the load features of different function areas, consisting of fan coil units (FCUs), fresh air units with liquid desiccant dehumidifiers, a varied refrigerant volume (VRV) system, and split units. In order to reduce the energy consumption of the distribution system, the chillers are small and are installed for each function area independently. The cooling water system is centralized.

Most areas of the building employ FCUs and a fresh air system. The supply temperature of the chilled water for the FCUs is 16 °C, as FCUs only undertake a sensible load. The moisture load is charged by the fresh air units with liquid desiccant dehumidifiers, and the indoor temperature can be controlled by the occupants. The

**Fig. 4.8** Entry and atrium

**Fig. 4.9** High-floor plan

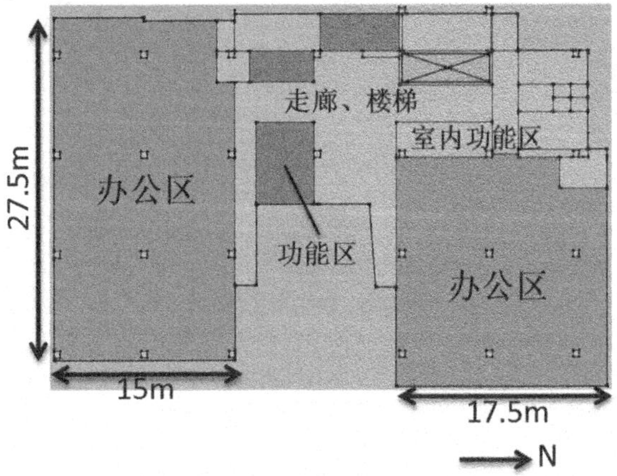

**Fig. 4.10** Outdoor platform of a high-floor office

**Fig. 4.11** Openable windows and outside sunshades

Fig. 4.12 Reflectors installed inside and outside windows for natural lighting

Fig. 4.13 Lecture hall with openable windows

lecture hall is serviced by separate chillers and fresh air units. The underground labs and other specific spaces (information technology [IT] rooms, control room, etc.) are also designed with separate air conditioning systems. In addition, a few split units are used for the expert apartments on the 11th floor and in other spaces.

### 4.5.2 Shen-du Modern Building in Shanghai

The Shen-du Modern Building is located at 1368 Tibet South Road, Shanghai, which is only 800 m from the Ningbo Exhibition Hall of the Shanghai World Expo 2010. The building has obtained Three-Star green building design certification (Tsinghua University Building Energy Research Centre 2014).

**Fig. 4.14** Outside stairs of the lecture hall

The building covers an area of 1106 $m^2$, and its building area is 6231.22 $m^2$ aboveground and 1069.92 $m^2$ underground, with a height of 23.75 m. The building is a commercial office building, with six floors aboveground and a one-floor basement, as shown in Fig. 4.15.

The basement includes a parking garage, equipment room, and other accessory rooms. The 1st floor includes a kitchen, a dining room, an exhibition area, an entrance hall, and a security equipment room. The 2nd to 5th floors are office rooms, occupied by design consulting companies engaged in design, consulting, engineering supervision, and project contracting; the 5th floor contains offices for company leaders. The 6th floor also contains offices, occupied by a real estate development company.

#### 4.5.2.1 Design Features for Energy Saving

*Energy-Saving Building Envelope System:* The building is L-shaped, as shown in Fig. 4.16. The east–west length reaches 17 m on the northeastern side, and the south–north length is 19 m on the southwestern side. The building faces south by east 10° and its shape coefficient is 0.23. The window-to-wall ratio is 0.67 on the east side, 0.66 on the south side, 0.08 on the west side, and 0.33 on the north side. The building information model (BIM) of the building is shown in Fig. 4.17.

The building envelopes were retrofitted according to the Design Standard for Energy Efficiency of Public Buildings in China. The outdoor walls were fitted with indoor and outdoor insulation with inorganic thermal insulation mortar (35 mm indoors and outdoors) and an average heat transfer coefficient of 0.85 $W/(m^2 \cdot K)$.

The roofing includes several types: a planted roof, a plane roof, and a metal roof with insulation material of spun glass wool (80/100 mm) and a resin composite plate (80 mm); its average heat transfer coefficient is 0.48 $W/(m^2 \cdot K)$.

**Fig. 4.15** Architectural appearance of the building (east)

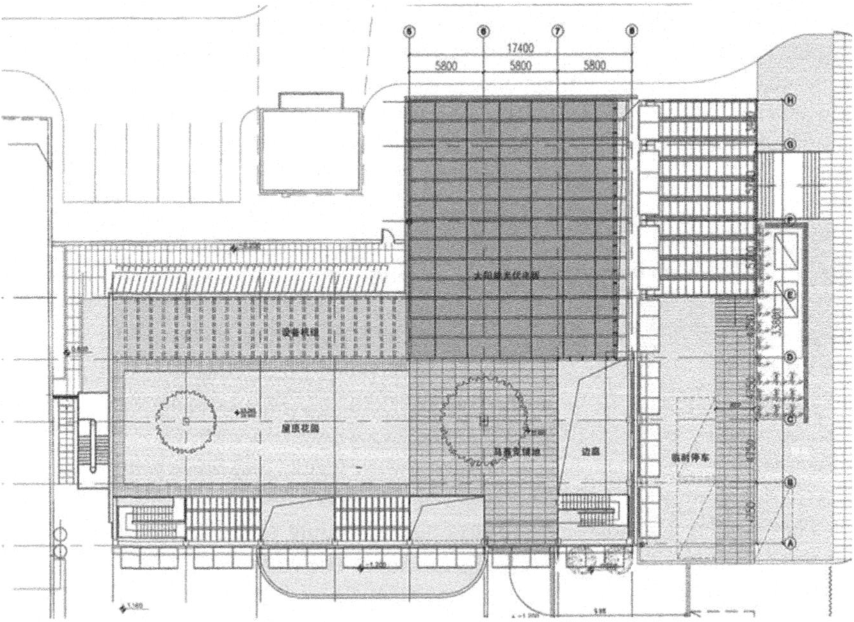


**Fig. 4.16** Overall plan of the building

**Fig. 4.17** Building information model (BIM) of the building

Considering the functions of heat insulation, shading, and lighting, insulated aluminum alloy windows with high-transmittance low-e insulating glass (6 + 12A + 6 shading type) was used. Its heat transfer coefficient is 2.00 W/($m^2$·K), and its overall shading coefficient is 0.594 with transmittance of 0.7.

#### 4.5.2.2 Passive Technologies for Energy Saving

The building makes full use of passive energy-saving technologies, consisting of natural ventilation, natural lighting, solar shading, etc.

*Natural Ventilation:* The building is located in a densely constructed area of the city, very near to surrounding buildings.

Although the surrounding environment of the building was unfavorable, several features have been incorporated into the scheme design, consisting of an atrium, windows, skylights, angled outside vertical sunshades, etc.

*Atrium Design:* There is a six-floor atrium in the building with a skylight at the top. The height of the atrium is 29.4 m, and the area of the atrium is 23 $m^2$, which is also a natural ventilation shaft. The ventilation shaft is 1.8 m higher than the roof, and the ratio of the height above the roof and the equivalent diameter of the atrium is 0.33, as shown in Fig. 4.18.

*Window Design:* The opening areas of the eastern and southern facades were increased by using mobile glass doors, because the increase in the opening area benefited the ventilation effect by wind pressure as the prevailing wind direction in Shanghai in the transition seasons is mostly southeasterly. The opening area ratio of the windows is 39.35%.

*Skylight Design:* The skylight is raised in order to enhance the ventilation effect by thermal pressure, as shown in Fig. 4.19. The opening faces the north, which is

**Fig. 4.18** Atrium of the building

**Fig. 4.19** Skylight of the building

located in the negative pressure zone to enhance discharge of wind. The opening area is 12 $m^2$ and the type is a top-hung window.

*Design of Outside Vertical Sunshades:* The eastern sunshades (vertical greening sunshades) are tilted by 30°, which is beneficial for ventilation.

*Natural Lighting:* The windows have been retrofitted, as shown in Fig. 4.20. The original windows in the building were of a traditional type, which was replaced by new windows outside the main function areas, and was reserved in equipment rooms, toilets and the northern facade. As a result, indoor daylighting has been improved and the operation of artificial lighting has been reduced.

The space of the entrance hall has been improved, as shown in Fig. 4.21. As the floor height of the 1st and 2nd floors is relative low, the main entrance was on the

**Fig. 4.20** Open large office areas (south)

**Fig. 4.21** Eastern entrance hall

northeastern side of the building and the spatial depth was relative large; direct light could not enter into the deepest part of the space, and the entrance hall was not wide. In order to solve the space requirements of the entrance hall, part of the floor slab of the 1st floor was cut through and solar light was brought in by outside windows in the entrance hall to improve the natural lighting environment of the northeastern indoor zone. In addition, windows rotating on a central axis were installed in the leisure and exhibition area on the southeastern side, which has increased the opening areas and types, as well as resolving the need for introduction of natural light.

As the building plan was L-shaped and the depth was large, a side-opening space was added on the south of the 2nd to 6th floors. The side-opening space includes vertical perforation of semi-outer space and has been enlarged floor by floor, which has not only formed a necessary slight transitional space but has also reduced the negative impact of direct light irradiation due to the large depth.

An atrium at the front of the elevator hall was added to the 3rd floor of the building, which improved the indoor environment of the deepest areas within the building by introducing natural light and ventilation, as well as satisfying functional requirements.

A sunken courtyard was added to the southeastern side of the 5th and 6th floors, which reduced the depth and width of the office areas. It improved the natural lighting effect in the indoor areas by introducing solar light effectively, as well as increasing the feeling of space.

Original office rooms separated by walls were changed into a large open space. Except for the 5th floor of small office rooms, the other floors were retrofitted as open large spaces without separation, which increased the space transparency. Thus, one-way natural light was turned into two-way natural light, improving the natural lighting effect in the inner space.

*Solar Shading:* Several methods of solar shading were put forward in the scheme design, considering the shading requirements in summer and heat gain in winter, as well as the influence on surrounding buildings.

The main applied methods were vertical sunshades and horizontal outward grating, applied on different facades (east and south).

*Vertical Sunshades:* Vertical sunshades slanted outward at 30° were applied on the eastern facade, as shown in Fig. 4.22, which satisfied the solar shading requirement in summer and had very little effect on the sunlight in winter. Besides, it could be greened to improve the microenvironment as well as enhancing the effect of the sunshades in summer and increasing the sunlight after the leaves fall in winter.

*Horizontal Outward Grating:* There was an outside corridor (the width was 3.9 m) on the southern facade, and the grating was designed for solar shading to improve the office environment.

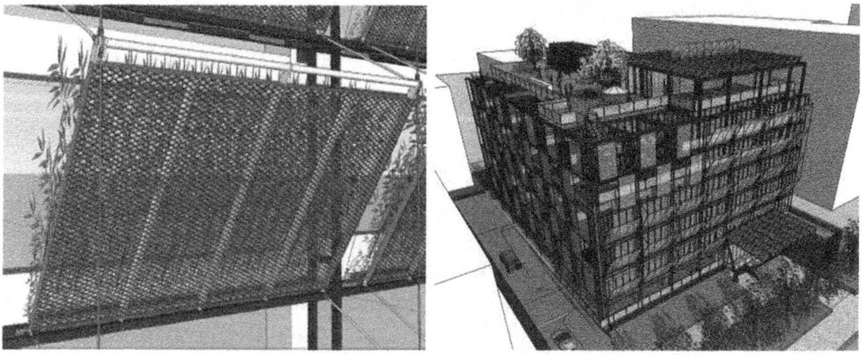

**Fig. 4.22** Effect of vertical sunshades

### *4.5.3 Tenio Office Building in Tianjin*

The Tenio Green Design Building is a project of architectural reconstruction, located at 17 Kaihua Way, Huayuan New Technology Industry Parks, Tianjin (Tsinghua University Building Energy Research Centre 2014). Before reconstruction, it was a five-floor electronics plant with an ordinary architectural appearance and no thermal insulation of the building envelope. Its natural ventilation was poor because of the large building depth, and its air conditioning system was a heating network with split units. After reconstruction in 2012, it became a six-floor office building with a total building area of 5700 $m^2$ and a height of 25 m. The 1st floor includes an exhibition area, meeting rooms, monument rooms, an equipment room, etc. The 6th floor includes a newly built gymnasium and office rooms. Figures 4.23 and 4.24 show the architectural appearance of the building before and after reconstruction, respectively.

#### 4.5.3.1 Design Features for Energy Saving

The energy-saving operation of the Tenio Green Design Building was the result of many factors, consisting of optimal design of the air conditioning system, with different strategies for different seasons, control modes for part of the time and parts of the space, air conditioning operation with natural ventilation and electronic fans, energy saving by human behavior, etc., which were introduced in the following ways.

**Fig. 4.23** Architectural appearance of the Tenio office building before reconstruction

**Fig. 4.24** Architectural appearance of the Tenio office building after reconstruction

*Design of the Air Conditioning System:* The cooling and heating source of the air conditioning system for the building is a combination of modularized ground source heat pumps (as shown in Fig. 4.25) and a water storage tank (as shown in Fig. 4.26). The heat pumps include two types: heat pump A, supplying chilled water of low temperature to the outdoor air handling units; and heat pump B, supplying high-temperature chilled water to radiant terminal devices. In this way, heat pump A is used to take off the moisture load, and heat pump B is used to remove the sensible load; therefore, they operate at high performance efficiency.

Since electricity prices in Tianjin vary tremendously, as shown in Table 4.8, water storage is used to save energy consumption costs. At off-peak times, cooling is stored in summer and heating is stored in winter; at peak times, the cooling or heating storage is employed for the air conditioning system.

Radiant floors and outdoor air handling units (AHUs) have been adopted as air conditioning terminals in the building. With vertical FCUs beside windows (as shown in Fig. 4.27) and ceiling fans, the air conditioning terminals have an effective mode of mixed ventilation.

*Operating Strategy:* As there are many methods used for thermal environmental control, such as natural ventilation, ceiling fans, free cooling, heat pump A, heat pump B, cooling or heating storage, etc., optimal combinations can be adopted in different seasons to reduce energy consumption. In the following sections, the operating strategies in the summer and winter of the year 2013 are introduced.

*Operating Strategy in Summer:* As it was not hot in early summer, natural ventilation was employed by opening windows in the first place. If natural ventilation

**Fig. 4.25** Modularized ground source heat pump

**Fig. 4.26** Water storage tank

**Table 4.8** Electricity prices in Tianjin

| | Peak time | | Ordinary time | Off-peak time |
|---|---|---|---|---|
| | Summer | Other seasons | | |
| Time | 1000–1100 hours | 0800–1000 hours | 0700–0800 hours | 2300–0700 hours |
| | 1900–2100 hours | 1600–1900 hours | 1100–2100 hours | |
| | | 2100–2300 hours | | |
| Price (yuan) | 0.9768 | 0.888 | 0.6025 | 0.333 |

**Fig. 4.27** Vertical fan coil unit beside windows

could not supply sufficient thermal environmental control, ceiling fans were used in combination with natural ventilation to enhance the heat transfer.

If ceiling fans did not work well enough, low-temperature water in underground pipes was used by heat transfer between the plate heat exchanger and radiant floor loops without opening the heat pump. In this way, the supply water temperature of the radiant floors was lowered by using free cooling.

When the outdoor air was very hot and humid, windows were closed and heat pumps were used to supply cooling with ceiling fans to achieve a good effect because the combination of natural ventilation, ceiling fans, and free cooling could not satisfy the need for indoor temperature and humidity control. Females and males experienced different thermal sensations from the radiant floors. As females had more tolerance of heat and also wore fewer clothes in summer, such as short shirts, short pants, and sandals, they would complain that it was too cold when males felt satisfied with the indoor temperature. In order to solve this problem, a combination of radiant floors and ceiling fans was adopted to improve the cooling effect. The indoor temperature was controlled at 26–27 °C, which satisfied the temperature requirement of females, and males also felt comfortable due to the mixed ventilation (as shown in Fig. 4.28 and Table 4.9).

*Operating Strategy in Winter:* In winter, radiant floors and FCUs were used as heating terminals. From 1100–0700 hours, heat pump A operated to store hot water, and during 0100–0700/0800 hours, heat pump B operated to supply hot water for radiant floors. In the daytime, heat pump A did not work and the hot water in the water tanks was used for FCUs on the 1st and 2nd floors during 0730–1700 hours, because the indoor temperature of the 1st and 2nd floors was relative low, thus it was necessary to raise the indoor temperature quickly by use of FCUs during office hours. The operating strategy of heat pump B depended on the weather. If it was sunny, heat pump B operated only during 1400–1730 hours to supply hot water for the radiant floors; if it was not sunny, heat pump B operated during 1100–1730 hours. As there was a lot of indoor heat generated by occupants, office devices, solar radiation due to the high window-to-wall ratio on the southern facade, heat storage of radiant floors operating during the night, etc., heating for the 3rd, 4th, and 5th floors was not required in the daytime (0800–1400 hours). The operating strategy in winter is summarized in Table 4.10 and Fig. 4.29.

It can be seen that the electricity consumption was reduced largely during peak electricity hours to reduce energy consumption costs (as shown in Fig. 4.30), as the hot water storage was used and heating for the 3rd, 4th, and 5th floors was not required in the daytime (0800–1400 hours).

**Fig. 4.28** Methods of controlling indoor temperature in summer

**Table 4.9** Operating strategy in summer

| Functions of ceiling fans | Shortened operating time of heat pumps | | Raised indoor temperature during cooling |
|---|---|---|---|
| Operating strategy | Natural ventilation and ceiling fans | Natural ventilation, radiant floors with free cooling, and ceiling fans | Heat pumps and ceiling fans |
| Operating time | Mid-May to mid-June | Mid-June to mid-July | August |
| Supply water temperature of radiant floors/ground surface temperature (°C) | 20/23 | | |

**Table 4.10** Operating strategy in winter

| Time | Strategy |
|---|---|
| 0100–0800 hours | Radiant floors supplied by heat pump B |
| 0730–1700 hours | Fan coil units on the 1st and 2nd floors supplied by water storage tanks |
| 1400–1730 hours (sunny) | Radiant floors supplied by heat pump B (utilizing solar radiation) |
| 1100–1730 hours (not sunny) | Radiant floors supplied by heat pump B |

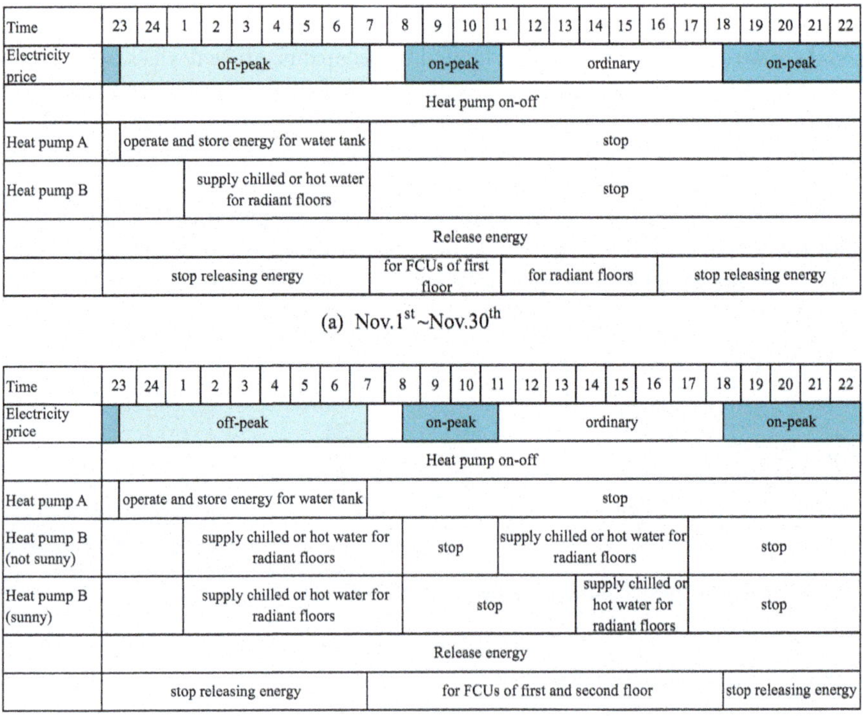


(a) Nov.1st~Nov.30th

(b) After December

**Fig. 4.29** Operating strategy for air conditioning devices during different periods in winter. **a** November 1–30. **b** After December

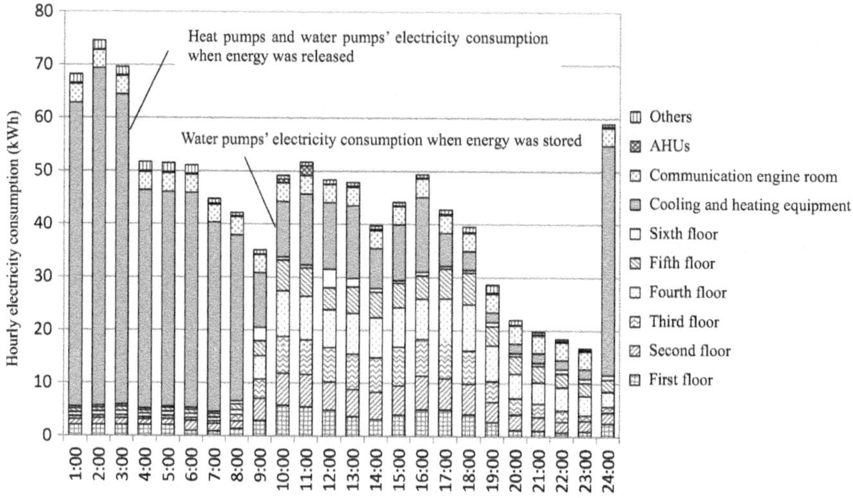


**Fig. 4.30** Hourly electricity consumption on November 25

## 4.6 Conclusions

This chapter introduces the Three-Star evaluation standards, energy performance of green office buildings, comparison of occupant satisfaction between green and common office buildings, and three typical green building cases in Shenzhen (a hot summer–warm winter zone), Shanghai (a hot summer–cold winter zone), and Tianjin (a cold zone).

The analysis of the energy performance of the green office buildings indicates that for building type A (mixed mode), it is only in the hot summer–cold winter zone that green buildings' energy consumption is statistically significantly less than that of common buildings. For building type B (mechanically conditioned), no statistically significant difference between green and common buildings is observed in any of the climate zones. It can also be seen that in the hot summer–cold winter zone, the average energy consumption values of building types A and B are 25% and 23% lower, respectively, than the upper limit values required by the Chinese standard.

In terms of the IEQ and BSP, the satisfaction of users in green office buildings is markedly higher than that of users in common office buildings in the aspects of thermal comfort, IAQ, facilities, operation, and maintenance.

## References


Boerstra AC, van Hoof J, van Weele AM (2015) A new hybrid thermal comfort guideline for the Netherlands: background and development. Archit Sci Rev 58(1):24–34

BRE (2016) What is BREEAM. Online [Available at] http://www.breeam.com

China Society for Urban Studies (2016) China green building. China Architecture & Building Press, Beijing. (in Chinese)

IBEC (2015) Online [Available at] http://www.ibec.or.jp/CASBEE/english/

Lin B, Liu Y, Wang Z, Pei Z, Davies M (2016) Measured energy use and indoor environment quality in green office buildings in China. Energy Buildings 129:9–18

MOHURD (2014) Evaluation for green building. China Architecture & Building Press, Beijing. GB/T 50378-2014 (in Chinese)

MOHURD (2016) Standard for energy consumption of building. China Architecture & Building Press, Beijing. GB/T 51161-2016 (in Chinese)

National Development and Reform Commission (2015) Enhanced actions on climate change: China's intended nationally determined contributions (http://www.sdpc.gov.cn/xwzx/xwfb/201506/t20150630_710204.html)

Newsham GR, Mancini S, Birt BJ (2009) Do LEED-certified buildings save energy? Yes, but…. Energ Buildings 41:897–905

Overview of the DGNB (2016) Online [Available at] http://www.dgnb.de/en/council/dgnb/

Pei Z, Lin B, Liu Y, Zhu Y (2015) Comparative study on the indoor environment quality of green office buildings in China with a long-term field measurement and investigation. Build Environ 84:80–88

Tsinghua University Building Energy Research Centre (2009) Annual report on China building energy efficiency. China Architecture & Building Press, Beijing. (in Chinese)

Tsinghua University Building Energy Research Centre (2014) Annual report on China building energy efficiency. China Architecture & Building Press, Beijing. (in Chinese)

US Green Building Council (2009) Foundations of LEED

US Green Building Council (2012) Foundations of LEED

US Green Building Council (2013) LEED v4—the next version of LEED

Xiao H (2011) Study on distribution features and influencing factors on energy use in office buildings by statistical method and survey. Tsinghua University, Beijing. (in Chinese)

Yi J, Xiu Y (2010) Electricity equivalent conversion method in energy analysis. Energy of China 32:5–11. (in Chinese)

# Chapter 5
# A Preliminary Comparative Study on Subtropical Ecological Community Indicators: EEWH-EC

**Ying-Ming Su and Shu-Chen Huang**

**Abstract** Rapid urban sprawl and resource exploitation have led to unprecedented challenges to the global environment. An ecological community, which is the integration of architecture, ecology, and society, includes various aspects such as people, construction, the environment, and overall development. The interdisciplinary integration of ecological cogitation and technology reflects harmonious coexistence between man and nature, and also becomes a development model for future sustainable settlement. In 2011, a green building ecological community assessment system, EEWH-EC (Ecology, Energy Saving, Waste Reduction, Health: Eco-communities), for subtropical regions, was promulgated in Taiwan. In this work, the ecological assessment systems used in the UK, the USA, Germany, Japan, and China are reviewed. The categories, items, content, evaluation criteria, and weightings of these assessment systems are studied and analyzed. The results show that the EEWH-EC indicators give more emphasis to ecology and to the living and natural environments. The weightings of the indicators have not been developed yet, therefore the degree of importance of each category and item cannot be determined. With the ecological community indicators established by different countries, social, economic, and humanistic indicators can be strengthened. With reference to the established international eco-community assessment systems, it will be more favorable for the overall development of ecological communities to formulate assessment systems better suited to subtropical regions.

**Keywords** Eco community · Sustainable city · EEWH-EC · Subtropical regions · Community sustainability assessment

Y.-M. Su (✉) · S.-C. Huang
Department of Architecture, National Taipei University of Technology, Taipei, Taiwan

Graduate Institute of Architecture and Urban Design, National Taipei University of Technology, Taipei, Taiwan
e-mail: ymsu@ntut.edu.tw; moximisty@livemail.tw



Z. Shen et al. (eds.), *Green City Planning and Practices in Asian Cities*, Strategies for Sustainability, https://doi.org/10.1007/978-3-319-70025-0_5

## 5.1 Introduction

### *5.1.1 Motivation and Objectives*

Urban environmental quality and ecological value have become victims of rapid urban expansion and nature resource exploitation (Ting-Fen, Liu and Ching-Nan Chen 2011). Because of the increasing effects of ozone depletion, greenhouse gases, and heat islands, the global environment is facing an unprecedented ecological catastrophe. The concept of sustainable ecology has gradually become employed in architectural design and urban planning. Cities can be integral parts of nature and also can be the key to solving urban environmental problems (Leitmann 1999). The objective of eco-community indicators is to construct green buildings that coexist with ecology, environmental protection, and sustainability to create ecological and sustainable cities. Connecting ecological communities among cities aggregates the foundation of the eco-city planning concept (Hsien-Te Lin 2008).

Although green building index systems such as LEED (Leadership in Energy and Environmental Design; USA, established in 1988) and BREAM (Building Research Establishment Environmental Assessment Method; UK, established in 1990) were developed much earlier and are therefore more mature than those in other countries, some of their indicators are not suitable for subtropical regions because those countries are located in temperate zones. Taiwan is a hot and humid island located in a subtropical region, with different climate conditions from those in the USA, UK, Germany, Japan, etc. Taiwan's green building assessment system, EEWH (Ecology, Energy Saving, Waste Reduction, Health) was established in 1999. It is the fourth quantization assessment system in the world, after the US, UK, and Canadian systems, and also is the first assessment system in a subtropical region. EEWH is an important reference for countries in subtropical regions to establish their own green building systems (Yi-Ping Tsai 2014). The purpose of this research is to encourage more effective and consummate development of these eco-community assessment systems.

In addition, the ABRI established an EEWH system family, which is capable of coping with various building types. The five individual systems and their corresponding evaluation manual revisions—including Basic (EEWH-BC) for general green building practices, Residential Building (EEWH-RS), Factory (EEWH-GF), Renovation (EEWH-RN) for existing buildings, and Eco-communities (EEWH-EC)—were all completed in 2012 (Fig. 5.1).

### *5.1.2 Scope and Methods*

By reviewing the relevant eco-community literature and theories, this research compares and analyses the indicators of EEWH-EC with those of BREEAM Communities (Building Research Establishment Environmental Assessment Method [BREEAM]

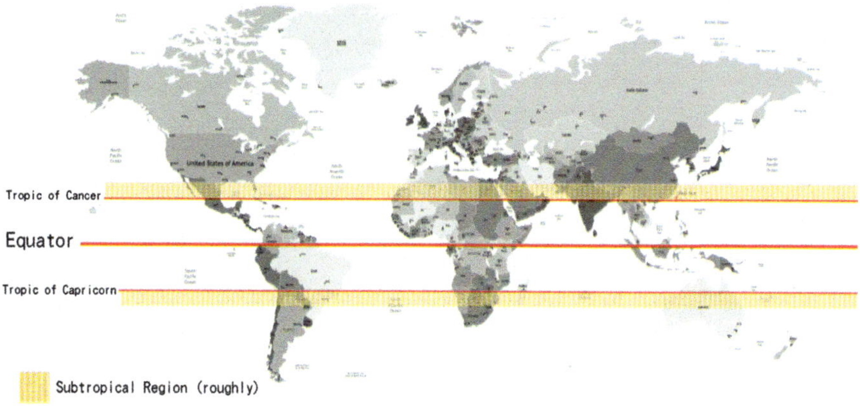


**Fig. 5.1** Subtropical region (*shaded area*). Source: (https://www.google.com.tw/search?q=%E4%B8%96%E7%95%8C%E5%9C%B0%E5%9C%96&rlz=1C1OPRB_enTW581TW584&tbm=isch&tbo=u&source=univ&sa=X&ved=0ahUKEwi14JHK78zXAhULtJQKHUZ5AxYQiR4IwgE&biw=1920&bih=974#imgrc=GaArC6EaxfZXRM:)

for Communities), LEED-ND (Leadership in Energy and Environmental Design: Neighborhood Development), CASBEE-UD (Comprehensive Assessment System for Building Environment Efficiency [CASBEE] for Urban Development), DGNB-NSQ (Deutsche Gesellschaft für Nachhaltiges Bauen: Neubau Stadtquartiere), and EIAS-GC (Evaluating Indicators and Assessing Standards for Green Communities). The structures of those six assessment systems are similar. The second-level indexes target sustainability issues. Points in the certification process accumulate through the fulfillment of index requirements. During an index-based comparison, the relation of the systems to the different aspects of sustainability becomes evident. The indexes contain an indicator of a combination of indicators (Hecht 2007). That main feature of the indicators is their ability to focus and summarize a complex issue into one measure (Singh et al. 2012).

## 5.2 Literature Review

### *5.2.1 Sustainable Development and Eco-community Assessment*

The significance of sustainability in the construction industry, especially in urban areas, has long been recognized (European Union 2007). Urbanization and its rapid increase are major concerns due to their detrimental effects on the environment (Register 2010; Jaeger et al. 2010). Cities play a key role in social economic development and sustainability technology development. In order to achieve

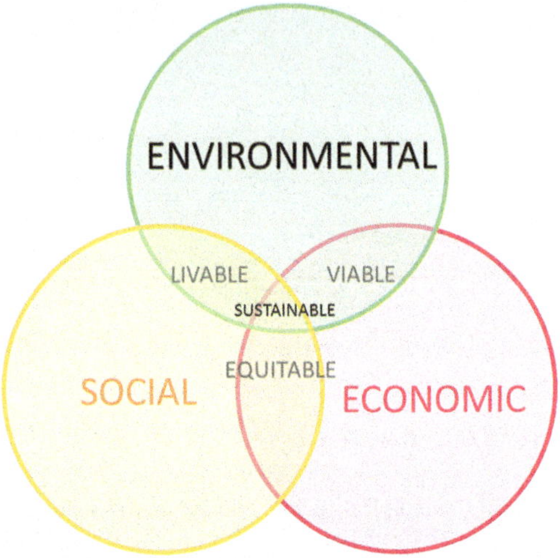


**Fig. 5.2** Pillars of sustainability (Tanguay et al. 2009)

sustainability of city goals, assessing the building alone is not enough. Sharifi and Yamagata (2014) and Choguill (2008) stated that if communities—the constituent units of a city—do not achieve their sustainability goals, the city is unable to achieve sustainable development. Neighborhoods are areas of cities that have their own architectural, cultural, and economic systems. They are limited to a specific area, and a common consciousness exists among their habitants (Berg and Nycander 1997). It is necessary to build a community assessment system to improve neighborhood sustainability, with consideration of their buildings, public spaces, and infrastructure (Luederitz et al. 2013).

The Brundtland Report (titled *Our Common Future*) from the World Commission on Environment and Development (WCED) in 1987 declared many definitions of sustainable development (Langeweg 1998). Though there are various interpretations, the main content always addresses three pillars: social development, environmental protection, and economic development (Fig. 5.2) (Tanguay et al. 2009). A neighborhood's sustainability assessment system is a tool that evaluates the sustainability performance of a given neighborhood against a set of criteria (Sharifi and Murayama 2013). According to Sharifi and Murayama's idea (2012), there are many types of urban sustainability assessment systems used as decision-making tools embedded into neighborhood-scale planning (e.g., HQE2R, Ecocity, SCR, EcoDistricts, SPeAR, One Planet Living, Cascadia Scorecard, EcoDistricts Performance and Assessment Toolkit). Efforts to promote sustainability in urban areas are present mainly in governmental programs, policies, and incentives (Shen et al. 2011). The importance of sustainable cities becomes more and more significant. A mandatory sustainability assessment system is considered a symbol of a developed country.

The concept of sustainable development germinated after the Second World War. Highly committed to economic development, industrialized countries abused natural resources. This triggered academic discussion (Caldwell 1994). Mollison (1989)

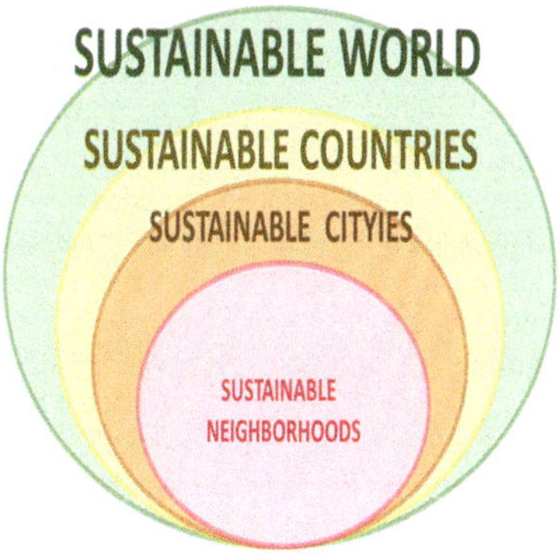


**Fig. 5.3** Sustainable development dimensions (A. Zeinal Hamedani, F.H, 2012)

constructed the concept of sustainable cultivation based on the idea of ecology and economics (Chin-Hsiung Hou 1996). In 1972, the UN Conference on the Human and Environment announced the *Declaration of Human Rights*. This caused people to pay much more attention to environmental problems. Since the WCED published *Our Common Future* in 1987 and the Earth Summit (United Nations Conference on Environment and Development [UNCED]) announced its *Agenda 21* in 1992, many countries have set up sustainable development indicator systems at different levels (Heng Chang and Chih-Hang Hsing 2004). Sustainable development ideas have been carried out at global, national, city, and community levels. The more advanced countries run sustainable development programs with more extensive dimensions (Fig. 5.3).

The main concepts of sustainability focus on extending the logic of capital and commercialization to survive ecological conditions (Stefan 1996). The combination of sustainability and a community plan would be tantamount to integration of ecology and capital reproduction strategies. With regard to the direction of sustainable development, the coordination of complete development and ecosystems can be achieved. The purpose of the ecological community is to introduce the concept of ecological living into remodeling and new construction. This brings new understanding of the living environment from the ecological aspect.

### *5.2.2 Principles of Sustainability Assessment Indicators for Eco-communities*

Although there are similar concepts, the scope of a sustainable city is more extensive and comprehensive than that of an eco-city (Chang-Lin Hua 1998). Not being limited only to a single building, evaluation of sustainable cities must be regarded as a complete system that includes a wide range of all environments, activities, and composition (Sharifi and Yamagata 2014). Therefore, such an evaluation system must include economic sustainability, social sustainability, and environmental sustainability (of natural resources and ecology)—the definition of the three pillars of

**Table 5.1** Definition of the three pillars of sustainability (Gillbert et al. 1996)

| Economic sustainability | Environmental sustainability | Social sustainability |
|---|---|---|
| This occurs when development, which moves toward social and environmental sustainability, is financially feasible | This requires that natural capital remains intact. This means that the source and sink functions of the environment should not be degraded. Therefore, the extraction of renewable resources should not exceed the rate at which they are renewed, and the absorptive capacity of the environment to assimilate wastes should not be exceeded. Furthermore, the extraction of nonrenewable resources should be minimized and should not exceed agreed minimum strategic levels | This requires that the cohesion of society and its ability to work toward common goals be maintained. Individual needs, such as those for health and well-being, nutrition, shelter, education, and cultural expression, should be met |

sustainability (Gillbert et al. 1996) (Table 5.1). An eco-city evaluation system is focused on economic, social, and special related indicators, based on the purpose of environmental sustainability (Chang-Lin Hua 1998).

## 5.3 Major International Eco-community Assessment Systems

### *5.3.1 BREEAM Communities (UK; 1990)*

BREEAM was the first evaluation system on earth, developed by the Building Research Establishment (BRE) in England in 1990. The system has become an important reference for other countries trying to develop assessment indicators. In order to reduce environmental impact, the system covers the ecological environment including buildings, natural resources, and land. The success of the assessment system depends on support from the UK government in terms of policy and funding. Moreover, government policies always pay much attention to the local economic system and welfare. The indicators related to social welfare and economic influence have been given the greatest weightings in the evaluation system.

### *5.3.2 LEED-ND (USA; 1998)*

LEED, the US green building rating system published by the US Green Building Council (USGBC) in 1998, is used for evaluating building performance through widely accepted standards and evaluation criteria. After 10 years of modification,

LEED has been developed as a series of assessment tools to adapt to different types and life stages of building construction. LEED-ND (Neighborhood Development) is an evaluation tool developed by the Congress for the New Urbanism (CNU), Natural Resources Defense Council (NRDC), and USGBC (Hurley and Horne 2006). It encourages innovation design practices and technology to resolve particular geographical environment, social justice, and public health priorities in line with local conditions.

### 5.3.3 DGNB-NSQ (Germany; 2007)

The DGNB evaluation system is sponsored by the private sector. Nonprofit organizations serve the environment and people as their own responsibility. The goals of the regulations are to promote environmental protection and to guarantee human health and social culture. NSQ (*Neubau* means new construction, reconstruction, or construction of a new development; *Stadtquartiere* means cities, near the local, regional, neighborhood, community or housing), broadly speaking, means "new town development." Unlike other countries' indicators, which are difficult to apply to large-scale urban or community developments, DGNB can be utilized from small-scale to large-scale building developments through the project life cycle. DGNB-NSQ focuses on the Life Cycle of Architecture (LCA) and Life Cycle Costing (LCC) as an assessment framework.

### 5.3.4 CASBEE-UD (Japan; 2002)

CASBEE is an assessment system by which the government of Japan, referring to the experience of implementing GBTool, links industry, government, academia, and research to accelerate the growth of domestic sustainable buildings. The development of CASBEE for Urban Development (CASBEE-UD) makes a broader assessment of the overall analysis for cities. Compared with other eco-community assessment systems, CASBEE-UD has much broader coverage. It emphasizes environmental performance efficiency. From environmental quality to environment load control and local management, the assessment indicators have detailed formulas and instructions.

### 5.3.5 EEWH-EC (Taiwan; 1999) and EIAS-GC (China; 1999)

EEWH was the first Green Building Assessment Manual and Green Building Label announced by the government of Taiwan in 1999. It has developed a family of EEWH systems (ABRI, MOI, Taiwan, 2012). The purpose of EEWH-EC is to account for the ecological quality and community function of a community. "Ecological quality" includes the content of species diversity and sustainable human survival. "Community function" includes the most basic living functions of convenience, health, comfort, efficiency, security, and culture (EEWH-EC Assessment Manual 2015).

Almost at the same time, EIAS-GC, declared by the Minister of Environmental Protection in China during the *Five-Year Plan* for 2001–2005, was used to promote grassroots participation in the establishment of ecological communities at all levels—from the nation to provinces to cities. The first edition of the official system was published in 2001. It was revised in 2007 and renamed *China Ecological Residential Technology Assessment Manual, 2007*. Due to the large regional differences, each country has its own evaluation criteria (Ecological Residential Community Building Guidelines 2001). Because the large land area of China includes five different climatic zones, uneven distribution of resources, and rapid expansion of building, the content of EIAS-GC focuses on the aspects of resources, energy, and construction.

## 5.4 Analysis and Comparison of Assessment Indicators

LEED: Neighborhood Development v.4 (2014), BREEAM Communities (2012), CASBEE: Urban Development (2014), DGNB: Neubau Stadtquartiere (2012), EIAS-GC (2007), and EEWH: Eco-communities (2015) are analyzed and compared within sections in this chapter. Particularly, the basic information, including the development agencies, development year, scoring patterns, and ranks of those assessment systems have been compared in this work.

By comparing and analyzing the category items of those assessment systems, we investigate classification factors for categorizing the indicators. We group indicators and indices of each neighborhood sustainability assessment system into topics based on the principles they follow. To make the huge number of indicators comparable, a common categorization system is required. For this purpose, we have aggregated the list systematically for nine classifications: buildings, community, ecology, economy, energy, infrastructure, location, resources, and transportation. Compared with the number of mandatory evaluation indicators of those systems, it helps us to figure out the relationship between environmental characteristics and assessment system design. After classifying optional indicators and comparing the weightings of new categories, the intentions of each assessment system are investigated (Fig. 5.4).

### *5.4.1 General Comparison*

Besides EIAS-GC, EEWH, and CASBEE, operated by governments, the other three evaluation systems are run by the private sector. Though BREEAM was launched by the UK government, it was subsequently executed by the private sector. The UK

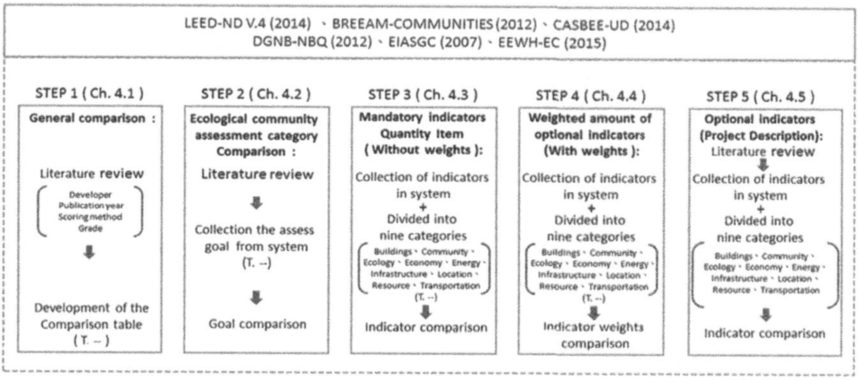


**Fig. 5.4** Five simplified analysis steps

government has become a strong supporter in terms of policy and funding. This could be a model for cooperation between the public and private sectors for evaluating system operation. Because people are generally unfamiliar with these eco-community evaluation systems and without or slightly government finical support, EIAS-GC, EEWH, and CASBEE, operated by governments, are not as widely used in practice as LEED, BREEAM, and DGNB. In terms of rating systems, CASBEE and EEWH have numerical formulas.

CASBEE determines the grade by the ratio between the environmental quality and the environmental impact. EEWH, with a formula for each evaluation item, calculates scores for all items following the formula, and adds up those scores to determine the performance grade. Although the evaluation process is relatively complex, relatively accurate data are calculated. LEED, BREEAM, DGNB, and EIAS, totaling all the points obtained from each item, are much simpler and clearer. These six evaluation systems divide performance grades into 3–5 levels.

Because BREEAM, CASBEE, and EEWH have smaller classes, they are more precise in grade determination (Table 5.2).

## 5.4.2 *Classifications of Eco-community Assessment Systems*

Due to differences in regional conditions and geographic locations, classifications are compared among evaluation systems (Table 5.3).

The classification of DGNB is the most extensive and comprehensive consideration, which covers the economy, environment, social affairs, and technology quality. Although it has only two categories, CASBEE emphasis the ratio of environmental quality and the environmental load in order to achieve maximum comfort improvement with a minimal environmental load.

The focus of LEED is innovation. In order to solve local geographical environment, social justice, and public health problems, LEED encourages innovative

**Table 5.2** Basal comparison of eco-community assessment systems

| | LEED | BREEAM | CASBEE | DGNB | EIAS-GC | EEWH-EC |
|---|---|---|---|---|---|---|
| Developer | USGBC | BRE Global Ltd. | JaGBC, JSBC | GeSBC | China Building Industry | Taiwan |
| Publication years | 2009 | 2000 | 2001 | 2008 | 2001 | 2011 |
| | 2014 (new) | 2012 (new) | 2014 (new) | 2012 (new) | 2007 (new) | 2015 (new) |
| Scoring methods | Point system | Point system | Special | Point system | Point system | Special |
| | Regional weighting ≧ target | Weighting of issues ≧ target score | Issue score ≧ special target score | Weighting of issues ≧ target score | Target score | Issue score ≧ special target score |
| Grades | Platinum | Outstanding | S = excellent | Gold | 3 stars | Diamond |
| | Gold | Excellent | A = very good | Silver | 2 stars | Gold |
| | Silver | Very good | B+ = good | Bronze | 1 star | Silver |
| | Certified | Good | B− = fairly poor | | | Copper |
| | | Pass | C = poor | | | Certified |

*BREEAM* Building Research Establishment Environmental Assessment Method, *CASBEE* Comprehensive Assessment System for Building Environment Efficiency, *DGNB* Deutsche Gesellschaft für Nachhaltiges Bauen, *EEWH-EC* Ecology, Energy Saving, Waste Reduction, Health: Eco-communities, *EIAS-GC* Evaluating Indicators and Assessing Standards for Green Communities, *LEED* Leadership in Energy and Environmental Design

**Table 5.3** Classification of evaluation systems

| System | Category | System | Category |
|---|---|---|---|
| BREEAM | Governance | CASBEE | Environmental quality |
| | Social and economic well-being | | Environmental load |
| | Resources and energy | | |
| | Land use and ecology | | |
| | Transport and movement | | |
| LEED | Smart location and linkage | EIAS-GC | Location and residential environment |
| | Neighborhood pattern and design | | Energy conservation and environment |
| | Green infrastructure and buildings | | Indoor environmental quality |
| | Innovation and design process | | Residential water environment |
| | Regional priority credit | | Materials and resources |
| DGNB | Ecological quality | EEWH-EC | Ecology |
| | Economic quality | | Energy conservation and waste reduction |
| | Sociocultural and functional | | Health and comfort |
| | Technical quality | | Service function |
| | Technological quality | | Crime prevention |

*BREEAM* Building Research Establishment Environmental Assessment Method, *CASBEE* Comprehensive Assessment System for Building Environment Efficiency, *DGNB* Deutsche Gesellschaft für Nachhaltiges Bauen, *EEWH-EC* Ecology, Energy Saving, Waste Reduction, Health: Eco-communities, *EIAS-GC* Evaluating Indicators and Assessing Standards for Green Communities, *LEED* Leadership in Energy and Environmental Design practice and technology use in the overall assessment. Although LEED, BREEAM, CASBEE, DGNB, and EIAS-GC cover location, socioeconomic, and environment categories, EEWH does not evaluate these items. This may be due to Taiwan's high population density, high building density, and close relationship between communities. The content of EEWH demonstrates the consideration of the quality, function, and ecology of life within communities.

### *5.4.3 Mandatory Assessment Indicator Analysis (Weightings Not Included)*

According to Fig. 5.5, the indicators of the six evaluation systems can be classified into nine categories. Although mandatory indicators are important parts of evaluation systems, only LEED, BREEAM, and EIAS-GC have mandatory items; CASBEE, DGNB, and EEWH do not. EIAS-GC consists of the largest number of mandatory indicators, particularly in the buildings, location, and energy categories. To address the rapid expansion of the population and buildings, China has the most restrictions in terms of buildings. Because of the topographic, geological, and restricted residence, location needs to be evaluated carefully. Having a large land area, uneven distribution of resources, and high population density, China considers energy reserves and renewal to be important items for energy supply.

Although EIAS-GC has the most mandatory items, it covers only six categories (buildings, ecology, energy, infrastructure, location, and resources). There is no mandatory restriction in the community, economy, and transportation categories, and there are the most mandatory provisions in the ecology and location categories in LEED. Located in the temperate regions, the USA has a wide variety of animals and plants. Location selection considers the coordination of the site and local ecosystem—that is, it does not destroy the balance between the local environment,

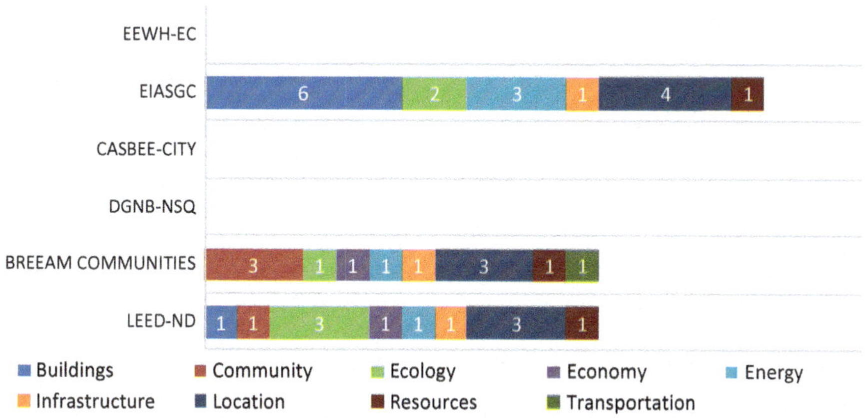


Fig. 5.5 Numbers of mandatory eco-community assessment indicators (weightings not included)

the ecology, and the plants. With the exception of transportation, LEED has mandatory items in the remaining eight categories (buildings, community, ecology, economy, energy, infrastructure, location, and resources). BREEAM is focused on the community and location. Even though there is no mandatory prerequisite for the building category, there are mandatory items in the transportation sector. The UK has a lot of land, flat terrain, buildings are scattered here and there, large distances between regions, and historical heritage; therefore, location selection and transportation between communities are considered carefully.

LEED and BREEAM are widely used in practice and are highly visible, since their mandatory items are distributed in categories much wider than those in EIAS-GC.

### *5.4.4 Weighted Numbers of Optional Indicators*

Weightings may show the focus and the relative care issues for each indicator. The topics and issues concerning each assessment system are shown in Table 5.4. After classification and analysis, our findings are as follows.

LEED gives greater weightings to buildings, infrastructure, and transportation categories. Because the evaluation system is concerned that buildings have a long life cycle and greater impacts on the environment, the buildings category and infrastructure category have greater weightings (35.94% and 15.99%, respectively), followed by the transportation category (13.65%).

BREEAM gives the ecology category the greatest weighting (25.25%). Since BREEAM was the first complete evaluation system to be developed, and has strong support in terms of government policy and funding, there is a lesser weighting in the energy category (0.35%).

**Table 5.4** Weighted numbers of optional indicators in each category

| Category | LEED (%) | BREEAM (%) | DGNB (%) | CASBEE (%) | EIAS-GC (%) | EEWH-EC (number of items) |
|---|---|---|---|---|---|---|
| Buildings | 35.94 | 9.11 | 12.1 | 11.51 | 31.65 | 2 |
| Community | 3.82 | 9.76 | 6.33 | 4.16 | 0.14 | 13 |
| Ecology | 8.87 | 25.25 | 20.39 | 23.84 | 9.17 | 14 |
| Economy | 2.87 | 16.07 | 13.07 | 0 | 0 | 1 |
| Energy | 6.12 | 0.35 | 7.12 | 6.99 | 15.6 | 2 |
| Infrastructure | 15.99 | 7.35 | 10.6 | 17.01 | 1.74 | 7 |
| Location | 7.96 | 12.66 | 21.02 | 12.66 | 2.66 | 13 |
| Resources | 4.78 | 14.35 | 5.29 | 14.35 | 33.3 | 3 |
| Transportation | 13.65 | 9.48 | 4.07 | 9.48 | 5.73 | 5 |

*BREEAM* Building Research Establishment Environmental Assessment Method, *CASBEE* Comprehensive Assessment System for Building Environment Efficiency, *DGNB* Deutsche Gesellschaft für Nachhaltiges Bauen, *EEWH-EC* Ecology, Energy Saving, Waste Reduction, Health: Eco-communities, *EIAS-GC* Evaluating Indicators and Assessing Standards for Green Communities, *LEED* Leadership in Energy and Environmental Design

DNGB assigns greater weightings to location (21.02%) and ecology (20.39%), and then economy (13.07%). Because Germany sets sustainable development as a goal and the three pillars (the economy, environment, and society) as the direction, the intention of the distribution of category weightings is apparent.

CASBEE assigns large proportions of weighting to ecology (23.84%) and infrastructure (17.01%). Based on good living environments and strict controls, there are good living conditions, community systems, and sustainable resources. A system focused on all factors may affect the environment.

EIAS-GC gives the resources category (33.3%) the greatest weighting, followed by the buildings category (31.65%). Due to the high population density, uneven distribution of resources, and increased building volume, the weightings of the resources category and the buildings category are substantial in order to maintain quality of life and equal distribution of resources.

EEWH is assessed by numbers of items, since no weighting is assigned. To take ecological quality and community functioning into account, the evaluation system gives the ecology category the greatest weighting (14 items) and gives community and location the second greatest weightings (13 items).

### 5.4.5 *Optional Indicators*

In this section, symbols are used to indicate the results of the assessment. There are four grades represented by the symbols: excellent, good, medium, and none. Following the classification in nine classifications, the items in the six eco-community assessment systems are reintegrated and reanalyzed again (Table 5.5). Based on the classification, the items in all evaluation systems are explored and compared.

After sorting and analyzing all the items in the six systems, this study finds that the content and the structure of an evaluation system reflect the initial attitude and purposes of the government.

1. Buildings

The content and the standards of the evaluation items for architectural quality are defined clearly in all assessment systems. Since those evaluation systems focus on the development of the whole region and the reuse of existing buildings, there is no item related to aesthetic quality and innovation. In LEED, BREEAM, DGNB, and EIAS-GC, items related to calculation of formulas and auxiliary policy are defined clearly.

2. Community

Except for EIAS-GC, all assessment systems have the most items related to enhancement of well-being and then items related to community resources/history and culture. Although LEED has provisions related to community links, it does not specify the qualification and the assessment standard for indicators. BREEAM has perfect standard norms and conditions of indicators for the community category.

**Table 5.5** Comparison of indicators in the six eco-community assessment systems

| | Assessment indicator | LEED | BREEAM | CASBEE | DGNB | EIAS-GC | EEWH-EC |
|---|---|---|---|---|---|---|---|
| Buildings | Building certification | ◎ | ◎ | ○ | ◎ | ◎ | ○ |
| | Aesthetic quality | – | △ | – | – | – | – |
| | Accredited professional design | ◎ | – | △ | △ | – | – |
| | Sustainable construction | – | – | – | ◎ | △ | – |
| | Architectural quality | ◎ | ◎ | ○ | ◎ | ◎ | ○ |
| | Innovation | ○ | – | – | – | – | – |
| | Reuse of existing buildings | ○ | ○ | – | – | ○ | ○ |
| | Density development | ○ | – | – | – | ○ | △ |
| Community | Community links | △ | – | △ | – | – | |
| | Community participation | △ | – | – | – | – | △ |
| | Community resources/history and culture | ○ | ◎ | ◎ | – | ◎ | ◎ |
| | Community management of facilities | – | ◎ | – | △ | – | ○ |
| | Lively neighborhood | ○ | ◎ | ○ | – | – | ◎ |
| | Preservation of local identity | ◎ | ◎ | – | ◎ | – | – |
| | Enhancement of well-being | ○ | ◎ | △ | ◎ | – | △ |
| Ecology | Ecology | △ | – | ○ | – | △ | ◎ |
| | Soil protection/use | ◎ | ◎ | ◎ | ◎ | ○ | ○ |
| | Minimum green area | – | – | ◎ | – | △ | ◎ |
| | Environmental impact/biological survival | – | – | ◎ | ◎ | △ | ◎ |
| | Local water conservation | – | – | – | – | – | ○ |
| | Enhanced site ecology | – | ◎ | – | – | – | ◎ |
| | Native biodiversity protection | – | ○ | △ | – | △ | ◎ |
| | Biological habitat | ○ | – | – | – | – | ◎ |

| | | | | | | | |
|---|---|---|---|---|---|---|---|
| Economics | Life-cycle cost | – | – | – | ◎ | – | – |
| | Monitoring | – | – | – | ○ | △ | – |
| | Governmental involvement | – | ◎ | ○ | – | – | – |
| | Quality control | – | – | ◎ | ◎ | – | – |
| | Maintenance | ◎ | ◎ | ◎ | ◎ | △ | △ |
| | Operating and maintenance (costs) | – | – | – | ◎ | – | – |
| | Affordability of residential houses | ○ | – | – | – | – | – |
| | Marketing | – | – | ○ | ◎ | – | – |
| | Local labor | ○ | ○ | – | – | – | – |
| | Business growth | – | △ | ○ | ○ | – | – |
| Energy | Energy efficiency | ◎ | – | – | ◎ | – | ◎ |
| | Energy performance | ◎ | ◎ | ◎ | – | ◎ | – |
| | Renewable energy technology | – | ◎ | △ | – | △ | ○ |
| | Energy monitoring | ◎ | – | △ | ◎ | – | △ |
| Infrastructure | Management, quality of service | – | – | – | ○ | △ | – |
| | Integrated/sustainable design | ◎ | ◎ | ○ | ○ | – | – |
| | Appropriate car parking | – | ◎ | ○ | – | △ | △ |
| | Quality of public spaces | – | – | – | ○ | △ | – |
| | Transportation infrastructure | – | ◎ | – | ◎ | △ | ○ |
| | Connection to the urban environment | – | – | – | △ | – | – |
| | Alternative car usage (electric vehicles, bicycles) | ◎ | ◎ | ◎ | – | – | |

(continued)

**Table 5.5** (continued)

| | Assessment indicator | LEED | BREEAM | CASBEE | DGNB | EIAS-GC | EEWH-EC |
|---|---|---|---|---|---|---|---|
| Location | Noise reduction | ○ | ◎ | ○ | ○ | ◎ | ◎ |
| | Lighting and illumination | ◎ | ◎ | ◎ | ○ | ◎ | ◎ |
| | Good air quality | – | ◎ | ○ | – | – | ○ |
| | Natural ventilation | – | – | △ | – | ○ | ○ |
| | Climate change | – | ◎ | – | – | – | – |
| | Disaster prevention | – | | ◎ | △ | – | △ |
| | Pollution and risks | – | – | – | – | △ | △ |
| | Emissions reduction | – | ○ | – | – | – | △ |
| | Natural disasters (flooding, hurricanes, etc.) | – | ◎ | – | – | – | |
| | Heat island effect | ◎ | – | – | – | – | ◎ |
| | Smart location | ◎ | – | – | – | – | – |
| | Sustainable site | – | – | – | – | △ | – |
| | Connection to green space | – | ○ | – | △ | – | – |
| | Sustainable site selection | ○ | – | – | – | – | – |
| Resources | Pollution control | – | – | △ | – | ◎ | △ |
| | Water efficiency and waste management | – | ○ | – | ◎ | ◎ | ○ |
| | Water consumption | ◎ | ◎ | – | ○ | ◎ | – |
| | Water recycling | – | – | ◎ | – | ○ | △ |
| | Rainwater discharge | ◎ | ◎ | ◎ | – | △ | – |
| | Waste reduction | – | ○ | ○ | – | △ | – |
| | Recycling | ◎ | – | △ | – | ○ | – |
| | Materials | ○ | – | ◎ | – | ○ | – |
| | Materials with low environmental impact | – | ○ | ◎ | – | ○ | △ |
| | Locally sourced materials | – | – | ○ | ◎ | △ | – |

| | | | | | | | |
|---|---|---|---|---|---|---|---|
| Transportation | Promotion of walking | △ | ○ | – | – | – | △ |
| | Promotion of cycling | – | – | – | – | – | △ |
| | Reduction of traffic | ○ | – | – | – | – | △ |
| | Green transport infrastructure (charging stations, bicycle parking) | ◎ | ◎ | | ○ | – | △ |
| | Quality of transit | ○ | ○ | ○ | ◎ | – | ○ |
| | Sustainable transport system | ○ | – | – | – | △ | △ |

*BREEAM* Building Research Establishment Environmental Assessment Method, *CASBEE* Comprehensive Assessment System for Building Environment Efficiency, *DGNB* Deutsche Gesellschaft für Nachhaltiges Bauen, *EEWH-EC* Ecology, Energy Saving, Waste Reduction, Health: Eco-communities, *EIAS-GC* Evaluating Indicators and Assessing Standards for Green Communities, *LEED* Leadership in Energy and Environmental Design

◎ Excellent: the assessment system has clear criteria, formulas, or provisions

○EGood: the assessment system only needs to satisfy basic conditions

△ Medium: the system assesses the project but has no clear or mandatory criteria

– None: no related items

3. Ecology

LEED, BREEAM, DGNB, EIAS-GC, and EEWH have evaluation items regarding soil protection/use, but CASBEE does not. Taiwan is small but rich in natural resources and species. In order to achieve conservation of biological diversity and sustainable development, which are the goals of the Convention on Biological Diversity (CBD) signed in 1992, EEWH has many assessment items concerning ecology. Furthermore, there are detailed evaluation formulas and standards developed by EEWH.

4. Economy

All six systems include economic evaluation items. The qualifications of the items in DGNB are the most clear and wide. The most important feature of DGNB is consideration of life-cycle costs and operation and maintenance costs to develop an overall sustainable plan. Although provisions related to this category are developed in EEWH, they are blurred and are not clearly defined or explained.

5. Energy

LEED and EEWH have many more evaluation items regarding energy. Because of the awareness of sustainability and the popularization of energy saving and green energy, the issues of natural energy and renewable energy have been considered in EEWH to develop evaluation items. Because Taiwan, an island nation, relies on imported energy and nuclear power, and has a heavy burden on its environment, EEWH has well developed regulations about energy monitoring and renewable energy technology. Evaluation items in CASBEE have been developed only for energy performance, because there are energy policies already in Japan.

6. Infrastructure

All six systems have well-defined limitations and requirements of the evaluation items for lighting and illumination. CASBEE, EIAS-GC, and EEWH have developed detailed formulas in evaluation items to control the lighting environment accurately and to avoid influence and loading on the environment. BREEAM has manifest limitations and standards for evaluation items for infrastructure facilities.

7. Resources

LEED, BREEAM, DGNB, EIAS-GC, and EEWH have evaluation items and regulations for water efficiency and waste management. CASBEE has strict and standard quantification formulas for this category. In China, resources are unevenly distributed, the population is large, energy supply is short, the degree of industrialization is high, waste output is excessive, and urban expansion is rapid. To mitigate the burden on the environment, EIAS-GC has assessment items for the resource category. The qualifications and standards for most assessment items are blurred.

8. Location

All systems have evaluation items for lighting and the acoustic environment. Only BREEAM has evaluation items for natural disasters, due to serious natural disasters (floods, tornadoes, fires, radiation storms, droughts, etc.) in the UK. According to a survey, nearly one fifth of the people live in fear of natural disasters (Daily Mail 2013). In order to avoid losses and harms caused by disasters, the UK government has developed assessment items regarding natural disasters.

9. Transportation

Except for EIAS-GC, all systems have transportation-related evaluation regulations. Owing to the development of modern transportation science and technology, changes in the social environment, development of economic activities, and augmentation of activities have a great impact on the environment and quality of life. In EEWH, although all the evaluation topics in this category are mentioned, only the evaluation items related to quality of transit are clearly defined; the other topics in the transportation category are rarely defined or are undefined. Therefore, the specifications and evaluation methods are weak for transportation topics.

### 5.4.6 Summary

In the evaluation process, weightings are used to evaluate the importance of the objective from different points of view. By analyzing the function of each evaluation factor in the evaluation process, we can understand the role and relative importance of each item in the system. For instance, though BREEAM has a larger number and specification of items in the community category, its weighting for community is only 9.76%. Therefore, the bigger the number of items does not mean the more important the category. Because ecological communities in Taiwan are divided into rural ecological communities and urban ecological communities, to understand how to define the weighting ratio between the two, study of diversification and localization is necessary. By reflecting the differences in the importance of different indicators, it is possible to build reasonable eco-community indicators and their weightings for subtropical regions. Since the norms and standards of the evaluation items in the economy and transportation categories are not clear, and the affairs of the economy and transportation are always administrated by government units, assessment systems do not draw up detailed specifications and indicators. CASBEE, BREEAM, LEED, and DGNB have put social environment, infrastructure, and population status into evaluation items in order to create a healthy, safe, and comfortable environment for residents. CASBEE pays much attention to gross local production and various matters concerning local economic efficiency. DGNB focuses on the life-cycle cost to the economy. In cultural aspects, LEED and DGNB bring innovative designs or concepts that combine cultural and urban texture into the ecological community, in order to protect coexisting city development and culture development. The indexes of EEWH-EC are currently more focused on ecology, the environment, and the natural environment, with less attention given to matters associated with the outside.

## 5.5 Conclusions

Today, sustainability development is the trend in urban planning. Different eco-community evaluation systems have varied index structures. The evaluation index systems, the weighting systems, the index reference standards, the result expression, and the conceptual understanding of the ecological community are diverse.

Although the eco-community evaluation systems developed in the USA and the UK, located in temperate regions, were developed earlier and are more mature, they do not suit the environmental conditions of the subtropical zone. Because of the diversity of life, natural environments, and cultures in subtropical regions, it is crucial to understand the concept of ecological communities to achieve the goal of sustainable development in subtropical regions.

It is recommended that the design of eco-community assessment systems in the near future should consider the structure of the weighting. Index weighting and evaluation criteria should take into account the actual needs of different regions and different communities with different living standards. Index weighting needs to consider diversification and localization, as well as the need to reflect the importance of the differences between different indicators. Evaluation criteria should differentiate between regions and between various community types. Evaluation criteria should take full account of the needs of old communities, use nonengineering techniques to approach ecological construction, and encourage the community to create sustainable ecology images. Referring to the community life-cycle assessment indicators in DGNB-NSQ, the life cycle of ecological communities can be established. Via analysis from raw material acquisition to product production to product usage, the environmental impact of the entire life cycle can be evaluated.

Based on the analysis of other systems' assessment indicators for the impacts of energy, the water environment, and the atmospheric environment during the community's life cycle, modifications and adjustments can be made to achieve the goals of a sustainable community. In recent years, Taiwan has continued to experience disasters (earthquakes, droughts, landslides, etc.). With reference to BREEAM's risk and natural disaster risk evaluation indicators, assessment indicators for natural disaster prevention and contingency measures will be developed.

Having been developed from 2011 to now, Taiwan's green building index focuses on the issues of the environment and natural resources. Though the rating indicators have been refined after several reviews and modifications, green building indicators are mainly energy conservation–oriented and lack consideration of the economy and social life style. In the social sphere, safety, convenience, and fairness have been put into consideration. The economic sphere is considered less. There is a big gap regarding eco-community development between Taiwan and other nations whose assessment systems cover a full range of health, finance, and the economy. Due to the lack of a clear standard addressing the economy and resource evaluation, Taiwan's green building index for eco-communities (EEWH-EC) is not yet perfect and comprehensive. The evaluation system should be improved and reintegrated to expand the scope and to develop new assessment items, and to make the evaluation system more widely used as a reference model for eco-community assessment systems for subtropical regions in the near future.

**Acknowledgments** This chapter represents part of the results obtained under the support of the National Science Council, Taiwan (contract no. NSC101-2627-E-027-001-MY3).

## References

Berg PG, Nycander G (1997) Sustainable neighborhoods – a qualitative model for resource management in communities. Lands Urban Plan 39(2):117–135

BREEAMINT (2012) BREEAM Communities 2012, an introduction for international use CASBEE for Urban Development (2007) Technical manual 2007 edition. Institute for Building Environment and Energy Conservation (IBEC)

Caldwell JH (1994) Photovoltaic technology and markets. Contemp Econ Policy. https://doi.org/10.1111/j.1465-7287.1994.tb00426.x

CASBEE for Urban Development (2014) Institute for Building Environment and Energy Conservation (IBEC)

Chang-Lin Hua (1998) Practice of ecological urbanism. A compilation of sustainable land development seminar. Chinese Culture University (in Chinese)

Choguill CL (2008) Developing sustainable neighbourhoods. Habitat Int 32(1):41–48. https://doi.org/10.1016/j.habitatint.2007.06.007

Chin-Hsiung Hou (1996) Envisioning a tourism Spot – a renewal project using sustainable development. J Outdoor Recreat Stud 9(4):51–62 (in Chinese)

Daily Mail (2013) Apocalypse Britain: terrifying images show Parliament wrecked by flooding and Edinburgh Castle hit by swarm of tornadoes. http://www.dailymail.co.uk/sciencetech/article-2393212/Perfect-Storms-TVshows-images-Britain-wrecked-natural-disasters.html#ixzz4zZH1Z6Dt

DGNB e.V (2012) Neubau Stadtquartiere: DGNB Handbuch für nachhaltiges Bauen DGNB e.V. and GeSBC. EPA (2013) EPA heat island impacts. Environmental Protection Agency. http://www.epa.gov/heatisland/impacts/index.htm

EEWH Assessment Manual for Eco Community 2015 edition (2015) Architecture and Building Research Institute (ABRI), Ministry of the Interior EIAS-GC 2007 v.4. Technical manual 2007. Edition

EIAS-GC (2007) v.4. Technical manual 2007. Edition

European Union (2007) "Leipzig Charter on Sustainable European Cities" document. http://ec.europa.eu/regional_policy/archive/themes/urban/leipzig_charter.pdf

Gilbert R, Stevenson D, Girardet H, Stren R (1996) Making cities work: the role of local authorities in the urban environment. Earthscan, London

Hecht JE (2007) National environmental accounting: a practical introduction. Int Rev Environ Resour Econ 1:3–66

Heng Chang, Chih-Hang Hsing (2004) A study on the concept of ecological community in community environment – taking Tainan County rural area as an example. J Architec Plan 5(1):29–47 (in Chinese)

Hsien-Te Lin (2008) Research on evaluation system of ecological communities. Ministry of the Interior Building Research Institute, Taipei (in Chinese)

Hurley J., Horne R. (2006) Review and analysis of tools for the implementation and assessment of sustainable urban development. In: Environmental Institute of Australian and New Zealand National conference, pp 65–174

Jaeger JA, Bertiller R, Schwick C, Kienast F (2010) Suitability criteria for measures of urban sprawl. Ecol Indic 10:397–406

Langeweg F (1998) The implementation of agenda 21 'our common failure'? Sci Total Environ 218(3):227–238

LEED (2014) For neighborhood development. US Green Building Council

Leitmann J (1999) Sustaining cities: environmental planning and management in urban design. McGraw-Hill, New York/London. ISBN-13: 978–0070383166

Luederitz C et al (2013) Learning through evaluation – a tentative evaluative scheme for sustainability transition experiments. J Clean Prod 169:61–76

Mollison PL (1989) Further observations on the patterns of clearance of incompatible red cells. Transfusion. https://doi.org/10.1046/j.1537-2995.1989.29489242803.x

Register R (2010) Ecocities, rebuilding cities in balance with nature, rev edn. New Society, Gabriola Island

Sharifi A, Murayama A (2013) A critical review of seven selected neighborhood sustainability assessment tools. Environ Impact Assess Rev 38:73–87. http://www.sciencedirect.com/science/article/pii/S0195925512000558

Sharifi A, Yamagata Y (2014) Major principles and criteria for development of an urban resilience assessment index. In: International conference and utility exhibition on: green energy sustainable development, p 63

Shen LY, Ochoa JJ, Shah MN, Zhang X (2011) The application of urban sustainability indicators – a comparison between various practices. Habitat Int 35:17–29

Singh RK et al (2012) An overview of sustainability assessment methodologies. Ecol Indic 15(2012):281–299

Stefan K (1996) Whose sustainability? Ecology, hegemonic politics and the future of the city: the case of Zurich. In: Keil R, Wekerle G, Bell D (eds) Local places in the age of the Global City. Montreal, Black Rose, pp 117–124

Tanguay GA et al (2009) Measuring the sustainability of cities: a survey-based analysis of the use of local indicators. http://www.sciencedirect.com/science/article/pii/S0195925512000558

Ting-Fen Liu, Ching-Nan Chen (2011) Comparison of green building evaluation system at home and abroad. Zhong Xing Eng J ZTE Eng Consult Assoc 113:87–94 (in Chinese)

Yi-Ping Tsai (2014) A study on visual preference of green building's appearance in hot-humid climate region. http://handle.ncl.edu.tw/11296/ndltd/53556138585551727022

Zeinal Hamedani A, Huber F (2012) A comparative study of DGNB, LEED and BREEAM certificate systems in urban sustainability. The sustainable city VII: urban regeneration and sustainability. WIT Trans Ecol Environ 155(1):121–132

# Part II
# Ecological Preservation and Restoration Planning by Greening the City

# Chapter 6
# Ecological and Intensive Design Tactics for Mountain Cities in Western China: Taking the Main District of Chongqing as an Example

**Feng Lu and Min Jiang**

**Abstract** To continuously increase urban space, it is quite unlikely for mountain cities to be able to expand their boundaries. With the lack of natural, ecological land resources and the unrecyclable nature of such resources, the only way out for them is to make the most of their existing resources and to find a new development mode that focuses on the internal renewal of those cities. As one of the largest mountain cities in Western China, faced with a series of grave problems in city development such as economic remodeling, functional updating, and resource reorganization, Chongqing is taking a development tactic focusing on optimizing and integrating resources, which may become a good model for other mountain cities in the same region. In view of the characteristics of the natural ecological and land resources in the mountain cities, the urban design tactics for these cities emphasize two aspects. The first is constructing hierarchical and systematic natural resource protection mechanisms, strengthening the dominant position and restrictive role of natural elements in the mountainous urban landscapes and maintaining the urban spatial texture of cluster growth; at the same time, in light of urban industry restructuring, exploiting the functions and tapping the potentials of the riverside belt as an urban landscape area open to the public, seeking a new model which is fit for the development of the riverside belts in cities. The second is taking resource integration as the main approach to efficient use of the land resources in downtown areas and by constructing three-dimensional public pedestrian systems in the cities, combining the cultural landscape and commercial facilities with urban street spaces so as to save these spaces from isolation; and by forming a compact, land-efficient, and multifunctional urban pattern, which is at the same time adapted for the various transport modes, shifting the mountain cities from an old pattern towards a new, intensive, resource-rich, and economizing pattern.

**Keywords** Western China · Mountain city · Resources · Integration · Urban design

---

F. Lu · M. Jiang (✉)
Architecture and Urban Planning, Chongqing University, Chongqing, China
e-mail: 20095705@cqu.edu.cn



Z. Shen et al. (eds.), *Green City Planning and Practices in Asian Cities*, Strategies for Sustainability, https://doi.org/10.1007/978-3-319-70025-0_6

## 6.1 The Main Characteristics of Spatial Growth in Chinese Mountain Cities

Mountainous areas account for about 70% of China and 68% of the western regions. Five western provinces face the challenge of more than 80% of their land areas being mountainous. Therefore, as the major carrier of the regional economy development, the mountain cities play a pivotal role in the western areas.

Having developed for more than 10 years after becoming a municipality, the super-large mountainous city of Chongqing in western China is currently facing two tendencies—namely expansion of boundaries and internal renewal, both of which may lead to diverse development modes; therefore, Chongqing's development mode is quite representative of western mountain cities.

### *6.1.1 Manifestation of Mountain Cities' Space Form Changes*

The redevelopment of urban core areas and extension of city boundaries coexisted during the development of Chongqing. Because of the urban economic transformation, lots of deserted industrial land along the riverside downtown became a major urban land resource and a new investment hotspot after the construction of riverside roads. However, the economic and social benefits from the expansion of the urban riverside space have not been fully realised due to the lack of coordination between developing sequences and objectives. In urban fringe areas, with the highway network taking shape, the medium-low-density-oriented urban fringe residential development mode has showed urban sprawl characteristics.

### *6.1.2 Negative Impacts of High-Intensity Urban Development Activities*

High-intensity development in the main district of Chongqing has altered not only spatial patterns but also the city skyline, which both took a long time to form and put greater pressure on urban natural ecological environment. With the decrease in land resources and rising urban construction demand, new urban areas have continued to occupy steep and ecologically sensitive mountains, breaking the balance between existing urban forms and the natural environment. In recent years, widely used building construction technology such as *heavy excavation* and a *high cut slope* have destroyed the original urban mountainous topography and resulted in numerous mountain disasters, thus intensifying the population–land contradiction. Additionally, increasing development of tall buildings have made mountain cities'

**Fig. 6.1** Aerial view of the urban design of the core area in the Jiefangbei area, Chongqing

integral landscape transform from a clustered "figure–ground relationship" to a natural–artificial mixed shape. Especially in the main district, tall building groups have already formed a new skyline and divided the urban green space into isolated patches, heavily damaging the scenery and the value of utilization of urban green spaces (Fig. 6.1).

### *6.1.3 Traffic Bottlenecks in Mountain Cities' Development*

With decades of development, the main urban area of Chongqing has formed a constellation–distribution structure with separate self-support subdivisions within the specific natural topography and environment. Along with fast economic growth, the traffic demand between centers has increased rapidly because of the polycentric structure. However, the strategies of the expanding urban traffic instead of making significant improvements, actually worsened the situation, and ended up forming new traffic-jam nodes. Therefore, mass rapid urban transport solutions began to receive attention. However, since it is still at an exploratory stage to integrate the urban spatial form and functional layout, the long-term effects have not surfaced. At the same time, with the lack of integral development strategies and effective coordination mechanisms, the function reorganization within each center has shown contradictions between high-intensity development and low-efficiency urban spaces.

## 6.2 Mountain Cities' Development Strategy Centring on Resource Optimization and Integration

Seen from the aspect of resource impacts on urban development in the long run, natural ecological resources and land resources are regionally monopolistic, non-renewable and unreplicated, which therefore become decisive elements for mountain cities' urban pattern and strategic development orientation. In the past 10 years, the fast economic development in eastern cities has offered successful examples for western mountain cities to find development directions. Under the general trend of urban ecological environment deterioration, mountain cities must transform from secondary industry which focus on energy-intensive industrial activities, to tertiary industry which focus on service industries, to avoid the passive situation of "polluting while developing." At the same time, mountain cities are not capable of infinite expansion due to the land resource limitation. To create more value per unit of land and obtain space for urban economic development in the long run, mountain cities must promote their urban function from low-level resource consuming mode to high-level resource integrating mode based on their original urban patterns. Therefore, to obtain a dominant position in urban development and achieve smart growth, mountain cities should take advantage of their unique geographic space and local cultures; develop a service industry system that can transform local resources into urban competitive advantages by combining the urban construction process with tertiary industry growth, meanwhile adopting an intensive urban form; help advantageous resources like commercial resources to accumulate and upgrade; and diminish the harmful effects resulting from shortage of land and human resources.

Urban design practice aimed at coordinated development is actually the process of reintegrating various urban resources. It tries to achieve optimal combinations of diverse resources performing well in economic, social, and environmental fields through certain mechanisms and control methods and then to create vitality for new development. The current research and practice of mountainous urban design, with regard for the idea of effective growth, ought to reevaluate cities' resources through an overall consideration of mountain cities' natural environmental conditions, urban layout characteristics, and long-term development goals. We need to find sustainable ways which represent their own value and urban characteristics through exploration and innovation in the protection of natural resources, efficient utilization of land resources, and revitalization of urban space, instead of superficial imitation of eastern urban development models.

Since 2004, the Urban Design Team at Chongqing University has finished two important urban design projects in the main district of Chongqing. One project is the urban design of the Riverside Region in the northern area of around 11 $km^2$. This project mainly analysed redevelopment modes and urban design strategies during industry transformation in areas along the Jialing River (refer to Fig. 6.2). The other project dealt with the urban design of central Jiefangbei area in Chongqing, covering around 0.92 $km^2$, involving modes of integrating urban public spaces and urban design strategies into a high-density urban texture (Fig. 6.3). These two projects helped develop both theory and methods for future mountainous urban development with vivid practical case studies.

**Fig. 6.2** Urban design of the riverside region in the northern area of Chongqing

**Fig. 6.3** Urban design of the central Jiefangbei area in Chongqing

## 6.3 Visible Landscape and Clustered Growth—Natural Landscape Resource Protection

In mountainous environments, natural restriction factors such as topography and water are crucial for urban textures to form and develop. As the main reference for positioning and orientation in urban areas, the outstanding natural scenery around mountain cities has gradually become a symbolic visual element in the local culture. From the larger perspective of cultural geography, traditional mountain cities are integrated into nature harmoniously as a whole, with their textures always representing local natural geography and dynamics. Hence, to protect the mountainous urban natural landscape patterns is important to protect and retain the urban contexts.

### *6.3.1 Protecting Natural Elements in the Mountainous Urban Landscape*

*The Ba Kingdom is floating and sinking like a leaf, while the precipitous Fotuguan is surrounded by two rivers.* This is a vivid reflection of Chongqing's urban form, which is hilly and surrounded by the Yangtze River and the Jialing River. The intersection of the two famous rivers created numerous mountainous areas, waterfronts, and open spaces. At the same time, it becomes a natural boundary for ecological environment protection. The integration of mountains, waters, and buildings brings distinction and diversity to mountain cities; it also offers a landscape framework for the public to recognize their cities. Therefore, for utilising natural mountainous landscape and creating attractive mountainous city image it is fundamental to make sure the natural landscape is integrated with the landscape framework in mountain cities' integral outline and as ecological restriction in urban space development, and to continue the co-existing mode between mountain cities and their environment.

#### 6.3.1.1 Hierarchical Protection Strategy

In mountainous urban design, natural landscape resources can be divided into three levels with specific protection strategies according to their position in the urban landscape. The first level is to protect the mountain tops or continuous mountain ridgelines which are iconic and can provide a landscape background for cities. This is essential to construct along the cities' natural patterns. Building skylines are forbidden to surpass the midheight of mountains or to damage the integrity of the mountain skylines. The second level is to protect the natural topography, which can form obvious boundaries such as riverside belts and cliffs, etc. These areas often become open spaces where construction is forbidden. The third level is to protect natural features including gentle slopes and terraces at different heights. Urban construction in these areas should adapt to the local topography without large topography changes and try to diversify the topography in accordance with the original one (Fig. 6.4).

#### 6.3.1.2 Systematic Protection Strategy

In urban design, maintaining continuity, complementarity, and networking of natural landscapes in aspects including vision, space, and usage in different regions or forms can promote natural elements' dominant position in cityscape. In the urban design of the waterfront in Jiangbei District, Chongqing, the natural landforms, which cross at different heights, including catchment areas, gullies, and valleys are classified as ecologically sensitive areas. The protected boundary has been clearly defined and these areas were intended to be turned into a continual green belt crossing three different elevations in the urban landscape to avoid complete disconnection caused by buildings between mountains and water bodies. Meanwhile, it promoted strategies to use the urban roads crossing different regions with green belts along them as connected green corridors. Thus, these green corridors together with vertical green systems can constitute the clustered mountainous urban landscape pattern (Fig. 6.5).

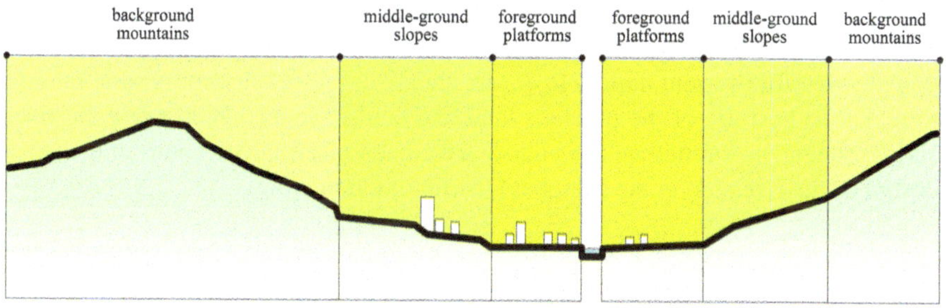


**Fig. 6.4** Illustration of hierachical landscape control in mountain cities

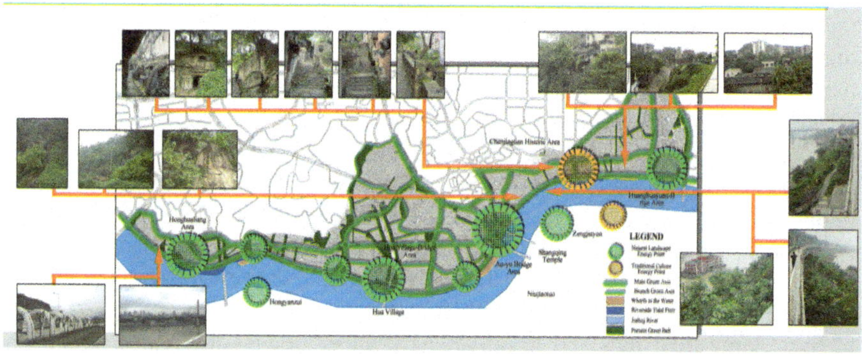

**Fig. 6.5** Green ecosystem networks

## 6.3.2 *Creating Diversity and Openness of Waterfront Landscapes*

In the main district of Chongqing, the long-term aggregation of urban industry, has made the riverside along the Jialing River the main area for transportation and production, with obvious problems such as monotone function and non-livability. Since 2000, as the urban economy has transformed, urban riverside areas have gradually been converted from production space into living space, improving land recovery and landscape value. The winding riverside are valuable land resources for urban spatial growth and also are able to make up for the shortage of public spaces and green land in the downtown area.

The current redevelopment of riverside areas in central Chongqing faces three bottlenecks. First of all, the former riverside roads paid too much attention to traffic and flood protection but ignored the urban landscape needs; consequently, these roads became major barriers between the city and the river. Secondly, the riverside tidal flats were not fully exploited and became the largest (though less efficient) natural green land because of a wrong perception to separate the riverside areas from the city, combined with a lack of basic physical infrastructure. Thirdly, the existing development programs alongside the two rivers are not good enough. Those programs are similar to each other; they fall short of unique architectural forms and urban public spaces. During the development, the influence of urban public traffic was not taken into consideration. Without field observations of traffic needs and public traffic route design, many riverside infrastructures stayed isolated from the downtown traffic system, failing to ensure steady consumes flow or profits. In order to make the best use of the precious land and landscape resources along the riverside, strategies to strengthen the vitality have been taken into account in the urban design of the riverside in the Jiangbei District of Chongqing.

### 6.3.2.1 Expanding the Landscape Potential of the Riverside

In order to increase the openness and accessibility of the riverside, and to allow more people to enjoy it, walkway under bridges, slopes and overpasses were incooperated in the design, according to the features of riverside areas with both embankments and viaducts. Meanwhile, the design also widened roads at certain positions in accordance with the original topography and connected them with bus stops to create viewing areas and thus to turn the riverside roads into riverside promenades with a variety of scenic areas.

As one kind of precious urban wetland, riverside tidal flats are irreplaceable in optimization of the urban landscape, preservation of biodiversity and flood prevention. Though they are influenced by seasonal floods, the broad vision and public spaces of the riverside tidal flats are still attractive to citizens living in downtown areas which are crowded by tall buildings. So it is important to make full use of these large tidal flats and negative space under bridges to create recreational spaces—such as platforms near the water, riverside boardwalks and even cliff parks, and to introduce temporary infrastructures such as mobile latrines and retail kiosks—to connect all the riverside parts to form an urban water-based leisure space with various functions and friendly public participation to promote special urban riverside activities (Fig. 6.6).

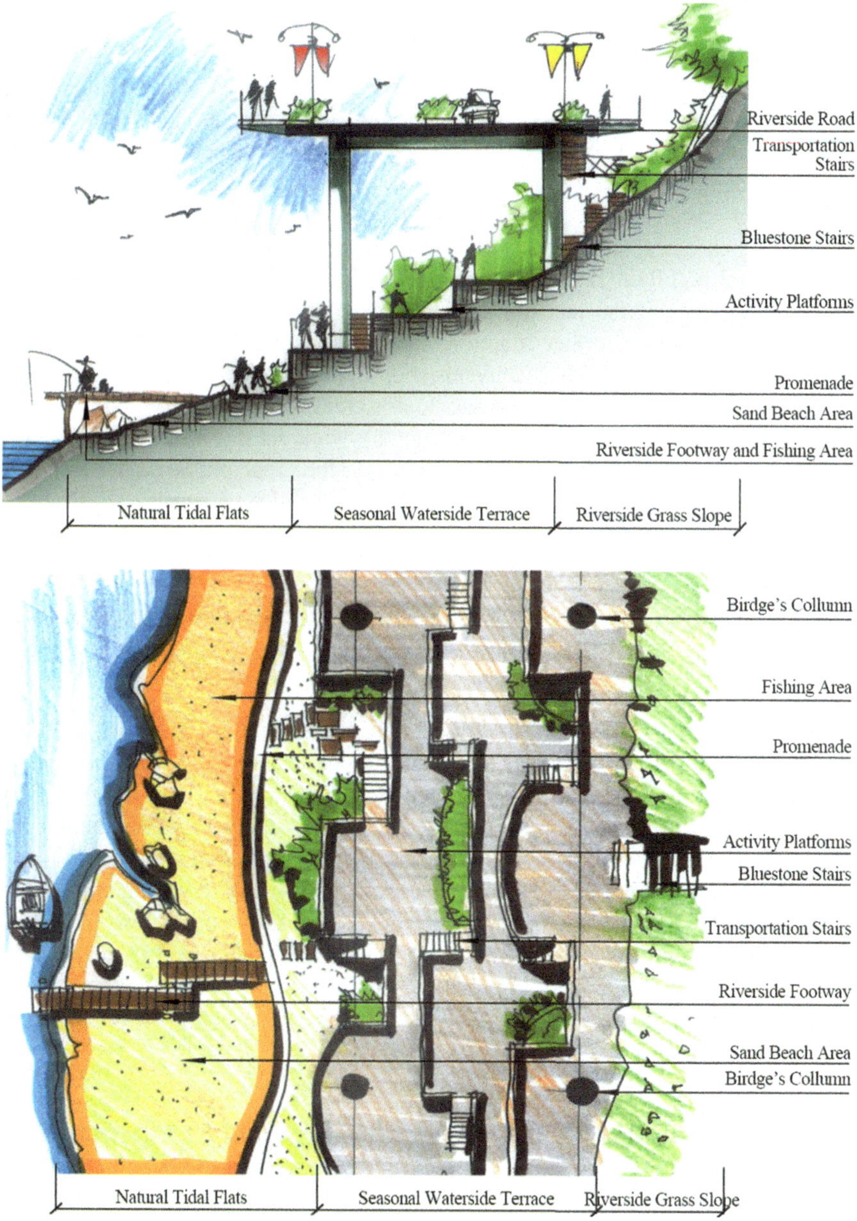


**Fig. 6.6** Make better use of riverside areas

#### 6.3.2.2 Urban Riverside Development Combined with Public Transportation

Multifunctional development can always achieve larger function agglomerations. The complementarity of mixed functions can lead to optimization and diversity of land use. Commercial developments in mountainous urban riverside areas are often designed into different terraces according to the original topography, and those near the riverside roads with a broader vision would be ideal for service industries, like tourism industry and leisure industry, for early development. But it would be difficult to form a convergence effect of the commercial service space and attraction for customers in a short time with a monotonous commercial category and sole dependence on car owners as the target customers. Therefore, the redevelopment of mountainous urban riverside areas should combine other goals including reinforcement of urban characteristics and promotion of urban tourism values. On the one hand, several highly distinctive comprehensive service areas should be promoted, based on features of the landscape nodes alongside the river and encouraging a mixed arrangement of service functions and complementary diversity—for example, with infrastructure such as souvenir shops, cultural displays, exhibitions, education and dining and recreation areas, to expand the service sector. On the other hand, given that the riverside area is too long for walking, it is important to connect public service areas and important scenic spots to the urban public traffic system to create a series of polycentric, convenient and interesting areas.

## 6.4 Complementary Resources and Compact Development—Intensive Usage and Organic Renewal of Land Resources

Downtown areas in mountain cities generally have multiple functions including retail, business offices, financial services, culture and entertainment, residence, administration, etc. The commercial accumulation effect after long-term development makes them leading areas in economy and social vitality. However, the scarcity of land resources, deteriorating transportation and stereotyped urban spaces become main obstacles in future development. Thus, urban design in mountain cities ought to make full use of favorable opportunities in urban spatial redevelopment, find the potential of existing urban space through resource integration and urban function reorganization, and, moreover, improve adaptability to new economic development demands by creating a new typology of urban space and development value.

### *6.4.1 Pedestrian System Construction–Oriented Urban Spatial Reorganization*

Street space is the most essential spatial element in mountain cities, which has multiple public urban functions. In the recent development of the commercial district in Chongqing, large-scale urban renewal projects changed the urban fabric dramatically, but it did not improve the vitality and capacity of the commercial space, as significantly as expected. On the contrary, isolated development resulted in more and more homogenous negative urban commercial spaces, especially in enclosed large commercial complexes whose closed facades weakened the diversity of city street activities and "destroyed" the vitality of the city streets. Therefore, the urban street space gradually turned into simple traffic passageways, resulting in the decline of urban public space (refer to Fig. 6.7). Although some old commercial plazas and pedestrian streets relieved the lack of public spaces in the old urban commercial district to some extent, but they failed to solve the problems of vehicle–pedestrian mixed traffic or homogenous pedestrian space, because little consideration was given to integrating the traffic network into urban areas or humanity design details. Improvement tactics are proposed as 6.4.1.1 to 6.4.1.3, based on investigation of existing urban public space conditions.

#### 6.4.1.1 Integration of Cultural Resources and Urban Architectural Heritage

Against the macro background of globalization of information and economy, competition among future cities will convert from physical space to city culture. Indeed it is a competition of city's features and characteristics which they derive from their unique local culture and environmental elements. Therefore, the attractiveness for international and cross-regional enterprises, tourism, commercial exhibitions, expositions and cultural communication can be improved by reinforcement of the cultural and regional uniqueness, as well as the irreplaceability, of mountain cities.

**Fig. 6.7** Fabric analysis of the central commercial area in Chongqing

A large number of historical buildings and particular places with rich culture are located in large numbers of mountain cities' downtown areas after long-term development, which are regarded as unique advantages when compared with newly built commercial districts. The main attraction for citizens is the cultural space of different levels and various supplies in old city commercial districts formed through long-term development. However, many isolated cultural places cannot survive or develop in the high-rent downtown areas. Though many old buildings bear great historical significance, it is quite difficult for them to form a complete urban context expression system because of insufficient maintenance and deterioration of the surrounding environment. Hence, combining cultural entertainment, information displays, and improvement of old buildings' surrounding environment with urban public open spaces, can enrich the cultural essence of urban public open spaces and lead to efficient utilization of urban cultural resources. Meanwhile, with selective rehabilitation or renovation to equip old buildings with new service functions such as businesses, culture, and entertainment in combination with their specific locations and space patterns to form a commercial street with special theme which not only enriches the typology of the urban commercial space and offers more choices to city residents and tourists, but also makes full use of old buildings' functions and space (Fig. 6.8).

#### 6.4.1.2 Enhancement of Spatial Cohesion of Urban Streets

In the city's core areas, to meet growing traffic demands only by increasing the density and capacity of motorways will inevitably lead to decrease in urban public space and degradation of spatial qualities. So it is essential for city core areas to manage intensive growth by constructing external traffic patterns with prioritising public traffic and internal traffic patterns that give way to pedestrian thoroughfares.

Firstly, interchange stations between motor traffic and pedestrian thoroughfares should be built on the edges of downtown areas to reduce the frequency of motor vehicles passing through the center areas. Meanwhile, the public transportation routes and station distribution should be optimized, with special emphasis given to enhancement of the supporting role that a high-capacity and high-efficiency express rail transportation plays in the downtown areas.

Secondly, the density of pedestrian thoroughfares should be increased, adding more connections between parallel main pedestrian thoroughfares and connecting lanes within closed sites to pedestrian streets nearby, so that the poor, unhealthy and messy state of backstreet areas can be optimized, more commercial frontages can be obtained, and the commercial value of sites and the surrounding environment can be improved (Fig. 6.9). As for large-scale commercial complexes, the main tactic is to, increase the number and width of internal shopping passages, improve artificial illumination to make the internal space look more like external streets, and combine them with the external urban pedestrian systems.

**Fig. 6.8** Rehabilitation of Luohan Temple block, Jiefangbei area, Chongqing

Thirdly, the green space condition of urban public pedestrian thoroughfares should be improved. Chongqing is located in a hot-summer and cold-winter district. In the hot summer and cloudy winter, outdoor activities become an important part of local people's daily life. Proper planting arrangements can overcome the disadvantages of local climate and provide resting places similar to the climate for people staying outdoors for long time. At the same time, by enriching the commercial and social functions of streets and organizing long-term regular folk activities, the relationship between urban residents and particular places can be nurtured and pedestrian space is more likely to keep people around therefore, it is able to rebuild the sense of place and the sense of belongingness among the residents.

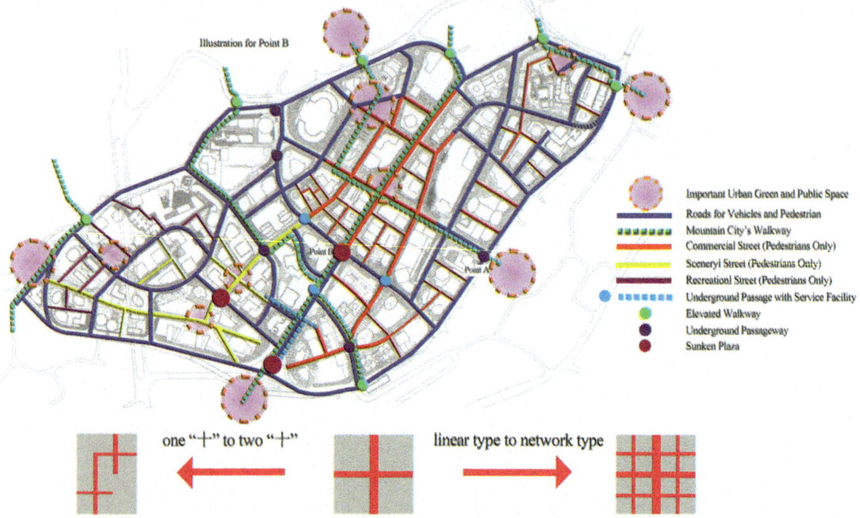


**Fig. 6.9** Networking urban pedestrian streets

#### 6.4.1.3 Emphasis on Integration of Urban Space and the Landscape by a Vertical Transportation System

The three-dimensional transportation space is one of the most important characteristics of traditional mountain cities. Latest urban transportation systems that perform smoothly include solutions dealing with topography such as continuous exterior stairs, gradient rail cable cars, escalators, elevators, and cable cars across the river are quite typical for mountain cities. However, because of the overemphasis on the development of horizontal traffic, vertical transport in mountain cities has not been fully exploited and lots of vertical pedestrian systems were demolished during the urban revitalization, which accelerated the extinction of unique urban space and left very less scope for the urban landscape to extend in depth. In mountain cities, vertical transport has many more advantages—in terms of efficiency, investment, and land occupation—than horizontal ones, which are costly and difficult to construct, and have adverse effects on the original urban topography. It is important for people to be able to move vertically in mountain cities, and this is also a crucial part of building connections between citizens and cities. Therefore, in future design of mountain cities, designers ought to recognize the efficient, land-saving, and ecological features of vertical transport systems, strengthen their roles in connecting and integrating the urban space structure, and turn them into an innovative space formation in mountain cities.

In the urban design of the riverside area in the Jiangbei District of Chongqing, attempts have been made to create continuous public open spaces among different lots, to organize public facilities in resting platforms, green spaces, and squares, and to provide scenic platforms in high-density buildings to build viewing corridors between the mountain and the water. In order to ensure the continuity of the vertical transport system, the altitude difference was used to form overpasses with urban motorways and provide pedestrian entrances on different levels (Fig. 6.10). These pedestrian spaces perpendicular to the contours constitute a three-dimensional spatial shape and a multidimensional landscape for mountainous residents to exercise, communicate, and trade. They are also ventilation corridors connecting the mountains and riverside green spaces at the same time, contributing a lot to improving the microclimate in small urban areas (refer to Fig. 6.11).

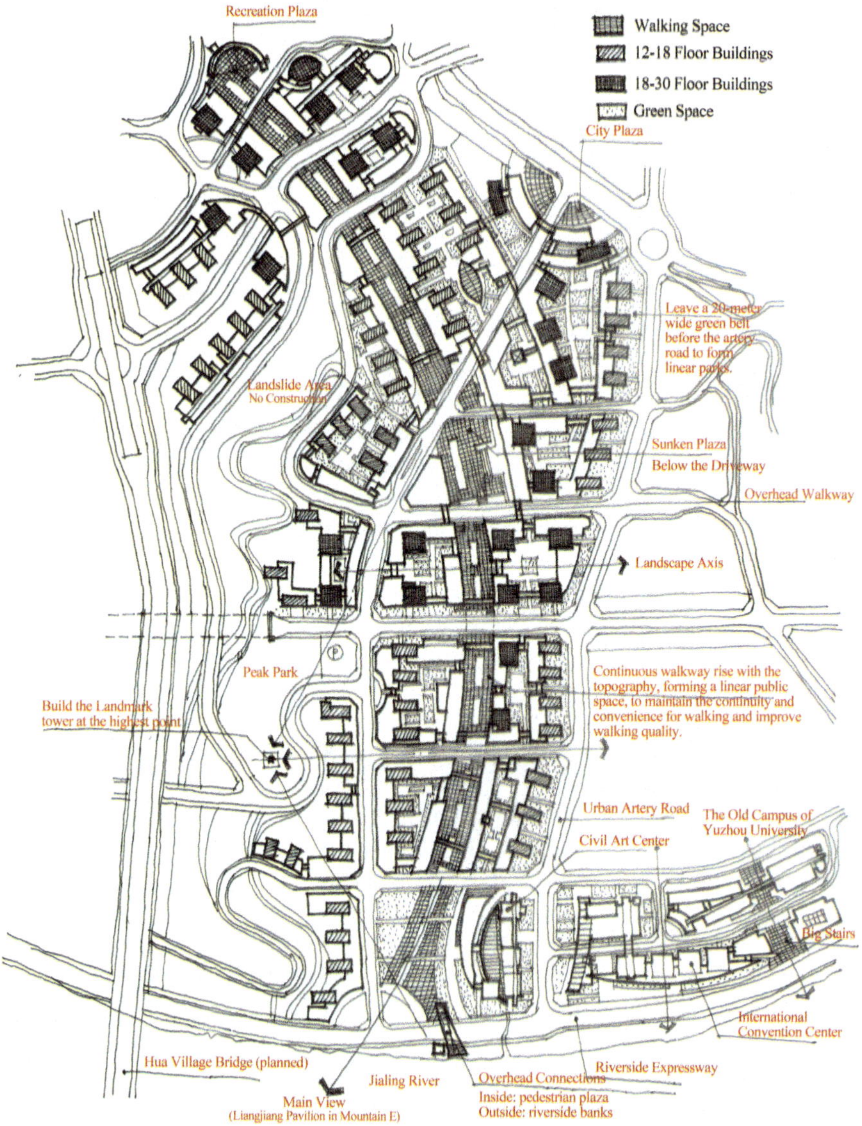


**Fig. 6.10** Multilayered and systematic urban pedestrian street network

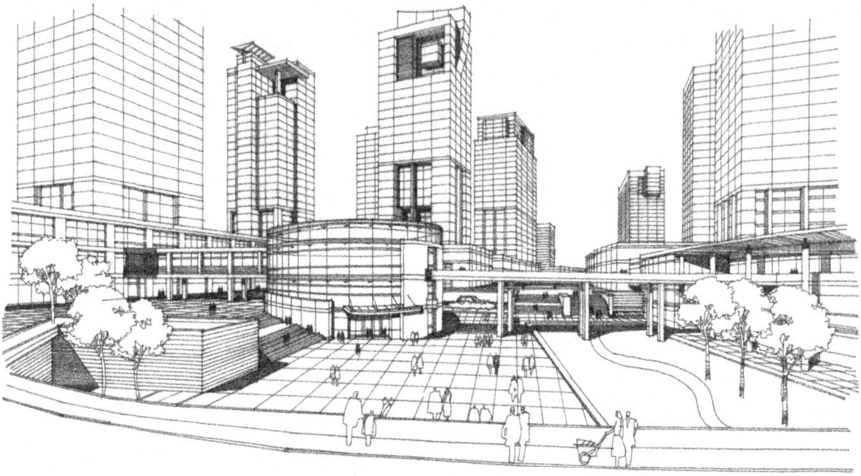

**Fig. 6.11** Three-dimensional public spaces in mountain cities

## *6.4.2 Mode of Intensive Land Use*

The high-density and high-capacity use of small pieces of land, which stems from the limited land resources as well as reasonable budgeting, is compactly structured and complementarily functional. Especially, the overlap of working and living spaces can relieve the stress of public transportation, improve working efficiency, strengthen citizens' relationships and sense of place, and create a diverse urban landscape. Meanwhile, the diversity of functions can also contribute to the flexibility of cities, which can better adapt to the complicated and changeable markets. Therefore, under the circumstance of a great shortage of land resources, it is of greater realistic meaning to raise the utilization ratio of urban space than expand it.

### 6.4.2.1 Intensive-Function Mode of Resource Sharing

After detailed field research on the commercial space in the core area of Jiefangbei, it was found that the operating conditions above the second floor of many isolated shopping malls, even those located in busy areas, were not quite beneficial because of the scale and the inefficient transportation organization. The amount of this kind of area accounted for more than 30% of the core area, resulting in a huge waste of urban architectural space as well as commercial resources. Under this circumstance and based on integration of spatial and commercial resources, strategies including pedestrian overpasses, arcades, and continuous overhead corridors of different classes can enhance the connection of commercial spaces above the second floor and create a connection to interior atriums, squares in the air, and vertical transportation to form a three-dimensional shopping network. To connect spaces of different shapes and functions can achieve a diversity of both complementary functions and shopping activities. The overhead corridor–oriented pedestrian transportation is economical and efficient, with no influence on city street patterns, and can provide more effective commercial spaces; it also reduces the waste of investment and land resources caused by repeated development, avoids disorderly competition of commercial spaces, and achieves resource sharing among urban service systems (Fig. 6.12).

#### 6.4.2.2 Mixed-Function Mode Combining Architectural Function with Urban Public Transit

Restricted by topography, the transportation networks in mountain cities often have some natural shortages such as uneven distribution, poor accessibility, and insufficient circular connections, which make them less efficient. In urban places where the gradient is more than 45 degrees, many pieces of land available for construction are between roads at elevations that are far apart, where transportation is difficult. To solve this problem in mountainous buildings, it is common to combine architectural functional spaces with continuous rotary driveways and city roads oriented in different directions and at different elevations. The continuous rotary slope can not only improve the transportation condition of different floors but can also adjust city roads' traffic flows, thus bringing a new remedy for traffic jams caused by the mountainous topography. City roads at different elevations can also offer the possibility for vertical function overlapping and entrances at different floors, becoming a special method of functional integration in mountain cities (Fig. 6.13). At present, this strategy has been applied to urban light rail stations, showing many advantages including flexibility, land saving, investment, efficiency, functional complementarity, and topographic adaption. It is of unique significance for use of intensive land resource in mountain cities.

## 6.5 Conclusion

mountain cities in China are typically cities with resource shortages. With limited ecological and land resources, it is difficult for mountain cities to achieve a continuous increase by expanding their outer boundaries. However, the mountain–water landscape pattern, profound cultural accumulation and particular mountainous public spatial shapes together provide great potential for internal renewal of mountain cities. As a result, the following mountainous urban design strategies are suggested based on the resource shortage situation:

1. Centering on the natural resources, defining mountainous urban development ecological boundaries, and then constructing three-dimensional ecological green networks and urban public open space systems within these boundaries

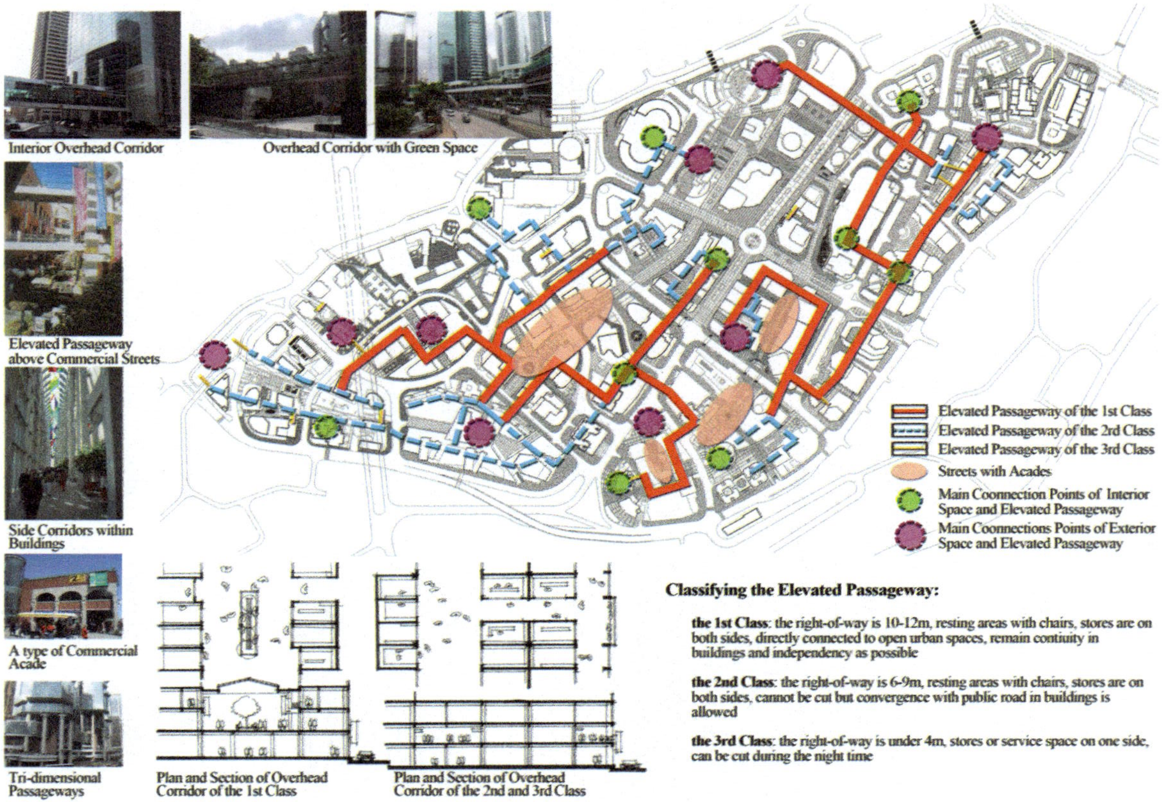


**Fig. 6.12** Elevated passageway system in a commercial center

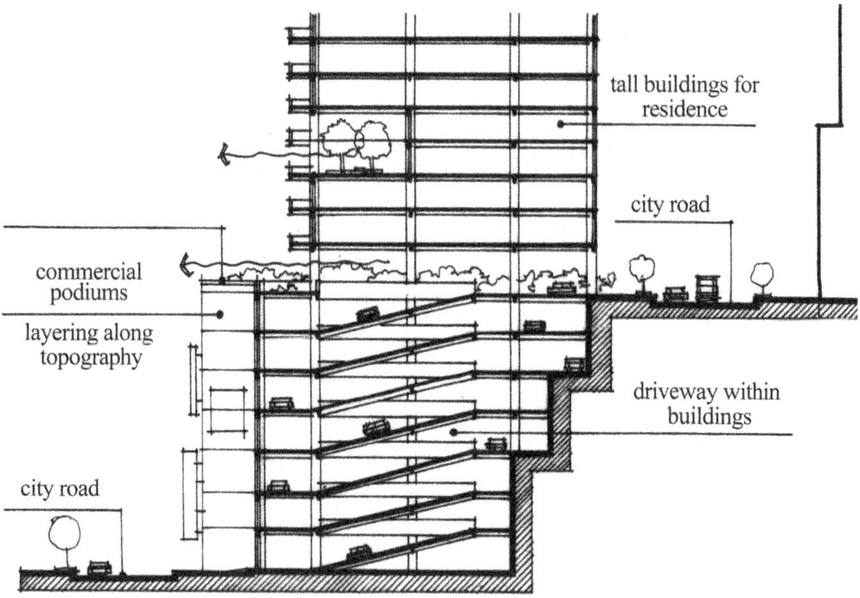


**Fig. 6.13** Combination of modern mountainous buildings and transportation space

2. Strengthening riverside areas' accessibility and landscape diversity in mountain cities, turning them into connections between public spaces in different districts
3. Constructing three-dimensional urban public space systems that give priority to pedestrians and public traffic mainly by optimizing the street network, therefore highlighting the cultural and historical characteristics of high-density urban public spaces

Mountainous urban design strategies oriented toward ecology and intensity can be helpful to guide mountain cities to overcome resource bottlenecks in the process of continuous internal renewal and finally become effective, intensive, and resource-conserving cities with their own advantages.

## References

Abbott C, Xu WT (1997) The metropolitan frontier: cities in the modern American West. Commercial Press, Beijing

Chen H-l (2004) Human geography. Science Publishers, Beijing

Chen G-J et al (2003) The report of mountainous area in China. Commercial Publishers, Beijing

Hou B-z (2005) City transformation: cycle, strategy and mode. Urban Plann Forum 2005(05):1–11. 16

Huang G-Y (2002) Mountain city science. China Building Industry Publishers, Beijing

Li Y (2006) Urban renaissance and experience from Cardiff, UK. Urban Plann Overseas 2006(02):23–28

McHarg IL, Rui J-W translated (1991) Design with nature. China Building Industry Publishers, Beijing

Mo T-W, Mo H-Z (2005) Market, bazaar, trading, the behaviour of living form: on the ability in bringing back living activity. Time Archit 2005(02):16–21

Qiu B-x (2005) Urban planning reform during the urbanization in China. Tongji University Press, Shanghai

Zhu D-J, Yi H (2005) Research on the basic theory of urban planning in China based on integrated courses. Urban Plann Forum 2005(01):21–23

# Chapter 7
# Ecological Wisdom in the Planning and Design of Ancient Chinese Cities and Its Contemporary Relevance

**Yan Shuiyu, Yang Huihui, Xu Lijuan, and Ren Tianman**

**Abstract** Ancient Chinese urban civilization has a long history. Many ancient cities were designed with ecological wisdom with regard to the site selection, the space patterns, and the urban forms of the cities, adapted to geomorphological agents, hydrological features, and climate states; thus, natural ecological services could be better utilized in urban activities to create comfortable urban environments. By means of city historical maps and local chronicle data, this chapter analyzes the relation between site selection, space patterns, and building clusters of typical cities in ancient China and the terrain, hydrology, and climate. This analysis shows that these cities have a good relation of adaptability to nature, so that the urban space coordination ecological wisdom of "skillfully utilizing natural ecological resources and creating coordination between nature and human beings" with appropriate street directions, building orientation, and building cluster forms are selected according to different climates to create comfortable temperature and wind environments. This ancient ecological wisdom needs to be blended into contemporary urban planning and design to realize the coordination between urban construction and natural systems. Urban land expansion and urban form planning and design need to conform to natural ecological process rules. Manmade landscapes and natural landscapes need to be blended to realize the balance between urban processes and natural evolution processes and basically maintain regional ecological stability.

**Keywords** Ancient Chinese cities · Ecological wisdom · Urban morphology · Adaptability to nature

Y. Shuiyu (✉) · Y. Huihui · X. Lijuan · R. Tianman
School of Architecture and Urban Planning, Chongqing University, Chongqing, China
e-mail: yanshuiyu@cqu.edu.cn



Z. Shen et al. (eds.), *Green City Planning and Practices in Asian Cities*,
Strategies for Sustainability, https://doi.org/10.1007/978-3-319-70025-0_7

## 7.1 Foreword

Ancient Chinese urban civilization has a long history. Since the time of the Qin and Han Dynasties (200 BC), hundreds of cities were constructed on the vast land of China, and the urban system was basically formed by the Tang and Song Dynasties. Although many cities such as Chang'an, Beijing, and Hangzhou as capitals and Lijiang, Suzhou, and Baoding as county cities experienced repeated declines due to war over a 2000-year period, they were basically located in the original places during peacetime and were prosperously constructed again according to the old shape, thereby displaying stronger vitality as all the cities are located in areas where a defense function is realized, the water and food supply is adequate, the ecological service resources are rich, and they are naturally and strongly supported. The space pattern adapts to geomorphological agents, hydrological features, and climate states. Natural ecological services can be better utilized in urban life activities; meanwhile, the ability to resist natural disasters is greater, and the safety and the operation efficiency are greater. The urban building cluster form can fully adapt to the local climate, and natural energy is utilized to create comfortable environments. All of these aspects display the adaptability of the ancient Chinese cities to the terrain, climate, water systems, and other natural elements, and the rich ecological wisdom contained in their planning and design. By researching the adaptability of the ancient Chinese cities to terrain, climate, water systems, and other natural elements, we can understand how the terrain, water systems, climate, and other natural elements affect and act on site selection, space patterns, urban textures, and building clusters, and how multiple services are provided for the cities by means of natural resources to meet people's life demands and ensure the sustainability of the urban ecological environment. Since the 1970s, a lot of research work has been done, mainly focusing on discussing the adaptability of certain cities to certain natural environments (Nianhai Shi 1996, 1995, 1999; Shiguang Zhu 1984, 1990; Lingfu Li 2009). Hangzhou has been sufficiently researched in aspects of the city history and the relation between the urban construction and the West Lake (Yulong Zhong 1982; Feng Zhou 1997; Zhengqiu Lin 2011; Kuan Yang 1993; Yeju He 1996). Sufficient research on terrain features, city origins, and the like, has been performed for Beijing (Renzhi Hou 1988, 2005, 2009; Fan Cai 1987; Bingjian Ma 1999), focusing on the adaptability of certain cities to certain natural elements, but it is short of overall exploration and lacks exploration of universal ecological wisdom. This chapter tries to comprehensively explore the ecological wisdom in the planning and design of ancient Chinese cities adapted to nature. As nature is a basic element of urban construction and an important component of the urban ecological system, how to handle the relation between cities and nature in the planning and design of the cities becomes one of the core problems in sustainable urban planning and design. At present, along with expansion and acceleration of urbanization, the pressure on nature caused by urban development and construction is increased, and the relation between cities and nature is becoming precarious, so adaptability to natural urban planning and design is vital for safe city construction, landscape construction, and harmonious human settlements. The ecological wisdom in the planning and design of the ancient Chinese cities adapted to nature explored in this chapter plays a large and enlightening role in contemporary urban planning and design.

## 7.2 Study Areas

Six typical ancient Chinese cities—Chang'an in the Sui and Tang Dynasties, Lin'an in the Song Dynasty, Beijing in the Ming and Qing Dynasties, and Lijing, Suzhou, and Baoding as county cities—are selected in this chapter to explore the ecological wisdom of ancient Chinese civilization in city construction. This group includes mountain cities and plain cities distributed in different climate zones and terrain areas in China, located in the north temperate zone, the warm temperate zone, and the subtropical zone, so they are typical of the construction of ancient Chinese cities. Chang'an in the Sui and Tang Dynasties, located in the Guanzhong Basin in northwest China's Yellow River Valley, has relatively flat terrain, has a relative height difference below 20 m, is surrounded by eight rivers, and has a warm temperate semihumid continental monsoon climate. Lin'an in the Song Dynasty, located in the south of the Yangtze River delta in China's southeast coastal area, next to Hangzhou Bay in the east and the Qiantang River in the south, leans on Fenghuang Mountain, Wu Mountain, and other ranges of Tianmu Mountain in the southwest. It is embraced on three sides by mountains, is positioned at the destination of the Beijing–Hangzhou Canal, and has a subtropical monsoon climate. The terrain inclines from southwest to northeast, and one side faces the West Lake, with a densely distributed river network. Beijing in the Ming and Qing Dynasties, located on the Yongding River alluvial fan and embraced on three sides by mountains in the northwest of the North China Plain, is next to the Wenyu River in the north and the Liangshui River in the south, stretches across the Gaoliang River, and has a typical north temperate zone semihumid continental monsoon climate. The Gaoliang River and the Tonghui River transversely penetrate through Beijing. Lijiang, located in the northwest of Yunnan Province, is taken as the transition zone from the Qinghai–Tibet Plateau to the Yunnan–Guizhou Plateau. It is in the Lijiang Basin central piedmont zone, embraced on the periphery by the Jinsha River, next to Shizi Mountain in the west and Xiang Mountain and Jinhong Mountain in the north, and connected with wide Pingpa in the southeast. It is in the gentle slope zone of 15.2°, is taken as a mountain city in which the water network is densely distributed, and has a warm temperate monsoon climate, with the terrain gradually decreasing from north to south. Suzhou is located in the slightly higher area in the middle part of the Taihu Plain, in a hilly plain zone, and has a subtropical monsoon climate, with crisscrossed rivers and lakes. Baoding is located in the middle part of the North China Plain, at the east foot of Taihang Mountain and on a flat low plain. It is surrounded by four rivers in the north and five rivers in the south, and has a temperate semiarid continental monsoon climate.

## 7.3 Research Method

First of all, historical references were consulted and local chronicles, maps, and other firsthand data on the ancient cities were obtained from the Local Chronicles Museum. These included the Chang'an Map (from the Northern Song Dynasty) and Xingqing Palace Map (from the Northern Song Dynasty) related to Chang'an in the Sui and Tang Dynasties; the Shuntian Chronicles (from the Ming Dynasty) and Beijing Palace Map (from the Wanli Reign of the Ming Dynasty) related to Beijing in the Ming and Qing Dynasties; the Lin'an Chronicles (Xianchun) related to Lin'an in the Song Dynasty; the Yunnan Chronicles, Brief Lijiang Chronicles, Lijiang History, Xu Xiake's Travel—Yunnan Tour Diary, Dali Chronicles—Geographical Chronicles, and Lijiang Chronicles—Private School related to Lijiang; the Suzhou Local Chronicles Museum–Suzhou Chronicles and Suzhou Records—Riverway related to Suzhou; and the Baoding Chronicles and Baoding Urban, Rural Construction Chronicles and the like related to Baoding. Through reading and analysis of these historical references and tracing of the graphics, the historical original appearances of the ancient cities were recovered, and the logical relation of space in the planning and design of the ancient cities and the logical relation between the cities and nature were searched for in the historical data.

Secondly, site investigations were carried out, with investigations of the natural environment, the surrounding mountain and water positions, the traffic, the functional divisions, the building layouts, the interior and exterior features of individual buildings, and the like, in the existing ancient cities, using note-taking and photographing methods and analysis by adopting an urban form analysis method. The maps of the urban site layouts were obtained from satellite maps, and the maps for the urban space patterns, urban textures, and building clusters, and other maps were obtained from the references.

Finally, the relation between the natural elements of the six ancient cities and urban construction was analyzed and elaborated: (1) respectively analyzing the adaptability, planning, and design of site selection, space patterns, urban textures, building design, ecological protection, and the like, of the six cities in the aspects of terrain, water systems, and climate, and searching their internal relation; (2) starting from the research for comparison, searching the common points of adaptability planning and design in the aspects of terrain, water systems, and climate of the capitals (Chang'an on the central plain, Hangzhou in the south, and Beijing on the northern plain) and the county cities (Lijiang, Suzhou, and Baoding), as well as the southwest cities (Chang'an and Lijiang), the cities in the south of the Yangtze River (Suzhou and Hangzhou), and the North China Plain cities (Beijing and Baoding), and summarizing the ecological wisdom in the planning and design of the ancient cities; (3) on the basis of the above analysis, exploring how the ecological wisdom in the planning and design of the ancient cities apply to modern urban planning and landscape design practice according to today's society, culture, and technical conditions, putting forward feasible sustainable planning strategies.

## 7.4 Planning and Design of Ancient Chinese Cities Adapting to the Natural Environment

### 7.4.1 *Chang'an in the Sui and Tang Dynasties*

With the advantage of convenience in drainage, Chang'an in the Sui and Tang Dynasties is located at the center of the Guanzhong Basin, is surrounded by mountains and water on the periphery, achieves a natural military defense function, is in the south of Longshouyuan on the south bank of the Wei River north of Qinling Mountain, and is on the highland on the plain, so as not to be threatened by flood water in the Wei River. The terrain is suitable for capital construction and gradually inclines from south to north. The Guanzhong Plain has a humid climate. Chang'an in the Sui and Tang Dynasties is rich in water resources; not only do the Wei River and the Jing River flow through Chang'an city from north to east, but also the Ba River, the Chan River, the Feng River, the Lao River, the Jue River, the Hao River, and the like, originating from Qinling Mountain, longitudinally penetrate through the southeast of Chang'an city, and water can be introduced from the southeast and the southwest simultaneously and be plentifully supplied to meet the demands for water resources. The small plain surrounded by the Feng River, the Hao River, the Jue River, and the Chan River has an east–west span of about 17 km and a south–north length of 40 km, is the widest area at the center of the Guanzhong Plain, and not only can provide enough sites for large-scale capital construction, but also can be expanded toward the four directions. Chang'an is located south of Qinling Mountain, receives plenty of sunshine, and is planted with exuberant vegetation. Qinling Mountain in the north can block the cold wind blown from the north, so Chang'an is not too cold in winter. In the aspect of site selection, Chang'an in the Sui and Tang Dynasties fully occupies a favorable natural foundation (Fig. 7.1).

Chang'an in the Sui and Tang Dynasties is located on the edge of hilly tableland, and is designed by a planner named Wenkai Yu in the Tang Dynasty by utilizing high and low terrain according to local conditions. Palaces, provincial offices, temples, and residences for emperors, royal relatives, dignitaries, and high officials are respectively constructed on six highlands transversely penetrating through the east and the west of Chang'an by fully using the base terrain. Thus, not only are the safety and the military defense of the Imperial Palace and the whole capital facilitated, but also flood control and drainage are facilitated. The six highlands have good ventilation effects and are smooth in drainage and plentiful in sunshine, so they are the best places in terms of the interior environment. Dwellings are constructed in flat places among the highlands. The natural valleys are utilized for storing water to form lakes and garden scenic spots such as Qujiang Pool in the southwest of Chang'an, Taiye Pool in the Daming Palace, Xingqing Pool in the Xingqing Palace. Lowlands among the highlands are constructed for recreation, and ditches are arranged on the lowlands among the highlands according to the terrain to meet the demands for urban water supply and drainage.

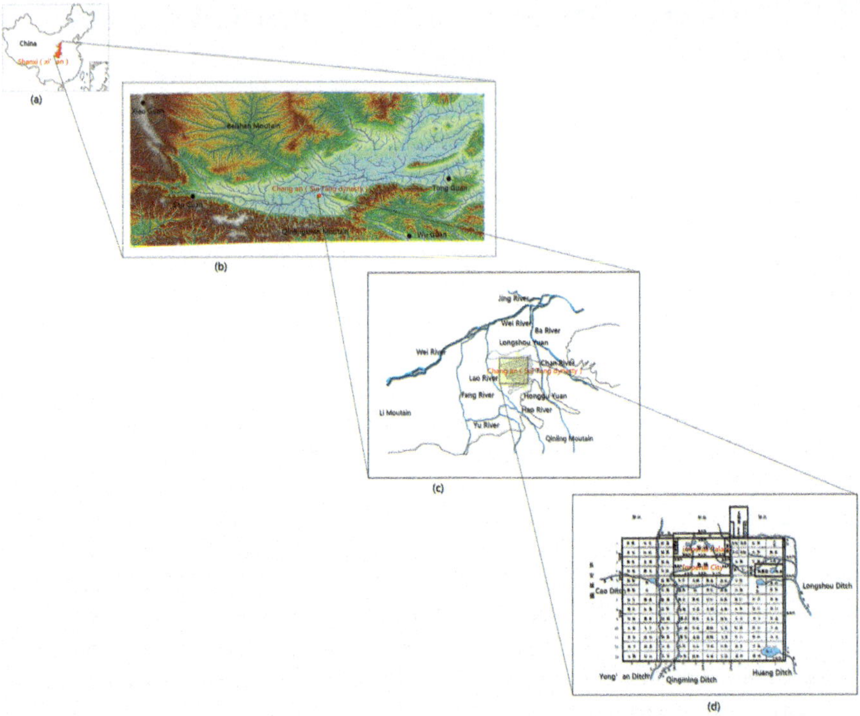


**Fig. 7.1** Examples of data sources on Chang'an in the Sui Tang Dynasty. (**a**) Study area. (**b**) Regional pattern. (**c**) Surrounding natural elements. (**d**) Urban space pattern and texture

In Chang'an in the Sui and Tang Dynasties, the "eight rivers and five ditches" water system pattern realizes the multiple functions of life, production, water transport, flood control, flood drainage, landscapes, and the like. Eight natural rivers outside the city wall penetrate through the periphery of Chang'an in a meandering way, with five ditches artificially dug inside the city wall, i.e., the Yongan Ditch, the Qingming Ditch, the Longshou Ditch, the Huang Ditch, and the Cao Ditch, introducing water from outside to inside, adapted to the geographical situations in route selection, and reasonably arranged along the lowlands among the highlands. The water in the ditches naturally flows through the whole city from south to north by skillfully utilizing the height difference, and pools and lakes formed by valleys among the highlands are used for water storage functions. Therefore, the rivers, canals, pools, and lakes form a complete water system which realizes the multiple functions of the water body.

Chang'an in the Sui and Tang Dynasties is high in the south and gradually and slightly declines from south to north, and the road network structure adopts a chessboard-type pattern mode. The Imperial Palace, the lanes, the city, and the city wall on the periphery are in regular rectangles and form a metropolitan city which is axis symmetrical and is regular in layout. There are 11 streets from south to north and 14 streets from east to west in Chang'an in the Sui and Tang Dynasties, which divide the city into 108 introverted lanes. The whole city faces south and is adapted to the local climate, fully absorbing the sunshine in the south, blocking the cold wind in winter, and creating a comfortable environment. Through the rich water system in Chang'an in the Sui and Tang Dynasties, the urban microclimate is improved, and the temperature of the ambient air in summer can be effectively reduced with the eight rivers flowing through the city. In Chang'an in the Sui and Tang Dynasties, rich water resources are fully utilized and green belts are planted, which are throughout the whole city and form a greening system together with the gardens and the water system, thereby creating a beautiful living environment.

The ecological wisdom in the construction of Chang'an is that the location on the edge of the hilly tableland facilitates flood control and drainage, and the "eight rivers and five ditches" water system pattern realizes multiple functions.

### 7.4.2 *Lin'an in the Song Dynasty*

Lin'an in the Song Dynasty is built near Fenghuang Mountain and is located in a place with relatively higher altitude at the foot of Fenghuang Mountain. Due to the high terrain, Lin'an is not easily threatened by tidewater, so the urban flood control requirements are met. Lin'an is next to the Qiantang River in the east. Due to gradual accumulation of silt, the Qiantang River gradually moves southward, so land space is provided for urban development. Meanwhile, the West Lake in the west is a freshwater lake, so plenty of fresh water is provided for the whole city (Fig. 7.2).

Lin'an is located at the destination of the Beijing–Hangzhou Canal, and is next to Hangzhou Bay in the east and the Qiantang River in the south, so as to form a traffic hub in the area in the south of the Yangtze River. Lin'an gradually expands from mountain to plain under the space pattern of mountains, lakes, and rivers.

Lin'an adapted to the surrounding environment in urban development, is limited by the West Lake in the southwest and the Qiantang River in the southeast, and is next to Fenghuang Mountain in the south, so a long and narrow irregular rectangle from south to north, in a waist drum shape, is presented. The construction of the city wall is also adapted to the topographic conditions. The watercourses in Lin'an mutually penetrate through, thereby forming a skeleton extending from south to north, and also forming a "three longitude lines and six latitude lines" grid-shaped space pattern together with three south–north and three east–west streets (Fig. 7.2).

In Lin'an in the southern Song Dynasty, a "south palace and north city" space pattern is formed, fully adapting to the complex terrain. Before site selection, the terrain and the climate are fully considered. Due to the higher terrain of Fenghuang Mountain, Wu Mountain, and other mountains, the south zone is the earliest land to be involved in the land-forming process of Lin'an, is far away from the dense water system and the damp air, and is smooth in ventilation. This plays an important role in cooling Hangzhou in the hot summer, which is rich in sunshine and water and provides comfortable living conditions, in which government offices and residences

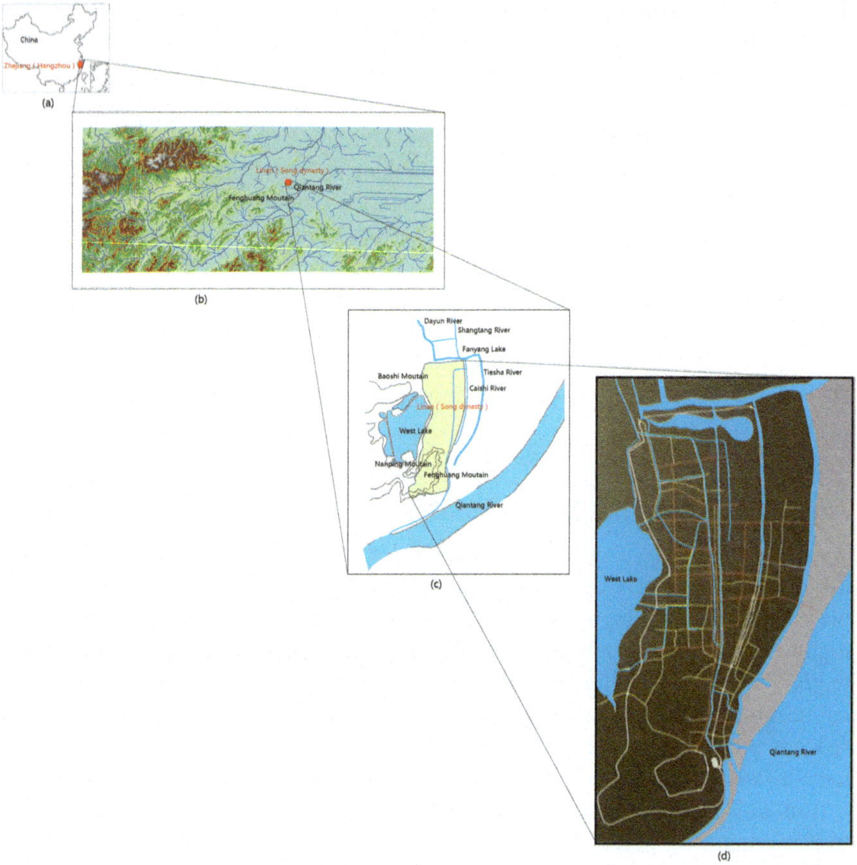


**Fig. 7.2** Examples of data sources on Lin'an in the Song Dynasty. (**a**) Study area. (**b**) Regional pattern. (**c**) Surrounding natural elements. (**d**) Urban space pattern and texture (Source: Hangzhou City Planning Exhibition Hall 2013)

are mainly constructed. In the north zone, the river network is densely distributed. The Beijing–Hangzhou Grand Canal penetrates through the north of the city, so the watercourse in the Canal becomes an important traffic line, and commerce develops along the two sides of the watercourse so as to form a dense berthing facility and concentrated commercial and industrial areas. The streets and the watercourse are crisscrossed to form the overall layout, and the commercial networks extend in a strip shape along the watercourses to form a mode of layout along the rivers and the streets.

Lin'an in the southern Song Dynasty is arranged by adopting a water and land road network (Fig. 7.2), and the roads are freely arranged according to the terrain features. The water lanes, formed by a smaller number of watercourses with larger flow rates, are superposed with main roads and streets and commonly form a main road network skeleton. The urban road layout is combined with the change in the city terrain and water system to form a free road network. Most of the streets adapt to the curve-shaped and fold line–shaped water system form and terrain. The urban space textures are formed by the river regime, water and roads are adjoining and parallel, the river in the middle is the main channel for cargo transportation, markets and warehouses are arranged on the two sides of the river to facilitate commercial development, and the main place for the common people's daily life is near the river.

Lin'an has a hot and humid climate, and the streets are arranged on the premise that good ventilation conditions need to be provided for buildings on the two sides of each of the streets. A free road network layout mode is adopted to meet the demands of the hot and humid climate in summer and the ventilation demands in multiple directions. The most important roads in Hangzhou are the three urban trunk roads parallel to the Shi River, the Yanqiao River, and the Caishi River, and are in the north–south direction. Thus, when the southwest wind prevails in summer, the direction of each of the three urban trunks is basically consistent with the southwest wind direction, so three air flues are formed in the city to enhance the interior ventilation and relieve the sultry weather in summer. Lin'an is adjacent to the West Lake; warm breezes in the local areas are formed by means of roads vertical to the West Lake, and cool air is conveyed from the West Lake to the inside of the city to facilitate ventilation.

The water system in Lin'an is regulated. The urban water supply problem is solved by making full use of freshwater resources. The Shi River, the Yanqiao Canal, the Xi River, and the Maoshan River, which crisscross the city, are connected with the Jiangnan Canal and the river and lake network in the whole Taihu Lake Area, wherein the Shi River reaches the Qiantang River toward the south, directly connected with the Jiangnan Canal and the river and lake network in the whole Taihu Lake Area toward the north, the north section of which communicates with the Wansha River. The Wansha River also communicates with the West Lake, and the water in the West Lake is supplied for these rivers and canals to form the urban water system, playing a role in flood drainage. The Longshan and Zhejiang water gates are built along the bank of the Qiantang River to restrain backward flow of saltwater, so the rivers and canals in Hangzhou are not disturbed by tides, and the original salty land gradually desalinates to adapt to urban development. The courtyard-style building is taken as the main residential building style in Lin'an, which has a high wall, is wide, and is penetrated through from front to back to form through flow, and natural wind can also facilitate ventilation in the residential buildings, adapted to the hot and humid climate environment.

The particular ecological wisdom in the construction of Lin'an is that the streets and the watercourse are crisscrossed to form an overall layout, and the commercial networks extend in a strip shape along the watercourses to form a mode of layout along the rivers and the streets.

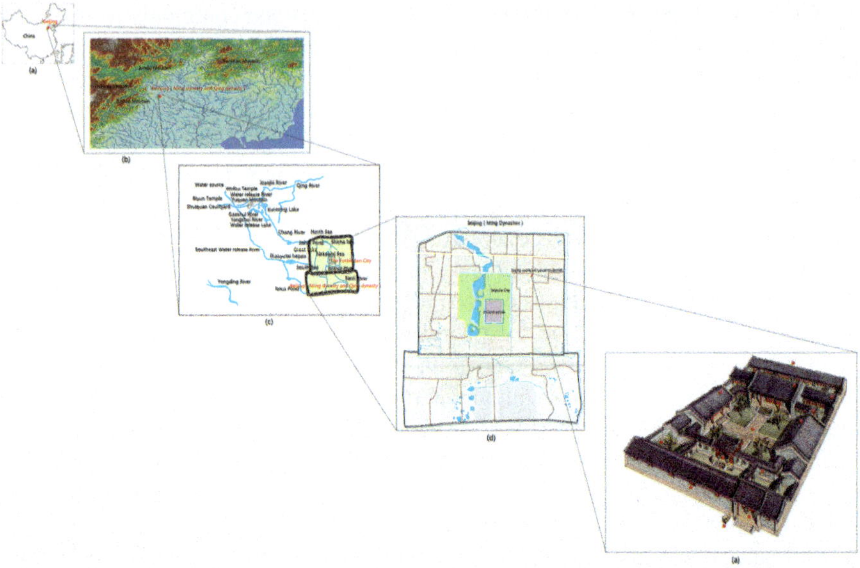

**Fig. 7.3** Examples of data sources on Beijing in the Ming and Qing Dynasties. (**a**) Study area. (**b**) Regional pattern. (**c**) Surrounding natural elements (Source: Hou 2009) (**d**) Urban space pattern and texture (Source: http://www.zgchb.com.cn/newspaper/show.php?itemid=2016) (**e**) Typical residential courtyard in Beijing (Source: Li 2009)

### *7.4.3 Beijing in the Ming and Qing Dynasties*

Beijing in the Ming and Qing Dynasties is embraced on three sides by mountains, so a natural barrier is formed. The Yongding River's water system originating from Taihang Mountain penetrates through the plain zone (Fig. 7.3). Beijing, with the advantage of convenience in drainage, is located on the highland at a distance away from the Yongding River, so that not only is the threat caused by flooding in the Yongding River eliminated, but also plentiful water is ensured, and its terrain gradually inclines from the foot of Xi Mountain to the southeast. Xi Mountain in the range of Taihang Mountain and Jundou Mountain in the range of Yanshan Mountain surround Beijing, so a "Beijing Day" expanding toward the southeast is formed; the warm and moist air from the southeast in summer is blocked by Yan Mountain and Taihang Mountain, is forced to rise, and is met with cold air to fall as rain. As the rainy area is formed on the windward slope in the southeast, Beijing is therefore the area with maximum rainfall within the North China Plain. Meanwhile, the spring water system formed in the mountains all passes through there, so the overall environment in Beijing is very comfortable.

Beijing in the Ming and Qing Dynasties is located on a small plain embraced on three sides by mountains. Its terrain is high in the northwest and low in the southeast, and is gentle and wide, so the urban construction foundation is good. The central axis of Beijing in the Ming and Qing Dynasties is closely related to the natural water body. The water is central to determining the urban pattern. In the urban construction, the lake in the Taining Palace is centered firstly, then three palaces are respectively arranged on the east bank and the west bank of the lake, and finally, the planning and the layout of the whole city are started after the positions of the three palaces are determined. The Imperial Palace is positioned in the center to the south, and its central axis is taken as the main axis of the whole city planning. As the central axis is just superposed with the east bank of the Jishui Pool, the east bank of the Jishui Pool is selected as the center of the plane layout of the whole city, and the Imperial Palace, the Taiye Pool, and the Jishui Pool are taken as the main bodies for urban construction according to a chessboard-type layout. The roads in Beijing are crisscrossed in the east–west and south–north directions, and a plurality of lanes parallel in the east–west direction are formed along the north–south trunk road, which are the roads through the residential districts; most of the streets in Beijing are in the east–west direction. Both the courtyards and the buildings face south and are adapted to the north temperate zone monsoon climate.

In Beijing, the terrain is high in the northwest and low in the southeast, the flowing directions of rivers and ditches adapt to the terrain, and the trunk ditch for drainage runs from north to south and is consistent with the topographical slope. For example, as there is a height difference of 1 m between the north and the south of the Forbidden City, the water in the Jinshui River is introduced into the Imperial Palace from the northwest direction, flows to the south along the west side outside the Imperial Palace, then turns to the east and finally flows out from the southeast corner of the Imperial Palace. Thereby the drainage is very smooth; after its establishment, no flood disaster occurs. The perfect drainage system in the inner city, the outer city, and the Forbidden City of Beijing in the Ming and Qing Dynasties is closely related to the rivers in the city. The three-level ditches—i.e., the Zhi Ditch, the Dou Ditch, and the Mao Ditch—are branched from the streets, from the urban drainage trunk ditch in the south–north direction, taking the Tongzi River as an axis, blending into the moat surrounding the inner city, and finally entering the Huitong River. The moat not only has the functions of urban flood shunting and urban sewage drainage but also is taken as an important water source for urban life, firefighting, and garden irrigation; meanwhile, urban sewage is discharged out of the city to form an organic urban water conservancy system. The urban wastewater supply system in Beijing in the Ming and Qing Dynasties takes the Gaoliang River water system in the suburban Yuquan Mountain as the urban water supply area; meanwhile, the spring water from Xi Mountain is introduced into the Kunming Lake to commonly recharge the Huntong River and the urban water supply watercourses. The Kunming Lake not only is used for urban water supply but also is a storage and drainage water control pivot, which has the function of farm irrigation, is used for shipping and gardens, and provides a reliable water source guarantee for the survival and development of Beijing.

The residential buildings in Beijing are represented by quadrangle courtyards. This building style adapts to the temperate climate in the north of China, where natural light and heat are fully utilized (Fig. 7.3) to meet the demands for sunlight in winter, courtyard width, and reasonable depth, size, and proportion; and to ensure that better sunlight and daylighting are obtained. The proportion of the house height to the eave is set by the laws of the sun angle, so that the eave can block a large quantity of sunshine from direct irradiation in summer and can also ensure sunshine duration in the house in winter. The quadrangle courtyards, which are completely adapted to the climate changes, are sealed in the north and opened in the south so as to block the cold wind in winter and face the wind for cooling in summer.

The particular ecological wisdom in Beijing is the quadrangle courtyard design, which adapts to the temperate climate in the north of China, with natural light and heat fully utilized to meet the demands for sunlight in winter with courtyard width and reasonable depth, size, and proportion to ensure that better sunlight and daylighting are obtained.

### *7.4.4 Lijiang*

In Lijing, the natural military defense city wall is shaped by utilizing the unique geographical environment. The Jinsha River is rapid in water flow, embraces Lijiang Bazi, and becomes a natural barrier for defending the ancient city. Lijiang is embraced on the periphery by mountains, so that a natural "city wall" is formed. Lijiang leans on the mountains in the north and the west, is located on the diluvial fan, and is in the phreatic water overflow zone, in which there is plenty of water. The terrain is high in the northwest and low in the southeast to facilitate diversion and drainage, and to reduce and avoid flood disasters. The mountains and the rivers on the periphery of Lijiang form a natural closed city, which forms a microclimate and has the characteristics of exposure to the sun, warmth, and cold wind avoidance in winter, with coolness in summer.

In Lijiang, the "river water first" principle is abided by in the urban development process. The terrain in Lijiang is high in the north and low in the south. The naturally formed Zhong River flows from north to south, and the Naxi people dwell along the river in the river governance process, so that a cone is formed. In the Yuan Dynasty, when the Tusi Government office is constructed, the water in the Zhong River is introduced from the Yuhe Bridge by means of the terrain height difference, then the Xi River is excavated and Lijiang expands along the Xi River. When the Liuguan Government is constructed in the Qing Dynasty, the water in the Zhong River is still introduced according to the demands of urban development, then the Dong River is excavated and Lijiang further expands along the Dong River. The Zhong River, the Xi River, the Dong River, and their branches are densely distributed in Lijiang, so a radial water system network is formed, which is taken as the basis of the urban space and building growth in Lijiang (Fig. 7.4). Both urban construction and water system optimization are developed synchronously to form the urban textures combining the water and the city. The terrain height decides the direction of the Zhong River, the Xi River, and the Dong River, and further decides the urban pattern and the road textures. The main roads extend along the main rivers, so a picture that the main roads are next to the rivers, and the lanes are adjacent

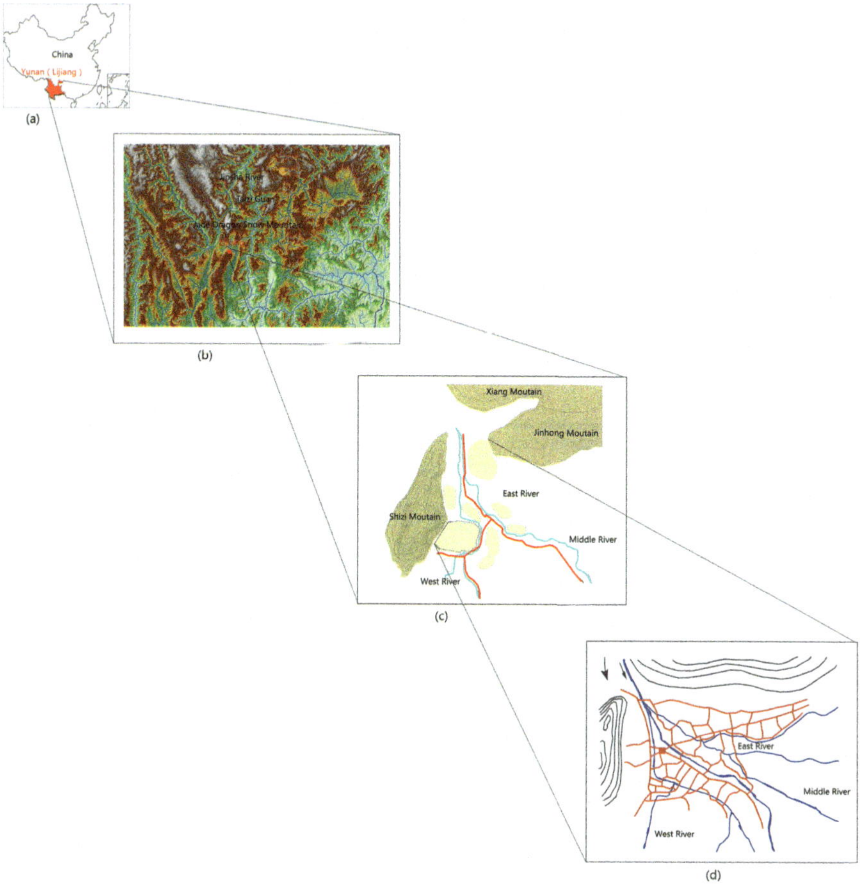


**Fig. 7.4** Examples of data sources on the ancient city of Lijiang. (**a**) Study area. (**b**) Regional pattern. (**c**) Surrounding natural elements. (**d**) Urban space pattern and texture

to the ditches, is formed; the required cobweb staggered road system is combined. The dense building textures in Lijiang adapt to the local plateau climate, and the high-density buildings maintain appropriate temperatures in Lijiang, with an altitude of 2400 m and a mean annual temperature of 12.6 °C, and face south or east to be conducive to obtaining ample sunlight. The traditional residential building for the Naxi nationality in Lijiang is the triangular courtyard enclosed by three squares and one screen wall, which is sealed inward and in which a small number of windows are opened, so that enough sunshine can enter, thereby avoiding the wind and ensuring warmth.

In the construction of Lijiang, the natural rivers are utilized to reduce flood disasters (Lin Zhou 2010). After the Tusi Government excavates the Xi River, the Xi River is higher than the Zhong River in water level, branches are formed at many places on the Xi River to a great extent, and flood control and flood discharge can be realized through frequent water diversion. After the Liuguan Government excavates the Dong River, due to the small gradient and slow water flow speed, sludge is easily deposited in the watercourses and the flood discharge capacity is weak; therefore the Xiehong Bridge is built to introduce flooding into the Zhong River for discharge. Based on this, the Dong River area is prevented from flooding, and the sound development of the water system pattern in Lijiang is ensured.

The ecological wisdom in the construction of Lijiang is that the natural rivers are utilized to reduce flood disasters.

### 7.4.5 *Suzhou*

Suzhou is located in a higher place at some distance from the Tai Lake on the Taihu Plain, and a group of mountains is spaced between the city and the Tai Lake. Suzhou is close to the Yangtze River in the north and is adjacent to the Tai Lake in the west, which not only is convenient for water usage but also can avoid direct impacts caused by floodwater in the Tai Lake, the terrain of which is flat. The slow water flow ensures excellent water navigational conditions in Suzhou, and the Grand Canal passes by the city, so Suzhou becomes a waterway transportation hub. The dotted lakes on the periphery of the city can play roles in dispersing the water when floodwater comes, and regulate the water flow so that flood disasters are avoided.

In Suzhou, the Imperial Palace is located in a southeast position on higher terrain and is rarely threatened by floodwater from the few rivers. Some administrative buildings are also located in a southwest position on higher terrain (Fig. 7.5). In order to meet the demands for military affairs, waterlogging prevention, traffic, and people's life, the Tai Lake water network is controlled, ditches are excavated outside the city, a crisscrossed watercourse network and inner and outer moats are constructed and excavated inside the city, and eight water and land double gates are constructed, so that the river and lake water system inside and outside the city mutually penetrate through. By utilizing nature, with the traffic system, which takes the river water system as a pivot and is supplemented by water and land parallelism, a "water and land adjacency and river and street parallelism" double-chessboard-type planning pattern is formed (Shengfang Yu, 206).

In order to adapt to the hydrological regime in which the rivers flow from west to east, with torrential water flow, a city wall is constructed. The northeast corner and the northwest corner are in a fold line shape, so the watercourse curvature is increased; therefore, the water flow is smooth, and sailing and drainage are also facilitated. The southwest corner of the city wall is in a convex arc shape, so the water in the Xu River and the ditch pass by the arc-shaped corner of the city wall; therefore, the water pressure from floodwater in the Tai Lake is reduced. The Pan Gate turns from southwest to southeast, so front flood peaks can be avoided. The water flows in through the Pan Gate and enters the watercourses in the city, so the pressure caused by floodwater at the Pan Gate is reduced. At the southeast corner, the water mainly flows to the southeast along the ditch, and the water flow is gentle,

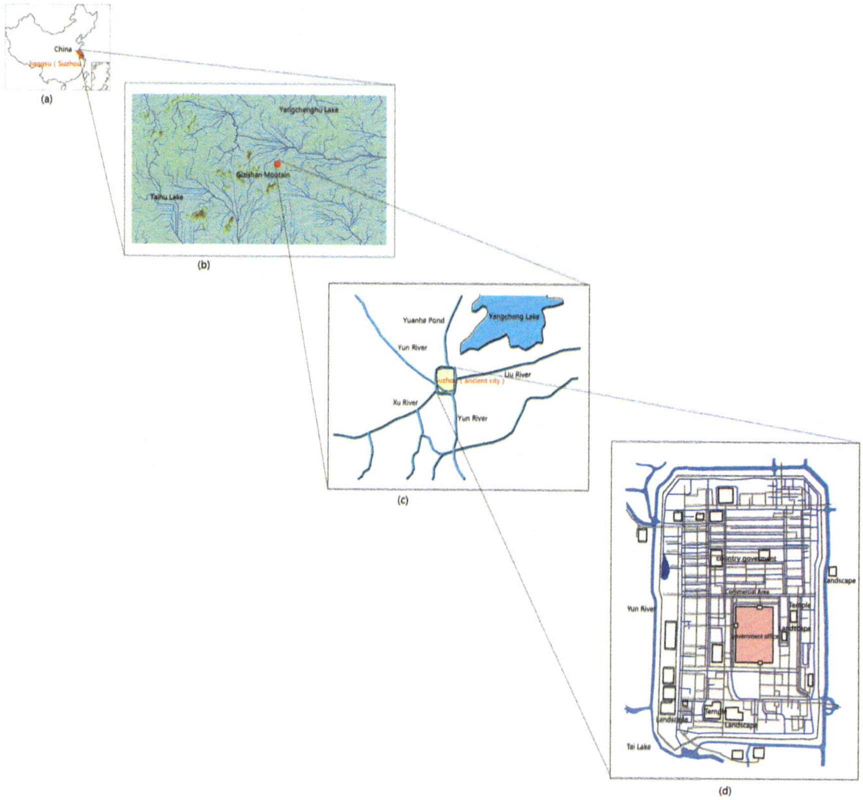


**Fig. 7.5** Examples of data sources on the ancient city of Suzhou. (**a**) Study area. (**b**) Regional pattern. (**c**) Surrounding natural elements. (**d**) Urban space pattern and texture (Source: He 1996)

so a right angle can be maintained (Shengfang Yu 1980). The terrain in Suzhou is high in the west and low in the east. The positions of the water and land city gates adapt to the flow direction and motion of the water and adopt an asymmetric layout. One city gate is arranged in a place close to the main watercourse, and no city wall or water gate is arranged in a position to suffer from frontal impact of floodwater. The water in the Tai Lake comes from the southwest, flows through the Chang Gate and the Pan Gate, is shunted in the city, flows out from the Feng Gate, the Lou Gate, and the Ji Gate, and finally flows into the sea. The positions of the city gates in Suzhou are specifically selected according to the hydrological characteristics, adapting to the water system environment features. Known from the "Pingjiang Map", the orientation of Suzhou is toward the southeast at an angle of 7°54′ (Shengfang Yu 1986), and the road pattern and the water system pattern cater to the prevailing southeast wind in summer, so a smooth ventilated corridor is formed.

Suzhou adapts to its water village features, with river and street-parallel water lane textures (Fig. 7.5), and the density of the water system determines the street size and the street space mode. There are three trunk rivers in the east–west direction and four trunk rivers in the south–north direction. Each of the four rivers in the south–north direction has a width of about 10 m, and their flow rates are basically consistent, so the water level throughout the whole city is ensured. Meanwhile, each of the three rivers in the east–west direction are communicated and regulated, with a width of about 6 m and channels for connecting all households. As the horizontal, flat, and vertical watercourses are longitudinally and transversely arranged, a parallel uniform water lane texture in the east–west direction or the south–north direction is formed. The watercourses and the lanes parallel to the watercourses form a water lane space that has a length of 200–4000 m. There are a large number of residential buildings and a small number of stores, handicraft workshops, or temples along the rivers and the streets. In Suzhou, each watercourse has a width of 6–8 m generally, and each street has a width of 3–5 m, which is hospitable to people. The city wall, the Hao Pool, the moats, the water gates, and the dike in the ancient city integrate a water conservancy system to block floodwater and water logging of drainage. In order to ensure that each household can conveniently use the watercourse, all the buildings are arranged along the rivers, most of them are in the south–north direction, and the courtyards are located in the east–west direction and are closely connected, so that a continuous residential group is formed. The residential courtyards are enclosed on four sides and are opened upward in the middle part to provide effective water collection, sunshine irradiation, ventilation, daylighting, and the functions of temperature reduction and humidity and moisture protection.

The ecological wisdom in Suzhou is that its location in a higher place at some distance from the Tai Lake avoids direct impacts of floodwater, and the Grand Canal passing by the city helps it become a waterway transportation hub.

### *7.4.6 Baoding*

Baoding, with an average altitude of about 20 m, leans on Taihang Mountain in the west and is adjacent to the Hebei Plain in the east. Its terrain experiences a 1% slope degradation from northwest to southeast. Due to geological factors and the terrain, the east foot of Taihang Mountain forms a diving (confined water) ascending spring group and is very rich in water. In Baoding, the water is introduced into the city according to the terrain conditions, so the problem of water shortages is solved and conditions are created for urban afforestation. Baoding is near mountains and water, and its urban pattern forms a natural barrier for urban defense (Fig. 7.6).

Baoding is situated at the east foot of Taihang Mountain, which exposes it to the sun, and is warm. It is not extremely cold in winter, as Taihang Mountain in the southwest–northeast direction blocks cold air from the northwest in winter. There are many rivers outside the city. The Baiyang Lake with its large water area is located to the east of the ancient city and has the effect of regulating the climate to a certain degree.

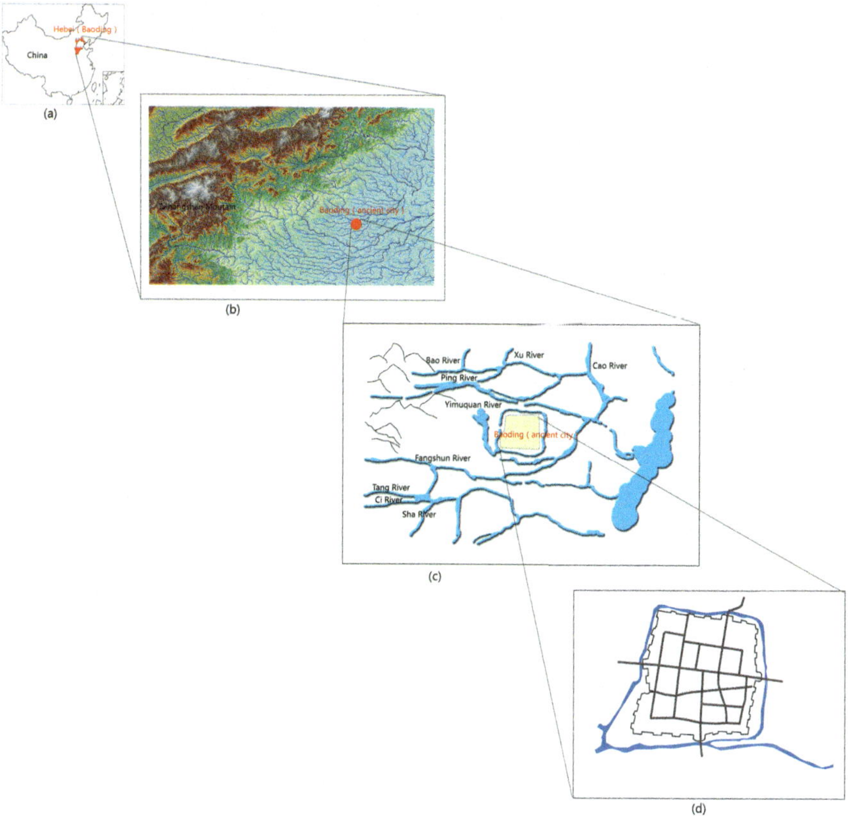


**Fig. 7.6** Examples of data sources on the ancient city of Baoding. (**a**) Study area. (**b**) Regional pattern. (**c**) Surrounding natural elements. (**d**) Urban space pattern and texture

At the beginning of urban construction, in order to meet the needs for military and grain transportation, water is introduced into the city and is shunted after entering the city, so an urban pattern that Jiju Spring surrounds the whole city is formed (Fig. 7.6). The city faces south by west, with roads at an angle of 22° from magnetic north and approximately parallel to the flow direction of the Fu River. As the flow direction of the Fu River is parallel to the prevailing southwest wind in summer, wind is favorably introduced into the whole city and the air temperature is reduced. At the beginning of the Jin Dynasty and at the end of the Yuan Dynasty, the water area accounts for four tenths of the urban area, so it plays a role in balancing the urban climate. The trunk road in the city adopts a north and south street dislocation way. The sharp wind blown from the north is blocked after entering the north streets and is dispersed and weakened, so the cold air does not rapidly disperse through the whole city. All south–north and east–west roads in the south of the ancient city are straighter and smoother and, in particular, the south–north roads can channel the wind in summer to reduce the temperature. The buildings in the ancient city have north and south–oriented quadrangle courtyards parallel to the roads, basically, and the height and the spacing of each quadrangle courtyard and the layout of each window are adapted to the cold and dry regional climate features in winter in the north.

The ecological wisdom in Baoding is that the city orientation is south by west, the roads form a 22° angle with magnetic north and are approximately parallel to the flow direction of the Fu River, and as the flow direction of the Fu River is parallel to the prevailing southwest wind in summer, the wind is favorably introduced into the whole city and the air temperature is thereby reduced.

## 7.5 Summary of Ecological Wisdom in Planning and Design of Ancient Chinese Cities

Through analysis of the planning and design of these six ancient Chinese cities, we discover that a good relationship of adaptability to nature exists among these cities, so that the urban space coordination ecological wisdom of "skillfully utilizing natural ecological resources and creating coordination between nature and human beings" in the planning and design of these ancient Chinese cities is embodied, which is concretely manifested as follows:

1. These ancient Chinese cities are located in places where a defense function is realized, the water resources are rich, and the land reserve is sufficient on the basis of fully analyzing the regional terrain pattern. Natural mountains, rivers, lakes, and other barriers are formed around most of the cities, which are fully utilized as defense installations in the ancient age to ensure city safety during wars. The cities are on the terraces of the rivers, are adjacent to the rivers, and keep a certain flood control distance, so that water is easily obtained while threats caused to the cities by floodwater are avoided. The cities can be expanded to be convenient for urban development.
2. The space layout of these ancient Chinese cities adapts to base terrain fluctuations; the urban land is arranged and the road network skeleton is designed for local conditions, and functions are achieved to facilitate city life. The plain cities generally adopt chessboard-type road networks, so the utilization efficiency of urban land is improved. The road networks of the mountainous cities are generally arranged along the contour lines, thereby presenting irregular forms. The cities in watercourse network areas adapt to the water system forms, and their urban road areas accompany the water system. The complete natural environment is retained and a state of harmony is maintained between natural forces and artificial construction, so that the services provided by nature can be accessed and the damage caused by fluctuations in nature can be avoided.
3. The water systems in these ancient Chinese cities are adaptively utilized. The regional water systems in most of the cities can be improved according to local natural hydrological features to ensure rich water resources and convenient water

transportation, and both flood disaster and drought disaster can be reduced through control of rivers and lakes. Meanwhile, the water systems inside the cities are created by skillfully using the water system layout to meet the demands for urban drainage, flood control, and landscaping, and to create suitable living environments.

4. The city textures of the ancient Chinese cities all try to adapt to the regional climate, with appropriate street directions, building orientations, and building cluster forms being selected according to the different climates to create comfortable temperature and wind environments. For example, the street direction must be parallel to the prevailing wind direction in summer as much as possible in hot and humid areas to ensure that the wind can be blown into the cities smoothly. The buildings face the south as much as possible to maximally utilize solar energy, and the building cluster forms also adapt to the different climatic environments to ensure comfortable temperatures and humidity levels in the courtyards and indoors.

## 7.6 Contemporary Relevance of Ecological Wisdom in Planning and Design of Ancient Chinese Cities

In order to solve the problems of urban environmental deterioration, urban disasters, habitability decline, insufficient sustainable development ability, and the like, in today's rapid urbanization process, the ancient ecological wisdom needs to be rediscovered in urban planning and design practice and blended into modern urban planning and design to realize greater coordination between urban construction and natural systems. Through analysis, this study has discovered the following relevance of the ecological wisdom used in the planning and design of ancient Chinese cities:

1. Urban land expansion and urban form planning and design need to conform to natural ecological process rules. On the premise that the natural ecological quality does not degrade, the urban land is selected, and the urban form is organized to realize internal coordination between artificial landscapes and natural landscapes, and the materials and energy provided by nature are utilized to blend more natural ecological services into the human life system. For example, in the urban expansion process, forests, wetlands, and other areas with rich biological diversity should be protected to the greatest extent, the functions and the services of the natural ecological system should be safeguarded to the greatest extent, the urban scale should be controlled within the carrying capacity range of the ecological system, and the health of the ecological system should be ensured. Urban forests, urban rivers, urban mountains, and other natural elements should be regarded as constituent parts of the urban ecological system and blended into the urban function system.

2. The planning and design of urban space layout need to adapt to natural ecological patterns to perform their functions. The urban layout should be designed on the basis of understanding regional ecological suitability, and the urban space structure should be constructed by taking the natural space as a base, on the basis of fully understanding the existing urban terrain and landforms, with the natural elements leading space growth. Both urban structures and natural structures should organically grow in a balanced way, and manmade landscapes and natural landscapes should be blended to realize a balance between urban processes and natural evolution processes to basically maintain regional ecological stability.
3. The adaptation of the planning and design of the ancient Chinese cities to the natural terrain are of crucial importance to the sustainable planning and design of the cities. The terrain decides the process and the pattern of urban nature; therefore the urban road system, the urban drainage system, the urban landscape system, and the like, must adapt to the terrain system and the base terrain fluctuations, so the urban land and its road network are arranged according to local conditions to facilitate city life.
4. The urban natural water system plays a huge role in the aspects of urban ecological system health and is an important object of landscape design and urban planning. The natural forces contained in the natural water system and the services it provides must be known in order to adapt to the space textures of these forces and services, and the natural landscape should be combined with the urban form in the design to create comfortable urban environments while achieving the functions of flood control and flood drainage.
5. Urban planning and design need to adapt to the regional climate. Through appropriate design of buildings, building clusters, and urban groups, the utilization of wind energy, solar energy, and other natural forms of energy can be enhanced to optimize the usage of the natural energy. For example, ecological buildings that adapt to the local climate should be used as much as possible in different climate areas. Creation of urban spaces capable of assisting urban ventilation and buildings that conserve heat should be emphasized in hot areas and cold areas, respectively, to reduce the use of artificial energy.

**Acknowledgments** The authors thank the Planning Exhibition Hall and the Planning Bureau for providing useful information.

## References

Cai F (1987) Study on Beijing ancient canal and urban water supply [M]. Beijing Publishing House, Beijing
Fu C et al (2009) History of China city's development [M]. Social Sciences Press, Beijing
He Y (1996) History of Chinese ancient urban planning [M]. China Building Industry Press, Beijing
Hou R (1988) Beijing historical map collection [M]. Beijing Publishing House, Beijing
Hou R (2005) About Beijing in Hou Renzhi's eyes [M]. Beijing Publishing House, Beijing

Hou R (2009) Life's imprint in Beijing [M]. Life/Reading/New Knowledge Sanlian Bookstore, Beijing
Li L (2009) Layout and geographical basis of Xi'an [M]. People's Publishing House, Beijing
Lin Z (2011) History of ancient cities in Hangzhou [M]. Zhejiang People's Publishing House, Hangzhou
Ma B (1999) Beijing Quadrangle Courtyard building [M]. Tianjin University Press, Tianjin
Shi N (1995) Historical evolution of Terrain in Xi'an [J]. Chinese Historical Geography, 1995, 04
Shi N (1996) Xi'an historical map collection [M]. Xi'an Map Publishing House, Xian
Shi N (1999) Longshouyuan and Chang'an in the Sui and the Tang dynasties [J]. Chinese Historical Geography 1999, 04
Yang K (1993) Study on history of ancient Chinese capitals [M]. Humanities & Social Sciences, Beijing
Yu S (1980) Masterpiece of planning and design of ancient Chinese cities—Suzhou map [J]. J Archit 1980(01):15–20
Yu S (1986) Discussion on water system problem in planning of Chinese cities from Suzhou [J]. J Archit 1986(09):37–42
Zhong Y (1982) About Hangzhou [M]. Zhejiang People's Publishing House, Hangzhou
Zhou F (1997) Hangzhou in the Southern Song Dynasty [M]. Zhejiang People's Publishing House, Hangzhou
Zhou L (2010) Comparison and study on diachronic growth and evolution and space form of ancient city of Lijiang [J]. Study Urban Dev 2010(07):106–112
Zhu S (1984). Macro geographical situations and micro geographic features in Chang'an [A]. Society of Chinese ancient capital. Chinese ancient capital research, vol II. In: Proceedings of the Second Annual Meeting of the Society of Chinese Ancient Capital [C]. Society of Chinese Ancient Capital, 1984: 13
Zhu S (1990) Chang'an–Beijing comparison and research on the most important ancient capitals in China [J]. Chinese ancient capital research. In: Proceedings of the eighth annual meeting of the Society of Chinese Ancient Capital, 1990, 08

# Chapter 8
# Green Space System Planning Practices of Multidimensional Network Construction in High-Density Areas Under the Coexistence Trend of Incremental Planning and Existing Stock Planning

**Jin Zuo, Chen Li, and Jing Dong**

**Abstract** Under the background of China's new normal, urban construction has entered a coexistence trend of incremental planning and existing stock planning, where green space in urban high-density areas are affected by an extreme shortage of land supply. Due to the lack of coordination of complex interests in traditional green space planning which can cause the huge transaction costs. This chapter takes two projects in the city of Tianjin as case studies and discusses existing stock space as a resource for land supply in order to reduce transaction costs for green space. The principle of "dividing levels, connecting networks, and building multi-paths", and the planning strategy that "turns stock into treasure, connects greens into a network, expands the green to three-dimensional" shows an innovative planning strategy for integrated open spaces of an urban green system, which concentrates on ecological conservation, public service, and community vitality. Through green space networks and three-dimensional and multilevel construction, new planning methods for urban green space systems in high-density areas have been explored for "urban repair and ecological restoration" in this work.

**Keywords** Urban high-density area · Transaction cost · Existing stock activation · Urban green space network · Three-dimensional green

J. Zuo (✉) · J. Dong
School of Architecture, Tianjin University, Tianjin, China
e-mail: zuojin0117@163.com

C. Li
Tianjin Urban Planning & Design Institute, Tianjin, China



Z. Shen et al. (eds.), *Green City Planning and Practices in Asian Cities*, Strategies for Sustainability, https://doi.org/10.1007/978-3-319-70025-0_8

## 8.1 Introduction

The speed of urbanization in China has increased more than 50% within less than 40 years, which is a highly compressed urbanization process (Zhang Jingxiang and Chen Hao 2010). Rapid economic development is accompanied by an urban spatial structure imbalance, and large numbers of city green spaces have been eroded, especially in old cities, which have high density and are highly established. Influenced by many complex factors, such as economic, social, historical, and land conditions, construction of an urban green space system is far behind the pace of urban development. Due to ignorance or avoidance of the contradiction of reality, traditional green space system planning cannot be implemented or permanent, creating a huge contrast between the form of green space planning and the negative diffused space of reality. Taking the Hedong District of Tianjin as an example, the area is about 42 $km^2$, the population density is 23,000 people/$km^2$, and the per capita public green space is only 2.4 $m^2$ (Tianjin Urban Planning & Design Institute 2016a), which is far behind the planning target value in the Tianjin master plan, which was set at 12 $m^2$ (Tianjin Urban Planning & Design Institute 2015).

Accompanied by China's new normal, and the transition of China's social economy, Chinese cities have entered a stage of development and structural adjustment in parallel under the construction goals of a livable urban environment. Chinese cities face a lack-of-green space problem, land shortages are severe in high-density cities, and there is no room to effectively expand. There are many reasons for the current predicament, but fundamentally it can be considered that resistance to the expansion of built-up areas is much greater than the pull of the creative value of expansion implementation. Although the green space in highly established areas is deficient, this cannot be entirely attributable to urban planning and development. This chapter attempts to reflect and explain, from the urban planning aspect, why the expansion of green space is blocked, and combines it with two practical plans in Tianjin: the Round-the-City Railway Park Planning in Tianjin and the Green Way Network Plan in Tianjin Hedong District. This work attempts to explore the reality of land shortages in urban high-density areas and determines how to use limited space resources to deal with the diversity of the space demands by fully cataloguing and using city stock space, moderately adjusting and perfecting the existing spatial structure of the city, and reshaping the backbone of the central urban ecological system. By deepening and expanding multilevel construction, the plane becomes a network, which allows the space to become three-dimensional green space, thus constructing a new ecological network system in an urban high-density area.

## 8.2 Problem Analysis

### *8.2.1 Difficulty of Land Requisition and High Transaction Cost*

From the perspective of urban management, urban green spaces cannot produce direct benefits for the main stakeholders. However, indirectly the premium can be increased for profit-oriented land around green space, such as residential and commercial land. In other words, the construction cost and opportunity cost of an urban green space should be covered by the appreciation of profit-oriented land. The benefits of urban green spaces for a city are firmly tied to profit-oriented lands surrounding them. The implementation of urban green spaces in isolated areas faces the financial paradox of "just investment and no output" (Su Ping 2013). Financial difficulties can directly and seriously weaken the motivation of government to create an urban green space system, which fundamentally limits the supply of urban green spaces and the implementation of urban green space planning.

Urban green spaces are important public resources, and traditional urban planning requires allocation of various types of urban green spaces to achieve construction of a green space system in a city. However, this is the work of the 1.0 stage of urbanization, which is established on the engineering fundamentals of incremental planning. Incremental planning implies a hypothesis meaning the entire city is a single property, and through the rationality of technology, the green system is arranged on a "white paper" to construct a highly ideal urban green space system structure. However, when entering into the current period of urbanization 2.0, especially in the high density of an old city, the space has already been filled with various color blocks. To construct a green space system by a conventional method requires land expropriation and housing removal to access the land use rights, which will certainly incur huge transaction costs and difficulties (Zhao Yanjing 2014). Therefore, before planning and implementation in urban high-density areas, a planner need to fully understand and respect the status quo of the city by re-excavating and organizing inefficient and abandoned stock space resources, looking for possible, relatively lower cost methods. It is necessary to design a reasonable benefit distribution method and allow premium income regurgitation feeding to implement urban green space, increase urban green space as much as possible, and promote the effective implementation of an urban green space system under coordination of interests among multiple subjects.

### *8.2.2 Blueprint Planning and Lack of Implementation Guidance*

At present, traditional urban green space system planning is still based on blueprint planning, which often resists action and is limited to the closed planning process within local city planning authorities. This chapter considers the following two aspects.

First, urban green space system planning is an important and indispensable part of an urban master plan, which arranges the goals, structures, and layout in a coordinated manner. However, such planning mainly focuses on a highly idealized green space form and lacks a specific coordinated process. During the compilation process of regulatory planning, which is influenced by conflicts of interest over real spaces, the boundaries of the master plan and many other aspects are indistinct, thus it is often difficult to achieve transformation from the "target" to the "indicators" (Liao Yuantao and Xiao Rongbo 2012). Second, traditional urban green space emphasizes the artistic and systematic form but lacks adequate and specific action guidance and tracking. The demands and interests of the grassroot levels of government, which is the main body involved in planning and implementation, are often ignored. Thus, green spaces are under pressure and boundaries are pushed back during different specific projects, which cannot meet the diversified needs of urban construction.

## 8.3 Methods: Activation and Remodeling

Under the trend that urban existing stock planning will be more and more important in urban high-density areas, rather than incremental planning, new opportunities for urban regeneration are raising development of urban green space systems. If planners fully dig the stock space resources, it is possible to greatly reduce the transaction costs and establish a new way to supply the land for green space. Under the principle of "dividing levels, connecting networks, and building multi-paths" and the action planning strategy of "reconstructing the skeleton, extending the meridians, and filling in the flesh," it is possible for green space to be treated as an accelerant to improve the public space environment, promote the regional urban space value, and optimize the urban functional structure in order to establish a positive interaction between urban green space and urban functions.

### *8.3.1 Turning Existing Stock Resources into Available Land and Reconstructing Green Skeletons*

In high-density urban areas, there are some existing stock lands that are not fully used or still have low value, such as abandoned railway lines that obstruct city traffic, influence the city's appearance, and will no longer be prosperous. Therefore, under the condition of unchanged property rights, it is possible for government departments and agencies to effectively reduce the transaction costs. The sites of existing stock lands can be further combined with urban rivers, landscape roads, etc., to form a new city skeleton for an urban open green space system. Taking the Round-the-City Railway Park Planning in Tianjin as an example, through "making stock land available and profitable," will not only form city-level green open spaces that combine "ecological conservation, public services, and green transport"; it will also preserve a large number of industrial heritage buildings, important public spaces (buildings), and a large number of communities for "turning from the back to the front." As this promotes the activation of stock properties and community infrastructures, it enhances the value of the area surrounding those existing stock urban spaces.

### 8.3.2 Combining Green Spaces Together for Green Networks

On the basis of reconstructing a green system skeleton, an urban green space plan must make various green spaces, including spots and lines, into a "network" in order that the ecological benefits of urban green space can be fully used. Taking the Hedong District of Tianjin as an example, the section of the round-the-city railway park in the Hedong District is treated as a premise, connecting to existing green "spots" and "blocks" in a series of line spaces, which can transform ecological areas, such as green river spaces, tree-lined roads, abandoned railway branch lines, and other stock resources. In this way, a crisscross "green park and road network system" can be formed, which extends the boundary surface of the green space, improves its accessibility, brings great kinetic energy for the city's economic, social, and ecological well-being, and maximizes the optimization of regional urban space (Wang Zhaolin and He Fang 2012). Meanwhile, complying with the will of district-level governments, detailed studies of land use conditions and actual demands are required. When driven by district-level governments, coordinated development in both the network system construction of green park roads and improvement of the spatial value of regional cities can be realized.

### 8.3.3 Expanding the Greening to Three Dimensions to Fill in the Flesh

Urban high-density areas face the reality of land shortages; thus, extending the plane of an urban green space network in a city requires upward expansion. Installation of green roofs on multidimensional surfaces will realize the three-dimensional extension of green spaces. At present, many cities in China, such as Guangzhou and Hangzhou, have a policy that green roofs can be converted into city green spaces. Moreover, special subsidies for greening roofs are available, which can actively encourage market forces for co-creation. From the technical perspective, the technology of fabricated light roof greening has the advantages of low cost and a light load, is easy to maintain, and has an especially high rate of resurrection and greening after winter (the rate can reach up to 90% or more in north China). Thus, fabricated light roof greening has the possibility to be popularized in northern China.

## 8.4 Results and Discussion

### *8.4.1 Reconstructing the Skeleton: The Round-the-City Railway Park in Tianjin*

Tianjin is the origin of the Chinese railway culture. As the cradle of northern industry, the railway has been the core pillar to support prosperous development of modern industry in Tianjin. Since the first railway started in 1888, Tianjin has built hundreds of industrial railway lines. Taking the railway as the core, a large number of factories, warehouses, and worker's villages have gathered along the line. In its heyday, one single railway branch line could collect hundreds of thousands of industrial workers, thus it became a "golden line" city for commuting and freight. However, in recent years, with the transformation of traditional industry and the "retreat into three" of the industrial structure adjustment in the central city of Tianjin, traditional industry has either declined or moved, and the industrial railway is no longer prosperous, thus it has gradually been neglected and abandoned to become a corner space of the city that is littered with trash, which hinders the development of the surrounding city areas and wastes land resources (Fig. 8.1). In order to promote land reuse in the Tianjin city center in the postindustrial era, a round-the-city railway park will be built for the purpose of retaining the overall historical pattern of the northern industrial base.

The Round-the-City Railway Park Planning in Tianjin proposal was submitted to the municipal government as a bottom-up volunteer status project in 2010, which proposed eco-transformation of the almost abandoned urban industrial railway to become China's high line park and would become the basis for the Round-the-City Railway Park Planning in Tianjin compilation in 2013. The Round-the-City Railway Park Planning in Tianjin is based on abandoned railway assets connected with urban rivers, green park spaces, scattered lands, and idle lands along the railway; forms the main green open space at the city level, which runs through seven urban areas of Tianjin, with a total length of 45 km and an average width of 100 m; and integrates the functions of ecological conservation, industrial culture, public service, etc. The round-the-city railway park in Tianjin includes the outer ring greenbelt and four rivers, which together constitute a "two rings and four lines" urban green space system (Fig. 8.2). This plan connects three city centers and ten key projects of Tianjin;

**Fig. 8.1** The plight of rail and its surrounding resources

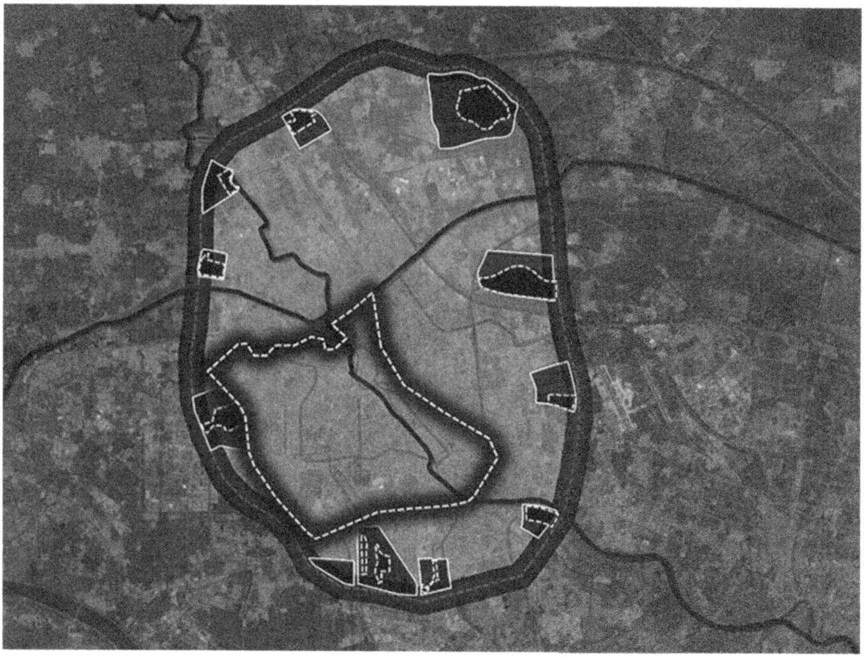

**Fig. 8.2** The urban green space system of "two rings and four lines" in Tianjin

retains a length of 22 km of railway lines for the city; integrates more than 30 industrial monuments; links about 18 parks, green spaces, and canals; extends to 15 greenway branches; connects 54 commercial blocks and complexes, 46 senior or higher-level educational resources, 35 medical facilities, and 22 pension facilities; and provides supporting amenities for the 69 surrounding mature communities and 40,000 urban residents. It is a rare city-level greenway resource in Tianjin and even in the nation, and thus is an important benefit implementation project of the Beautiful Tianjin action of 2013 (Tianjin Urban Planning & Design Institute 2013).

The planning fully respects the original ecological environment in the region, takes the abandoned railway landscape reconstruction as a starting point, combines urban industrial cultural heritage sites, and revitalizes urban stock assets. By reconstructing the ecological green space system of the main city and constructing the skeleton of an urban green network, it will turn negative space into positive space to guide city life, and will form an urban green growth economic belt. The "turn back to front" in such a space maximizes the value of incremental land use in the region, improves the value of land on both sides of the railway, and brings new opportunities for surrounding urban redevelopment and community upgrading.

### 8.4.2 Extending the Meridians: The Greenway Network Plan in the Hedong District of Tianjin

The green network in the central city of Tianjin must be implemented and extended by taking the administrative district as a unit under the overall framework of the round-the-city railway park. The goal of constructing a vibrant, livable city, as proposed by the Hedong District Government of Tianjin in their recent development action plan of 2016, provides a historical opportunity for green space network construction to change from being recessive to being dominant in the Hedong District. The planning puts forward the general idea of construction of a Hedong green space network under the incremental and existing stock parallel development pattern. On the one hand, it establishes a new view of space resources—from incremental expansion to existing stock promotion, and from functional construction to vitality creation. This attaches importance to top-down guidance, takes the round-the-city greenway as the main axis, and fully excavates urban elements (such as abandoned railway branch lines, rivers, boulevards, etc.), integrates surrounding vitality points, and forms an urban green network and an economic vitality belt at low cost. On the other hand, it establishes a new concept of subject method—from blueprint planning to action planning, and from government-led to share and co-construct. An "overall planning–recent construction plan–annual implementation plan" is composed by attaching importance to bottom-up power, launching action planning in a planned manner, and forming the implementation system of the construction of a green space network. Furthermore, the plan encourages enterprises, artists, and residents to build multiparty green spaces in order to achieve strong participatory planning.

The round-the-city railway park in Tianjin is about 11 km across the Hedong District, and contains a river and two lines composed of a crescent river, an abandoned small third line railway, and the Jingshan line railway green belt. It is the longest and most abundant component of the round-the-city railroad park in the Tianjin district administrative unit. In addition, the Hedong District of Tianjin is located on the east side of the Haihe River with abundant river resources. Its two secondary rivers have a total length of about 14 km, and it also contains more than ten unequal areas of urban parks in a scattered distribution (Fig. 8.3). The planning studies have conducted in-depth examination of the land use and existing conditions on the basis of the present green space system. The premise of minimizing conflicts of interest, and combining the round-the-city railroad park in Tianjin and the greenway network system, has strong implementation prospects, based on the unchanged

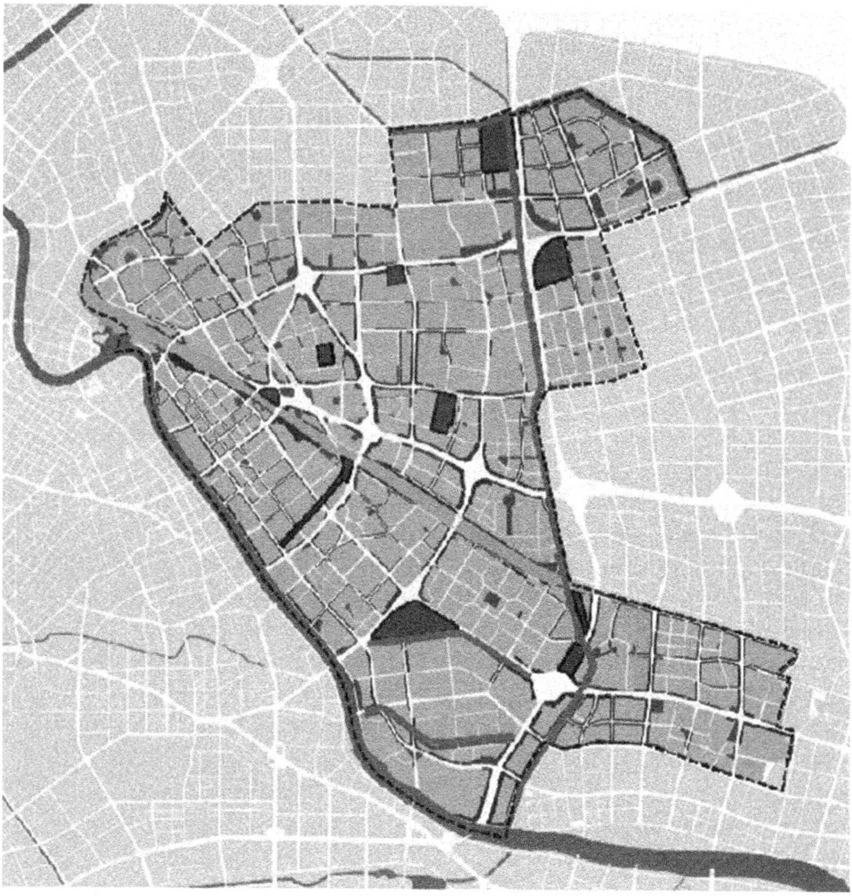

**Fig. 8.3** The green space system situation in the Hedong District of Tianjin

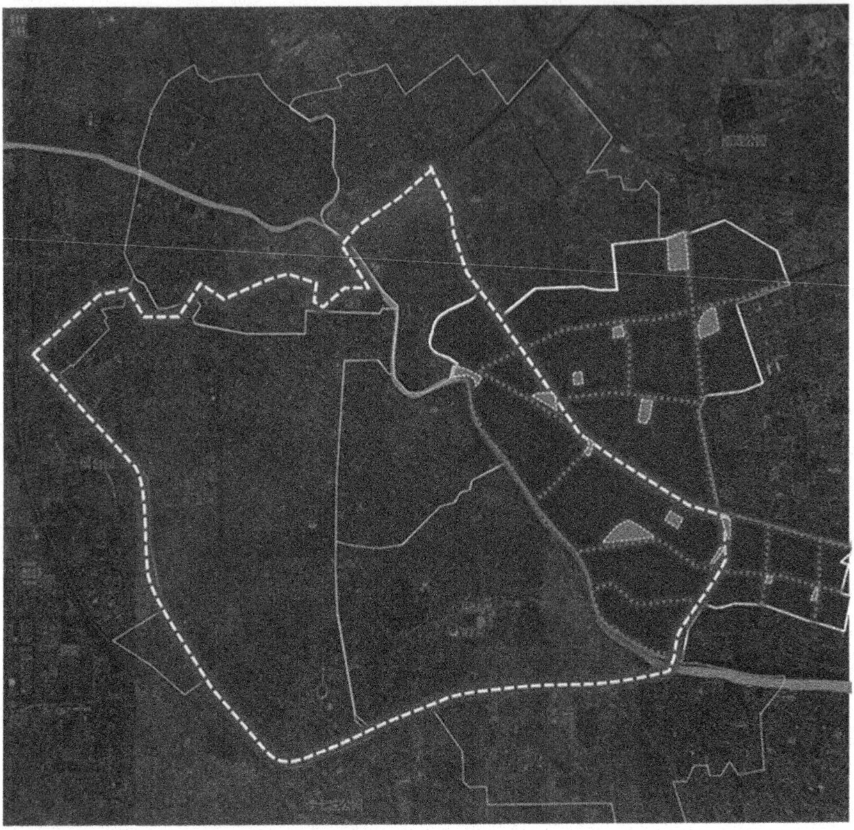

**Fig. 8.4** The greenway network system in the Hedong District of Tianjin

basic principle of land property rights (Fig. 8.4). The planning puts forward effective and enforceable greenway network plans according to local conditions by integrating space types, space resources, land usage, and other factors. The plan includes waterfront greenways, railway greenways, and boulevard greenways (Fig. 8.5), and integrates them to form the meridians of Hedong's green space system. The planning brings parks, green spaces, featured landscapes and slow traffic system into the greenway system; and forms a green network covering the whole region. Revitalizing the urban stock by taking urban management as the concept and urban redevelopment as the method will promote optimal allocation, flow, and function of the surrounding space resources for realizing industrial transformation (Tianjin Urban Planning & Design Institute 2016b).

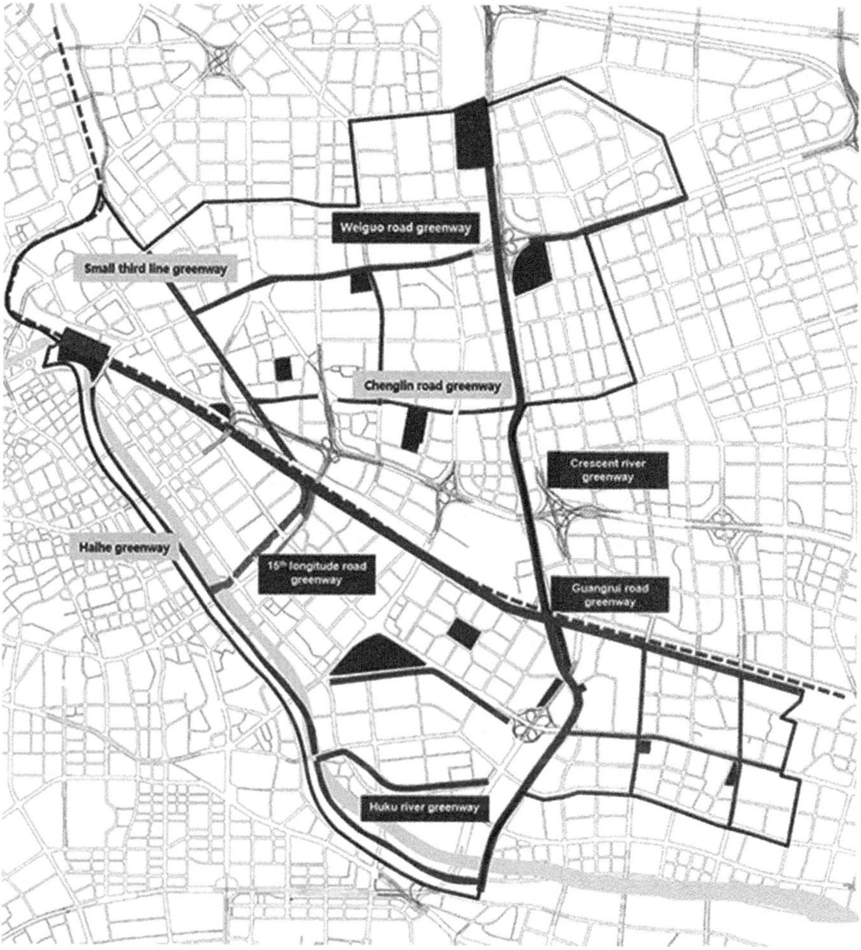


**Fig. 8.5** The greenway plan in the Hedong District of Tianjin

The greenway network in the Hedong District planning method is as follows:

1. Sorting out and utilizing existing blue and green corridors, carrying out ecological remediation and ecological environment promotion of the rivers (such as the Haihe River and the crescent river), making full use of the waterfront space of the river for planning and design on multiple levels and from multiple aspects, providing spaces for the composite functions that will be carried into the future, and activating the waterfront region will give the blue and green corridor multiple spatial meanings.
2. Activating the linear stock space and constructing a green corridor based on the unused railway, such as the 15th longitude road and the small third line railway, will transform the railway into a public space. At the same time, taking full

advantage of such stock resources with clear property rights and strong implementation, using it as a space carrier, taking the region's cultural characteristics and industry as the content, and carrying out multilevel and diversified activation of the existing stock space will give full play to the characteristics of the various types of stock resources and promote the development of surrounding areas. For example, the 15th longitude road area planning builds a vital, musical theme block along the old railway on the east side of the Haihe River by taking railway memories as the axis and music and fashion as the media.

3. The existing boulevard will be strengthened, with reliance on boulevards with excellent greening foundations, such as Guangrui Road, Weiguo Road, Chenglin Road, etc. The construction of an urban slow traffic system will be combined with expansion of leisure space, So as to optimize road landscapes, and construct green corridors associated with traffic.

This plan organically redevelops the green space of the Hedong District, based on the summary on current problems, by using an urban redevelopment technique. Meanwhile, combining the greenway network with the land use situation and the maintenance of unchanged land property rights will avoid the fundamental problem of high collection costs incurred by compulsory land acquisition for green land construction, reduce the transaction cost of green system organic renewal, and guarantee the implementation of a greenway network system. Through the integration of rivers, railways, roads, green spaces, and parks, revitalizing stock space resources will increase the use efficiency of stock spaces, endow ecological spaces with multiple functions, enhance space vitality, and increase the value of surrounding land, thus leading the surroundings into gradual redevelopment.

### *8.4.3 Filling in the Flesh: The Three-Dimensional Construction of Green Space*

Presently, the Hedong District of Tianjin has a shortage of green land resources in situations of scattered distribution. Moreover, with the acceleration of urbanization, the cost of urban surface greening has risen sharply. The premise of insisting on less compulsory land requisition for green land construction and ensuring the implementation of green space system construction will strengthen green corridor construction combined with a greenway network system. If possible, it is necessary to strengthen the green quantity and green quality improvement of blocky and dotted green lands. Under the tight constraint of green land use, it is necessary to "stretch up" and actively promote urban three-dimensional greening construction (roof greening systems with multidimensional elevation), as it is an important, enforceable means of increasing the urban green quantity and improving urban ecological environments. Three-dimensional greening, especially roof greening, has great social, ecological, and economic benefits, as it can alleviate the urban heat island effect, improve building energy efficiency, reduce the load on urban rainwater treatment systems, beautify city landscapes, enrich species diversity, etc. (Chun-Xia Zeng 2014).

In recent years, urban three-dimensional greening has shown a trend of globalization, diversification, and legalization. Foreign three-dimensional greening has entered a mature stage after decades of rapid development, which has a sound legal regulatory system and corresponding incentive policies and measures. While our country began in the 1910s, with its three-dimensional greening ideas presented at the Shanghai World Expo, some cities have successively started to allow more social forces to participate in the construction of three-dimensional greening through special subsidies, green rate conversion, and other policies. Beijing, Xi'an, Shanghai, Shenzhen, etc., have set up special subsidies for three-dimensional greening, and Hebei, Chongqing, and other cities have formulated greenery coverage encouragement policies. The greening rate indicator encourages policies in Beijing, Hefei, Hangzhou, and other cities, where local governments specify that green roof areas can be classified as green areas to participate in improving the greening rate calculation. In contrast to the way of encouraging promotion, Guangzhou is the first city to put forward mandatory provisions that institutionalize three-dimensional greening (mainly roof greening) in green building control requirements at the level of statutory planning. Most of our cities are actively promoting the construction of three-dimensional greening, with different levels of coercive measures (a legal push, regulatory guidance, planning limits) and incentive measures (target incentives, special subsidies, technical support), and have achieved certain results.

The planning in the Hedong District constructs greening and leisure facilities on multistory residential roofs and commercial annex roofs by renovating old residential areas in Tianjin. To find solutions to many problems of traditional roof greening (with regard to load-bearing, waterproof and drainage, thermal insulation, and other aspects), project teams must select a new type of mobile-assembly, lightweight, three-dimensional green products (Fig. 8.6), with the advantages of container type,

**Fig. 8.6** The real scene of new mobile-assembly, lightweight, roof greening products

assembly type, mobility, light load, no leakage, low cost, ease of maintenance, etc. In the Hedong District project, the multistory residential roof area is 4.22 million square meters, and the roof area of commercial annexes below 24 m is 1.09 million square meters (Figs. 8.7 and 8.8). Referring to the indicator that "greening rate of public roof will be 40% in 2017" for the 13th Five-Year Development Planning (2017) in Tianjin, The planning will create a roof greening area of 2.12 million square meters. Meanwhile, a green coverage rate of 16.5%, which is an increase of 3.9% over the planned green space ratio, enriches the greening level and changes negative space into energetic space (Fig. 8.9).

Combining the current transformation of old communities and changing spaces to green spaces, the planning reduces the transaction costs of a green space system, ensures implementation of the organic renewal plan, makes green space into three-dimensional greening space, and further promotes intensive land utilization. In addition, the comprehensive benefits of three-dimensional greening products will

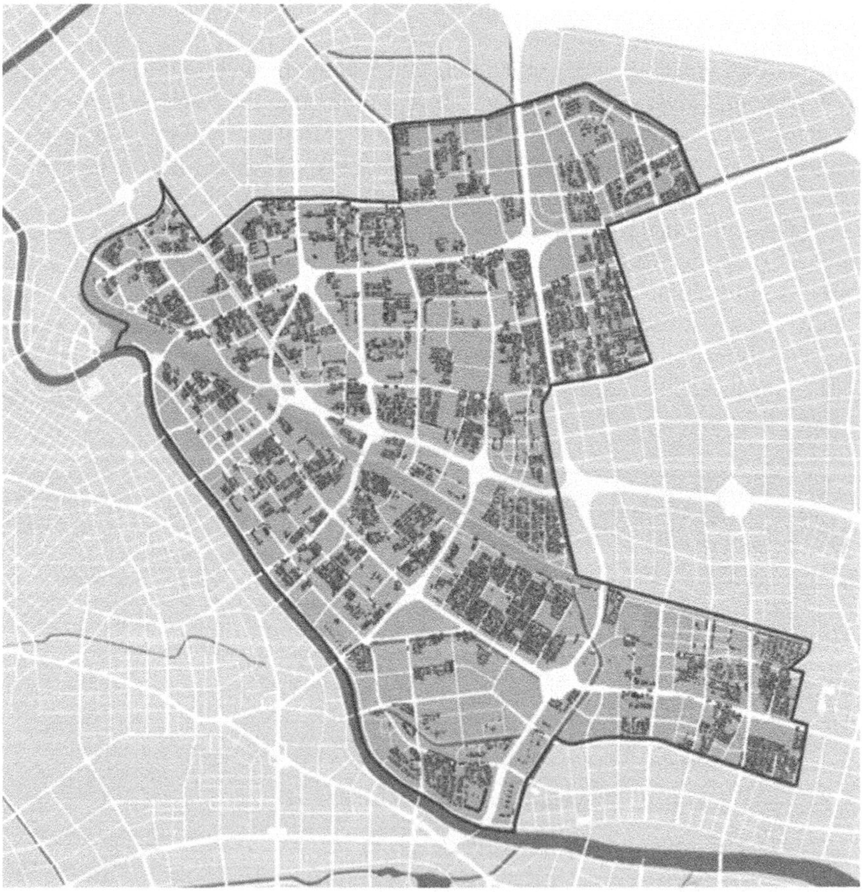

**Fig. 8.7** Distribution of multistory residential buildings in the Hedong District of Tianjin

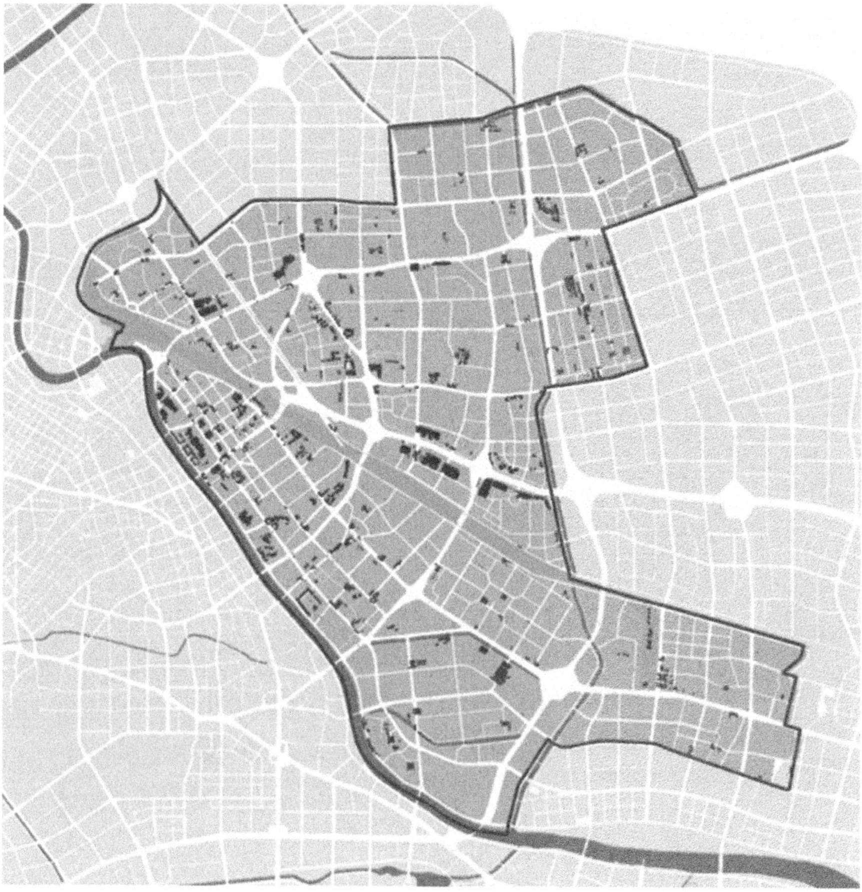

**Fig. 8.8** Distribution of commercial annexes in the Hedong District of Tianjin

improve initiatives for enterprises, property owners, and other implementing stakeholders to participate in three-dimensional greening construction, in order to "achieve multi-investment, multi-win situations and allow the government to share and co-construct."

## 8.5 Conclusion

Under the coexistence trend of incremental planning and existing stock planning, based on the concept of urban regeneration by fully using existing stock space resources, and applying a reasonable mechanism for benefit distribution under serious shortages and fragmentation of green spaces in urban high-density areas, transaction costs can be greatly reduced and a new way to supply land for green spaces

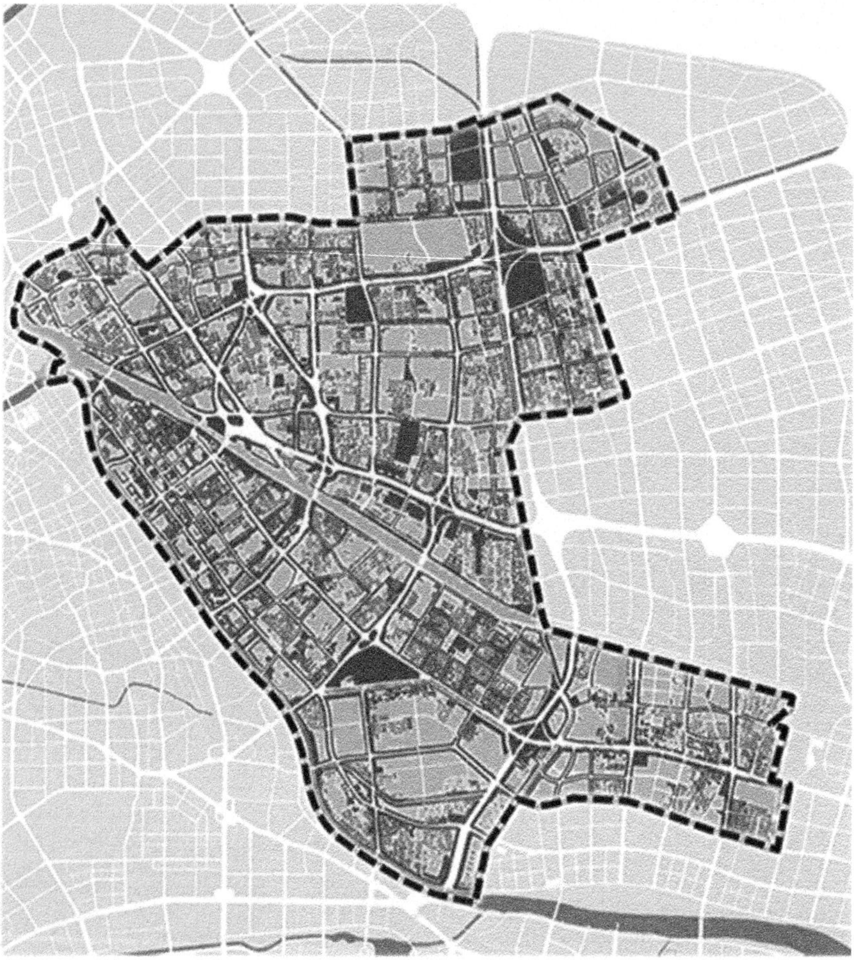

**Fig. 8.9** Layout of three-dimensional greening planning in the Hedong District of Tianjin

can be established. Under the principle of "dividing levels, connecting networks, and building multi-paths" and the planning strategy of "reconstructing the skeleton, extending the meridians, and filling in the flesh," high-density urban areas can be explored to establish a multilevel green space system. Summarizing the planning practice in Tianjin, this chapter has reviewed the gradual implementation of the Round-the-City Railway Park Planning in Tianjin and other special planning projects, including the Green Way Network Plan in Tianjin Hedong District, Old Communities Updating Plan in Tianjin Hedong District, and the Organic Regeneration of The 15th Road District In Tianjin. Existing stock land is gradually turned from idle to active, and the circle railway line forms the main skeleton of the city green space, where the total length is 45 km and the average width is 100 m.

Meanwhile, in Tianjin's Hedong District, by using existing stock resources—such as rivers, railway branch lines and roofs—the green coverage rate will increase to 16.5% in the district. The planners hope to build an integrated urban green open space system that can concentrate on ecological conservation, public service, and community vitality, in order that the city space can be flipped from back to front and improve its value. Tianjin is actively exploring renewal methods for urban repair and ecological restoration.

**Acknowledgments** This work was supported by the National Key Research and Development Program of China (2016YFC0502903).

## References

Chunxia Z (2014) New thinking and exploration of three-dimensional greening construction. Planner 7:148–152

Jingxiang Z, Hao C (2010) China's compressed urbanization and urban planning responses. Urban Plann Forum 6:10–21

Ping S (2013) Predicament of urban space management: urban design in transitional China. Urban Plann Forum 3:106–112

Tianjin Urban Planning & Design Institute (2013) The round-the-city railway park planning in Tianjin [EB/OL]. http://www.yangqiu.cn/zhongshun98/679572.html

Tianjin Urban Planning & Design Institute (2015) The master plan in Tianjin [R]. Tianjin

Tianjin Urban Planning & Design Institute (2016a) General urban design in Tianjin Hedong District [R]. Tianjin

Tianjin Urban Planning & Design Institute (2016b) Recent development action planning for Hedong District in Tianjin [R]. Tianjin

Yanjing Z (2014) The theory and practice of the urban stock planning. Beijing Plann Rev 4:153–156

Yuantao L, Rongbo X (2012) Establishing urban green space planning system. Planner 3:46–54

Zhaolin W, Fang H (2012) Discussion on urban greenway planning interacting with city. City Plann Rev 10:36–39

# Chapter 9
# Impact of Roof Greening on the Ecological Environment of the Green Building, Exemplified by the Roof Garden of the Mingde Building in Fujian Agricultural and Forestry University

**Tingting Hong, Xiaogang Wu, Yanhong Chen, and Xiaoshan Lin**

**Abstract** With the rapid development of urbanization in China, ecological problems in cities are increasingly alarming. In order to test the ecological benefits of roof greening on reducing energy consumption, cooling, humidifying, and controlling rainwater runoff, this chapter takes a roof garden at the Fujian Agriculture and Forestry University as a case study. This work verifies the effects of cooling, humidification, and lower energy consumption achieved by roof greening. Also, rainwater in the roof garden was monitored to test the utility of the rainwater collection system. Humidifying and cooling were effected significantly in the roof garden, where the temperature dropped by 1.8–4 °C and relative humidity increased by 5–10%. There was also a significant indoor temperature drop after the roof was greened. Furthermore, roof greening can have a significant insulating effect and is able to reduce energy consumption effectively. The monitoring results show that the roof garden also plays a significant role in controlling rainwater runoff.

**Keywords** Roof greening · Temperature and humidity · Ecological benefits

T. Hong (✉) · Y. Chen
School of Architecture, Fuzhou University, Fuzhou, China
e-mail: fjolive@qq.com

X. Wu · X. Lin
School of Landscape Architecture, Fujian Agricultural and Forestry University, Fuzhou, China



Z. Shen et al. (eds.), *Green City Planning and Practices in Asian Cities*, Strategies for Sustainability, https://doi.org/10.1007/978-3-319-70025-0_9

## 9.1 Introduction

The contradiction between land demands of urbanization and construction of urban parkland sharpens day by day. It is increasingly difficult to develop ecological parkland horizontally in urban areas, which forces us to seek new possibilities to explore potential spaces for greening in a city to alleviate conflicts and to offset building space. The utilization of vertical space is one possible solution. This chapter takes the roof garden in the administration building at Fujian Agriculture and Forestry University as a case study, and we explore the effective mechanisms of the roof garden in the green building landscape and its ecological effects via monitoring of the ecological environment before and after its construction.

### *9.1.1 Concept of the Roof Garden*

The concept of the roof garden is divided into a broad-sense concept and a narrow-sense concept. In a broad sense, the "roof garden" is a general term for all afforestation and garden construction activities on walls, bridges, rooftops, terraces of buildings or construction sites, or on top of huge man-made rockeries (Xu Feng et al. 2008). Narrowly speaking, the term "roof garden" refers to taking roof greening as a basis and applying techniques of ground garden design to roof garden construction to create a rich roof landscape and provide a space for sightseeing, resting, and sheltering (Wei Shiheng 1994). This chapter mainly relates to the narrow concept of the roof garden.

### *9.1.2 Research on the Roof Garden*

The "roof garden" was first described by Le Corbusier in his "five elements of modern architecture" (Le Corbusier 2004). He took roof garden design as the embodiment of humanism in modern city construction, which satisfies various needs such as living, working, leisure, and sightseeing, and deserves wide application. This is widely recognized. Landscape architect T. Brown once stated that "The roof of the building is supposed to be a garden in the air, providing a space full of sunshine and fresh air for residents and an outdoor terrace for rest, dining, and sightseeing" (Wang Haiou 2007). The famous architect Frank Lloyd Wright also regarded the roof as an extension of the building space and combined the design of the roof space with the whole building design (Zhang Feng 2009), such as the roof gardens of the Larkin Building and the Tokyo Imperial Hotel. There are other relevant theories, such as Yang Jingwen's "vertical greening" theory, the "garden platform" theory of Charles Corea, and Kurokawa's theory of "symbiosis roof" courtyards, etc. (Wang Shizhang 2007). These reveal their intense interest in the landscape aesthetics of the roof garden. Therefore, it has been an inevitable trend for eco-development of modern architecture to test the feasibility of vertical greening in building construction projects and to invent effective roof garden design.

## *9.1.3 Functional Benefits of the Roof Garden*

### 9.1.3.1 Landscape Function

Roof garden construction replaces the gray concrete and black asphalt platform of the traditional building roof, which eases people's visual pressure and creates a harmonious environment for citizens to emerge in the forest in a building. Furthermore, the roof garden, with combined landscape elements, provides users with a nice environment for leisure and entertainment. For people who live in a high-rise building or people who climb high to enjoy the view, they can feel the embrace from the natural beauty brought by a green garden by overlooking the ground or by looking up overhead. The roof garden enriches the urban green landscape system.

### 9.1.3.2 Ecological Effects

1. *Temperature cooling and heat insulation*: In summer, the green roof can cool indoor temperatures via transpiration. As Drefahl's test showed in 1995, the top floor of an apartment with a green roof was almost the same as that of the apartment which is without a green roof covered with vegetation can effectively avoid overheating on the top floor (Drefahl 1986) and reduce the indoor temperature. Through this, reduced use of air conditioning and enactment of energy conservation and environmental protection can effectively be achieved.
2. *Air purification*: Compared with ground plants, air purification by roof plants is far more widespread owing to its higher position, which makes it a "filter" of a multilayer distribution system in an urban space. Roof plants can exert a stronger regulatory function beyond the reach of ground plants. It is estimated that each 10 million square meters of green roofs can produce 750 tons of oxygen and absorb 100 tons of carbon dioxide, which satisfies 300,000 urban citizens' needs in terms of the carbon and oxygen balance. In addition, some plants, being able to emit volatile substances that are effective in killing bacteria, can play a significant role in the construction of ecological and environmental cities.
3. *Increasing air humidity*: A roof garden is effective in reducing the "dry island" effect and increasing air humidity in cities. In cities, owing to the higher sunlight reflectivity of a greenhouse than that of a dark cement roof, and the shading and assimilating effects of green plants, the net heat radiation of a green roof is much lesser than that of a non-green roof. Meanwhile, the transpiration from the green

roof is tremendously increased due to the transpiration effect of the green plants, which humidifies the air nearby.

4. *Storing water*: Normally, almost all of the rainwater on a roof drains into the underground drainage pipes, with 80% rainwater drainage from a non-green roof versus merely 30% from a green roof, which hugely decreases the amount of rainwater that drains into the pipes. Green roofs are not only helpful in increasing green plant coverage but also reduce the amount of roof rainwater discharge to one third. The other two thirds of the rainwater is stored on the roof and gradually evaporates into the air to regulate the microclimate in the city.
5. *Improving building performance and extending building life expectancy*: In most cases, damage to the building roof structure layer is caused by thermal stress from the roof's waterproofing layers. Temperature changes cause expansion and contraction of the roof structure, which finally leads to cracks in the building and infiltration of rainwater. By taking advantage of the evaporation and photosynthesis from the plants, the green roof construction can reduce the impact of solar radiation on the roof. In summer the temperature gap between a green roof and a non-green roof reaches 6–8°, while in winter the gap between the highest temperature and the lowest temperature is merely 1°, whereas that of a non-green roof reaches 5–12° (Wang Shizhang 2007). Therefore, the function of heat insulation in summer and temperature maintenance can effectively moderate extreme temperatures, hence protection of the roof structure and extension of its life expectancy are possible.

Briefly speaking, the roof garden plays an important role in reducing greenhouse effects, economizing land use, and achieving energy conservation and emissions reduction. It can increase the green coverage of the city in a limited urban space and can effectively retain, collect, and filter the urban water resource. To take the roof as the "fifth facade" of a building is to take roof greening as a new trend of utilizing the fifth facade in urban construction (Xue Feng 2016).

In this chapter, the roof garden in the administration building at Fujian Agriculture and Forestry University is taken as a case study example. In the following sections, we briefly introduce the design concepts of this roof garden and explore its ecological effects via monitoring of the ecological environment before and after the construction. On the basis of our research data, this chapter concludes with discussions on the principles of and approaches to roof garden design.

## 9.2 Roof Garden Used for Case Study

As one of the key universities in Fujian province, Fujian Agriculture and Forestry University, with a history of 77 years, has been committed to promoting the development of an ecological campus for many years. After many years of afforestation, the overall virescence of the campus has been raised with high green coverage. The green layout of the campus is reasonable, and the landscape works well for education and research. Endowed with the advantages of agriculture and forestry, the school is rich in species needed for teaching and research via careful importation and cultivation. Various plants and animals have been carefully introduced so that the species are rich in abundance. To date, the school has created several ecological landscape areas and has largely accelerated the construction of an ecological network and a campus water system. A series of achievements have also been made in line with educational promotion.

The roof garden at the Fujian Agriculture and Forestry University is located on the 13th floor of the Mingde Building in the southwest area of the campus. The total green landscape area is 960 $m^2$, with a 600 $m^2$ square green area, built around a tea room. The roof garden is mainly freely acccssible for internal office staff and students' practice. Beyond tedious work, it has become an ideal place for regulating brains, alleviating fatigue, and lifting spirits. The roof garden design efficiently collects rainwater and realizes the aim of green energy saving.

## 9.3 Ecological Landscape Design and Effect Measurement

### *9.3.1 Roof Green Configuration*

The design load on the 13th floor of the Mingde Building is 3.0 kN/$m^2$, the configuration of which consists of a substrate layer, a waterproofing layer, a drainage layer, a barrier layer, a protection layer, another waterproofing layer, and a floor layer (Fig. 9.1). In order to reduce the weight of the building materials and to reduce the load pressure on the roof, small trees have been planted. The base of the soil employs a substrate with a light inorganic medium, the capacity of which is 400–450 kg/$m^3$. The waterproofing layers use nonwoven soil cloth. The drainage layer includes 10–20 mm thick-PVC drainage boards, and the peak water value is 4.5 kg/$m^2$. The resistance layer consists of a polyethylene barrier membrane. The protective layer uses 50 mm thick fine stone concrete to protect the floor. The rainwater piping system in the garden is made of plastic materials.

### *9.3.2 Plant Deployment*

The plant deployment adopts a natural design, forming an arbor–shrub–grass compound planting mode. There are other varieties of planting combination modes, such as an arbor–shrub–ground plant–grass mode, an arbor–shrub–grass mode, a shrub–ground plant mode, a tree–shrub mode, and a shrub–bush mode. In total, 34 kinds of plants have been chosen for the garden, including 13 kinds of shrub, 10 kinds of tree, six kinds of herb, three kinds of bamboo, one species of climbing plant, and one kind of aquatic plant. This combination forms a variety of plant

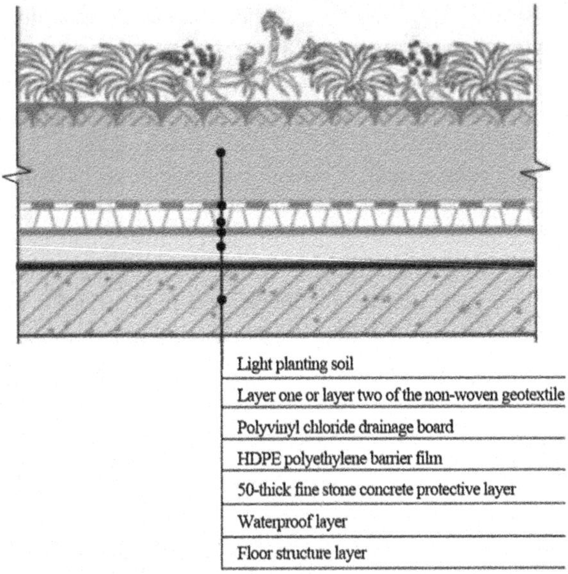


**Fig. 9.1** Green roof configuration

landscapes, such as tree views, leafy views, and flower views. The plant deployment fully considers the ecological characteristics of the species, with drought-tolerant tree species on the sunny side and shade-tolerant plants such as the arum, the octagon gold dish, and southern bamboo under the eaves. The types of plants used are listed in Table 9.1.

### 9.3.3 View Design

The roof garden on the Mingde Building is designed to meet visual psychological needs, with different views in different places. At no place in the garden can visitors see the whole landscape, which increases its mystique and arouses the visitors' interest. The Mingde Building is 13 floors high, from different triangles of which the campus view can be fully enjoyed. To the north of it is Guanyin Lake, to the east of it is the wetland park, to the south of it is the teaching building, and to the west of it is the Chinese botanical gardens (Fig. 9.2). The views in these four directions are completely different.

The roof garden is configured in different ways, depending on the needs of the four directions: for the north and west, the main observation surfaces are reasonably set aside. As for the south of the building, a three-dimensional plant configuration has been deployed to obscure the view. In the space transition area, trees are used to separate the views and low shrubs are arranged on both sides of the walkway to extend the view (Fig. 9.3).

**Table 9.1** Plant styles used in the plant types template

| Plant type | Plant name |
|---|---|
| Trees | *Osmanthus fragrans* var. *semperflorens* |
| | *Osmanthus fragrans* var. *thunbergii* |
| | *Lagerstroemia indica* L. |
| | *Camellia japonica* L. |
| | *Armeniaca mume* |
| | *Ficus microcarpa* L. f. var. *pusillifolia* |
| | *Pinus parviflora* |
| | *Podocarpus macrophyllus* (Thumb.) D. Don |
| | *Sophora xanthantha* |
| | *Prunus persica* |
| Shrubs | *Hibiscus rosa-sinensis* Linn. |
| | *Rhododendron simsii* Planch. |
| | *Bougainvillea glabra* |
| | *Cordyline fruticosa* (L.) A. Cheval. |
| | *Alpinia zerumbet* 'Variegata' |
| | *Loropetalum chinense* var. *rubrum* |
| | *Nandina domestica* |
| | *Rhapis excelsa* (Thunb.) Henry ex Rehd. |
| | *Fatsia japonica* (Thunb.) Decne. & Planch. |
| | *Cycas revoluta* Thunb. |
| | *Michelia figo* (Lour.) Spreng. |
| Ground cover | *Nephrolepis auriculata* (L.) Trimen |
| | *Dianella ensifolia* White Variegated |
| | *Cynodon dactylon* (L.) Pers. |
| | *Sedum lineare* Thunb. |
| | *Arachis hypogaea* Linn. |
| | *Alocasia macrorrhiza* (L.) Schott |
| Bamboo | *Phyllostachys nigra* (Lodd. ex Lindl.) Munro |
| | *Bambusa ventricosa* McClure |
| | *Bambusa vulgaris* Schrad. ex Wendl. 'Vittata' |
| Climbing plant | *Pyrostegia venusta* (Ker-Gawl.) Miers |
| Aquatic plant | *Nelumbo* sp. |

View from the north View from the east View from the west

**Fig. 9.2** Landscape surrounding the roof garden at the Fujian Agricultural and Forestry University

**Fig. 9.3** Roof garden at the Fujian Agricultural and Forestry University

## *9.3.4 Methodology Used for Measuring Temperature and Humidity*

### 9.3.4.1 Choice of Observation Points

The comparisons of temperature, relative air humidity, and wind speed were made through observations at test points. The first part of the test was a comparative analysis done by monitoring the temperature, relative humidity, and wind speed on the roof before greening and after greening.

For the second part of the test, the task was to observe the changes in the green layer and in the lower layer of the roof under normal daylight before and after the roof greening. Test measurements were made before and after the green roof was completed. The measurements included the temperature and humidity at the top and at the bottom of the roof. The tests were repeated to conduct a comparative analysis.

The third part of the test relied on selecting four points in the four directions, and testing the temperature, relative humidity, and wind speed at those points. Measurements were also made at a fifth test point located indoors on the 12th floor of the building. This test was also repeated to do a comparative analysis (Figs. 9.4 and 9.5).

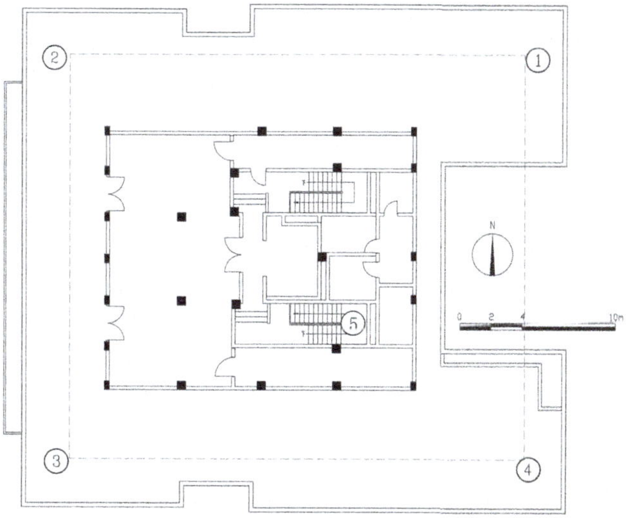


**Fig. 9.4** Test points before construction of the roof garden. Test point 1 was located at the northeast of the roof garden, test point 2 was located at the northwest of the roof garden, test point 3 was located at the southwest of the roof garden, test point 4 was located at the southeast of the roof garden, and test point 5 was located indoors on the 12th floor of the building

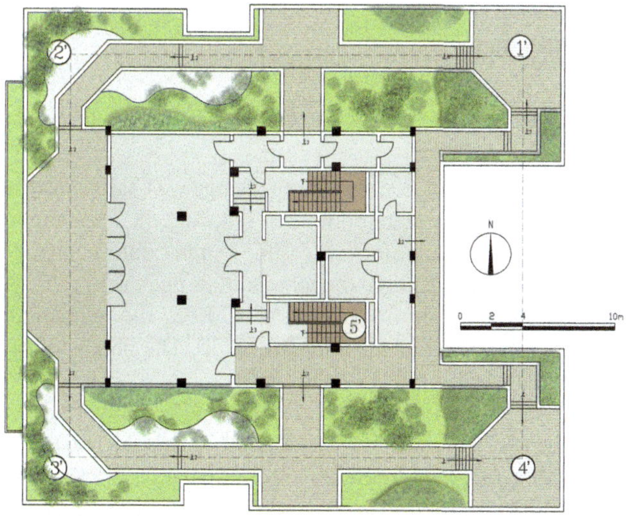


**Fig. 9.5** Test points after construction of the roof garden. Test point 1′ was located at the northeast of the roof garden, test point 2′ was located at the northwest of the roof garden, test point 3′ was located at the southwest of the roof garden, test point 4′ was located at the southeast of the roof garden, and test point 5′ was located indoors on the 12th floor of the building

#### 9.3.4.2 Observation Methods

The measuring instruments were installed on a stationary tripod at an observation height of 1 m. The observation days were clear, hot, and slightly windy. The time interval was half hourly from 0800 to 1700 h.

#### 9.3.4.3 Experimental Apparatus

The experimental apparatus included Tes1341 hot-wire anemometers (with precision of 0.1 °C, a relative humidity measurement range of 0–95%, and accuracy of 0.1%) and a handheld temperature tester (with precision of 0.1 °C and a temperature measurement range of –20 to 550 °C).

### *9.3.5 Rainfall Observation*

The rainfall observation was conducted by monitoring the amount of rainfall on the roof garden from January to May, 2017, and recording the rainfall and the amount of rainwater drainage during rainy days.

## 9.4 Data Analysis

### *9.4.1 Ecological and Cooling Effects of Roof Greening*

As shown in Fig. 9.6, the average temperature change was basically in a "unimodal" distribution. The peak appeared at about 1200 h during the measurement period of 0800 to 1700 h. Within the time coverage, the temperature changed sharply and varied greatly. From 0800 to 1200 h, the temperature increasingly went up with the intensive solar radiation, while from 1400 to 1700 h, the temperature gradually decreased with the weakening of the solar radiation. The photosynthesis and transpiration of plants themselves had a cooling effect.

In Fig. 9.7, it can be seen that the average humidity daily changes of the roof before and after the roof greening showed a trend of first decreasing, then increasing, and finally balancing. The humidity reached a peak at around 1300 h and a trough at around 1200 h. The humidity varied greatly at different times. Among them, from 0800 to 1200 h, the moisture declined most rapidly, and the humidity remained stable from 1300 to 1700 h. The humidity of the green roof was higher than that of the roof before the greening. Therefore, it was obvious that the air humidity was increased because of the photosynthesis, transpiration of the plants, and evaporation of soil water. Hence the humidity-increasing effect of the roof greening was obvious.

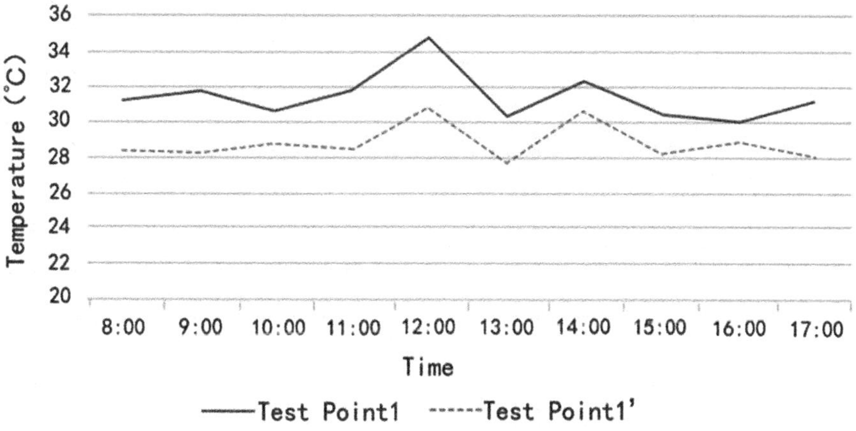


**Fig. 9.6** Temperature changes on the 13th floor of the building before and after the roof greening (from May 1 to 31, 2017)

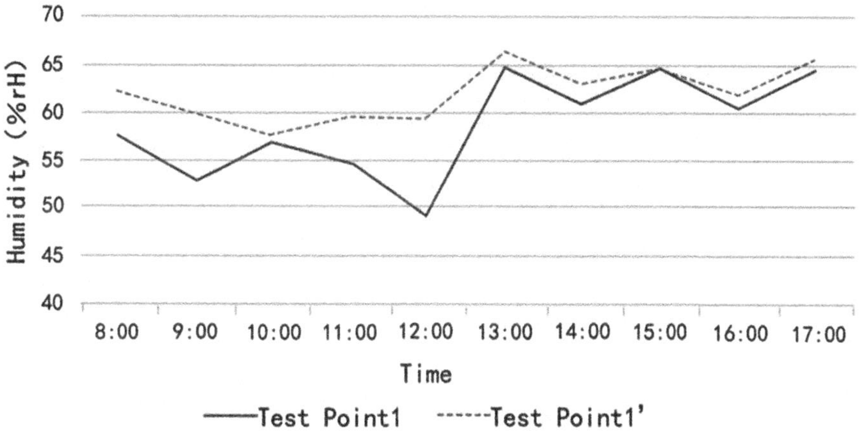


**Fig. 9.7** Humidity changes on the 13th floor of the building before and after the roof greening (from May 1 to 31, 2017)

### *9.4.2 Ecological Effect of Roof Greening in Different Directions*

Figure 9.8 shows that the temperatures on the four-azimuth roof fluctuated throughout the day. Among these, the temperatures at test point 3′, located in the southwest part of the field, were limited by the influences of the mild northeast wind and low green coverage. The highest peak for the day was achieved at around 1200 h. (Test points 1′, 2′, 3′, and 4′ were, respectively, in the northeast, northwest, southwest, and southeast directions of the construction after the roof greening.)

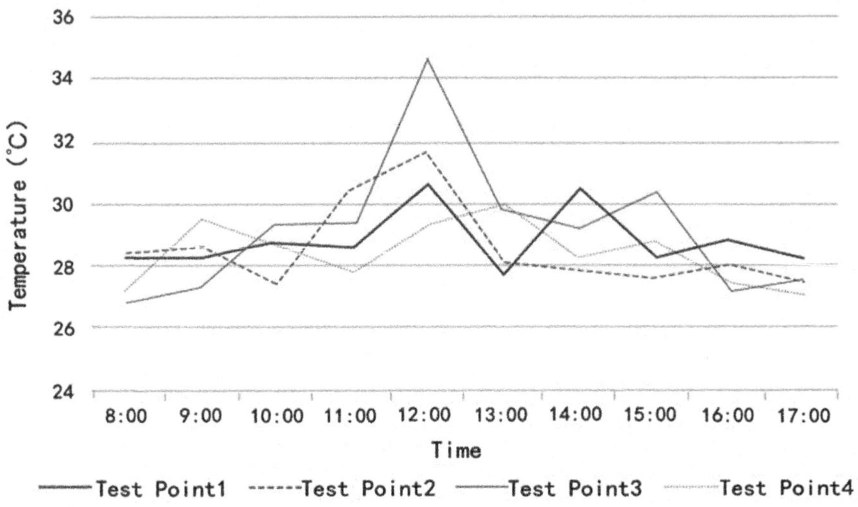


**Fig. 9.8** Temperature changes of the roof green layer in different directions (from May 1 to 31, 2017)

The humidity of the roof green layer was basically in a fluctuating state all day. Test point 3′, located in the southwest of the site, showed the most extreme values because of the northeast wind at the site and the sparse vegetation (Fig. 9.9).

### *9.4.3 Energy Efficiency Effects of the Roof Garden on the Body of the Building*

In this section the influences of the temperature and humidity effects from the roof greening on the 12th floor are explored (test point 5 was in the roof before the construction, while test point 5′ was in the roof after the construction of the roof greening).

Through Ecotect software analysis of the indoor temperature changes on the 12th floor of the administration building, the temperature changes before and after the roof greening on the day when the temperature reached its peak (the highest point) were analyzed in detail. By comparing Figs. 9.10 and 9.11, it is obvious that the indoor temperature on the 12th floor within a single day after the construction of the green roof was clearly below the simulated line of the outdoor temperature, while the interior temperature of the roof was higher than the outside temperature. As a result, a conclusion could be drawn that the roof greening had an obvious cooling effect on the floor below (the 12th floor).

Through field observation and exploration, Fig. 9.12 shows that the roof temperature changes on the 12th floor showed a trend of firstly going up and then going down. The temperature peak, like that on the 13th roof, occurred at around 1200 h.

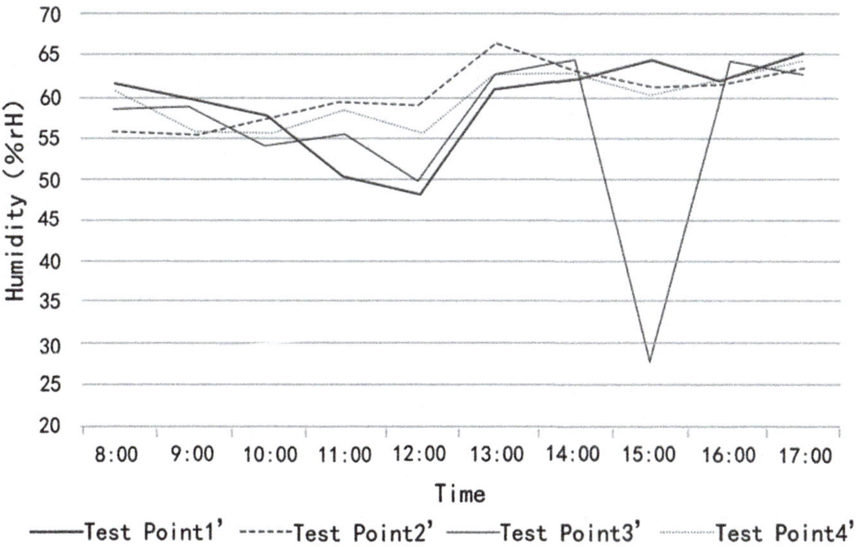


**Fig. 9.9** Humidity changes of the roof green layer in different directions (from May 1 to 31, 2017)

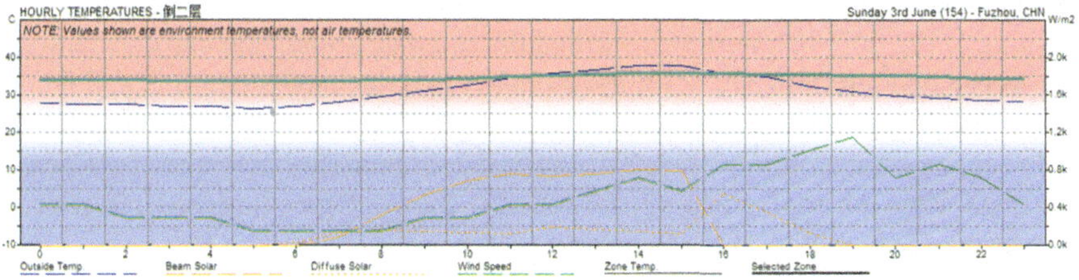


**Fig. 9.10** Interior temperature changes on the 12th floor of the building before the roof greening (from May 1 to 31, 2017)

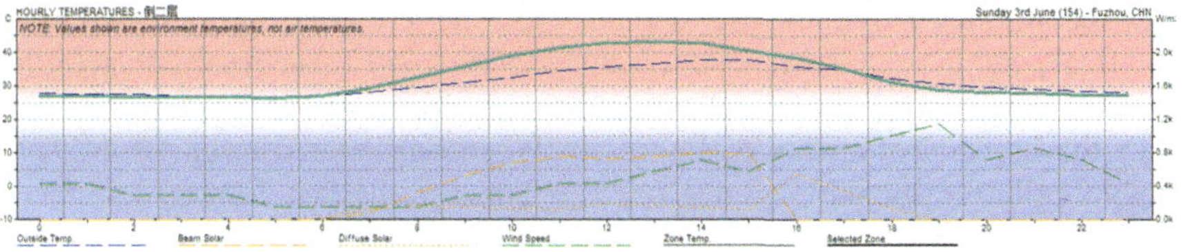


**Fig. 9.11** Interior temperature changes on the 12th floor of the building after the roof greening (from May 1 to 31, 2017)

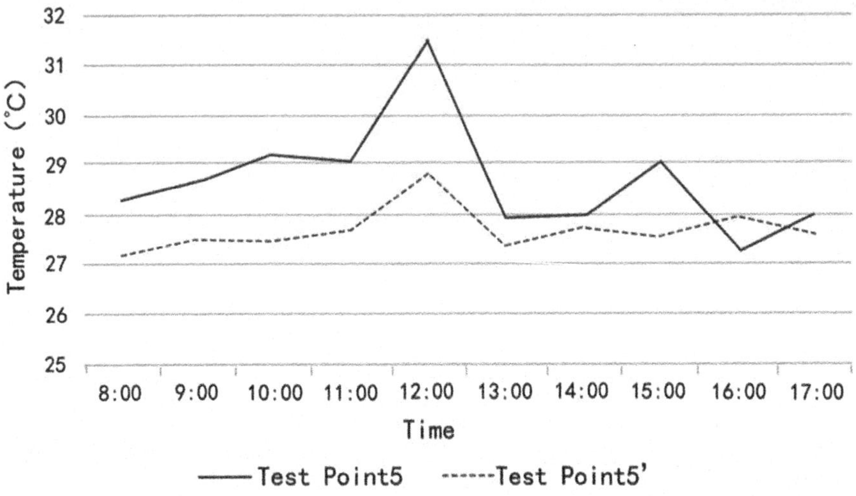


**Fig. 9.12** Interior temperature changes on the 12th floor of the building before and after the roof greening (from May 1 to 31, 2017)

Obviously, the indoor temperature was much lower after the roof greening than before the construction. Therefore, the roof greening had obvious effects on the cooling of the floor below, which verified the software simulation and calculation.

As can be seen from Fig. 9.13, the humidity first declined, then rose, and finally reached a balance on the 12th floor before and after the roof greening. The lowest value appeared at around 1200 h. The lower indoor humidity before the roof greening than after the roof greening showed that the roof greening was effective in increasing the humidity of the floor below.

### *9.4.4 Effect of the Roof Garden on Building Rainwater Runoff*

The runoff amount from the roof-draining pipe after each rain was observed under different conditions of rainfall from January to May, 2017. The results showed that in the first half hour of the rainfall, when the rainfall amount was 0.1–4.9 mm within 12 h, the discharge of water was almost zero. When the rainfall reached 4.9–14.9 mm within 12 h, the amount of water was very small in the first 20 min, and then it gradually increased. When the rainfall reached more than 14.9 mm within 12 h, the discharge of water within 15 min was very small. These results showed that the roof garden on the Mingde Building had a good effect in controlling rainwater runoff.

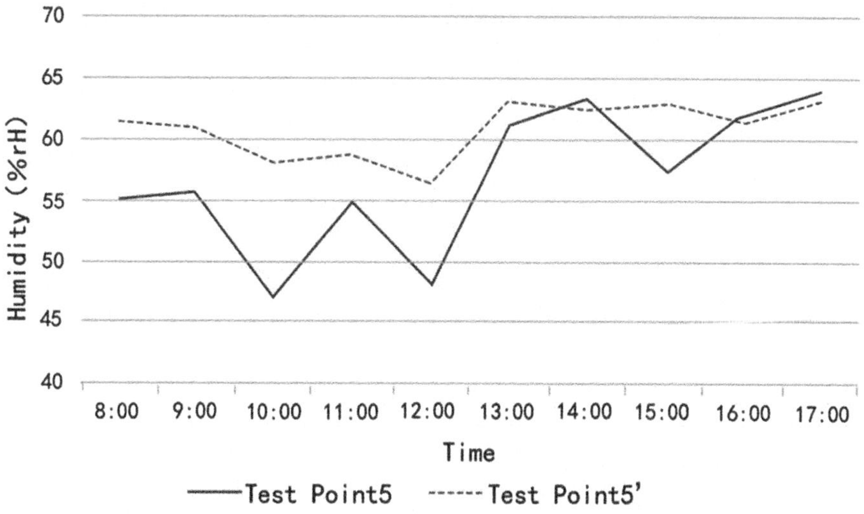


**Fig. 9.13** Interior humidity changes on the 12th floor of the building before and after the roof greening (from May 1 to 31, 2017)

### *9.4.5 Research Conclusions on the Ecological Mechanism of the Roof Garden*

By comparing the temperature and humidity of the roof greening layer before and after construction, a conclusion could be drawn that the temperature changes of the two conditions followed a unimodal trend, with the highest values occurring at 1200 h. The humidity first showed a trend of declining and then rising, with the highest value being observed at 1300 h. From the data, it was clear that the roof greening construction was effective in cooling the temperature and increasing the humidity.

Comparison of the temperature and humidity of the roof green layer showed that the temperature and humidity changes of the roof before and after the greening were basically in a balanced fluctuation, with some of them having peaks for particular reasons.

Through Ecotect software analysis and field research, it could be seen that the interior temperature of the green roof was much lower than that of the non-green roof. The interior humidity of the roof was much higher than that of the non-green roof. Therefore, a conclusion could be reached that the roof greening also had an obvious effect on the floor below in terms of cooling its temperature and increasing its humidity.

The monitoring of the rainfall showed that the roof garden had a good effect in controlling the rainwater runoff by way of rainwater collection.

## 9.5 Conclusion and Discussion

This chapter describes a comprehensive investigation of four aspects: roof configuration, plant deployment, landscape arrangement, and ecological effect measurement of the green roof on the administration building at the Fujian Agriculture and Forestry University. The results showed that the roof greening had a positive effect on the ecological environment.

As a result, how to design a roof garden according to local characteristics and how to exert the greatest effects are sure to be hot issues in the near future. In the light of this research, some techniques and strategies for roof garden design are as follows:

1. *Safety design:* The particularity of the roof garden lies in its special location and its function of serving people. It is a premise and foundation of roof garden design that the garden interface enclosure needs high-security protection. Therefore, the first consideration of the roof garden design is its design load. The construction of the roof garden especially is prone to undermining the roof's waterproofing layers (Osmundson 1979). This should be given special attention during the landscape deployment. Another consideration is the safety around the roof. A high-protection design is needed to prevent accidents caused by objects falling from the roof.
2. *Landscape design:* Compared with garden design on the ground, the roof space and the choice of the elements in the landscape configuration are limited. The landscape configuration mainly relies on plants, with a minimum structure size and a minimum quantity of landscape decorations. In this way, both capital costs and the roof load can be reduced. To make sure of the ornamental and economical effects, we should employ the principle of local-adaptive design, using the principles of "different views with moving steps" and "the winding path leads to beauty" from traditional garden design aesthetics to design a roof garden and create an ideal ecological leisure park for residents (MacHarg 1969).
3. *Plant disposition:* The design of the roof garden makes it hard to change its terrain and to create a man-made lake. For that reason, the plant deployment has a huge influence, though, with many limitations on the roof space, the requirement for landscape views is never reduced. In terms of plant disposition, high trees are always avoided because of the strong wind on the roof. Instead, species with strong trunks and flexible branches will be at the top of the list. The species of shrubs used are often those that are immune to the cold and are not selective with regard to soil quality. Generally speaking, it is a wiser choice to employ more rural species of trees and to pay attention to the combination of various types of plant to achieve ornamental and ecological values.
4. *Ecological design:* Different microecological effects are produced by different designs of roof greening. Configuring the density of plants in different directions to change the light and temperature in the building space below is one example. It is feasible to set up a "green fence" as a barricade against cold air, which can ensure gentle airflow in winter. Also, deploying plants with strong absorption

capability to absorb dust and harmful gases is an effective way of improving the air conditions in the roof garden. During the design of the roof garden, practical conditions must be fully taken into consideration. According to the functional needs of users, the design should apply humanistic–ecological principles and serve the requirements of modern ecological building construction.

In view of the above design methods and principles, the roof garden is a new landscaping form aimed at realizing ecological functions. Roof greening is an effective means of solving ecological and social problems caused by the present urbanization, and it is an effective way to counteract the natural environmental damage caused by human transformation. Roof gardens will be applied more and more widely to future urban design and for environmental modification. The government and the relevant construction institutions should pay attention to the study of roof greening, encouraging, guiding, and issuing related laws and regulations to promote reasonable application of roof greening.

## References

Le Corbusier (2004) Vers une architecture. Shaanxi Normal University Press

Drefahl J (1986) Sloping-roof covering

Feng Z (2009) Study on design of roof garden. China Central Academy of Fine Arts

Feng X (2016) Analysis of new trend of green building development in China from the perspective of the architects. Ind Constr 46(12):1–5

Feng X, Lei F, Ziyi G (2008) Roof garden design and construction. Chemical Industry Press, Beijing

Haiou W (2007) Research on the design of roof open space. Dalian University of Technology

MacHarg I (1969) Design with Nature. Wiley, New York, pp 23–45

Osmundson T (1979) The changing technique of roof garden design. Landscape Architecture

Shiheng W (1994) Exploration of Chinese natural aesthetics. China City Press, Beijing

Shizhang W (2007) Design and research of roof garden. Huazhong University of Science and Technology

# Chapter 10
# An Innovative Planting System for a Vertical Green Wall

**Shang-Jen Chen, Wen-Cheng Shao, and KuangHui Peng**

**Abstract** Under the influence of high building density, the heat island effect resulting from urban microclimates increases the heat load of the environment. Through evapotranspiration and shading effects of plants, the addition of green walls to exterior building walls can effectively lower the ambient temperature and satisfy greening needs. Most existing green wall systems use chemical foams and have a short service life. To effectively enhance the benefits of green walls, this study combined green planning, building–human coexistence, use and reuse of rainwater, and green materials to develop a green wall system that delivers sustainable greening. The proposed novel system is innovative and feasible, enables green walls to deliver long-lasting effects, uses the exterior structure of buildings for energy conservation, and facilitates sustainable development for eco-city.

**Keywords** Green planning · Vertical green wall · Symbiosis · Synergy

## 10.1 Introduction

The urban heat island effect has become increasingly apparent in highly industrialized cities with rapid economic development and high population density. Most cities feature considerable man-made structures and asphalt surface areas, which change the thermal parameters of the urban substrate. The materials used to construct these man-made structures often exhibit high heat absorption and low heat capacity, which renders the climatic environment warmer than that of neighboring

S.-J. Chen
Doctoral Program in Design, College of Design, National Taipei University of Technology, Taipei, Taiwan

W.-C. Shao
Department of Architecture, National Taipei University of Technology, Taipei, Taiwan

K. Peng (✉)
Institute of Design, Taipei University of Technology, Taipei City, Taipei, Taiwan
e-mail: khpeng@ntut.edu.tw



Z. Shen et al. (eds.), *Green City Planning and Practices in Asian Cities*, Strategies for Sustainability, https://doi.org/10.1007/978-3-319-70025-0_10

regions. This phenomenon is known as the urban heat island effect. Demand for sustainable building designs has increased over the years as a response to climate change, and the ratio of exterior walls to a building's total surface area has also gradually increased. If the thermal parameters of buildings' exterior walls can be lowered, the urban heat island effect can be effectively reduced and buildings can attain satisfactory energy conservation. The use of green walls is a common method used for these purposes. Through the evapotranspiration and shading effects of plant leaves, the ambient temperature can be lowered and environmental remediation can be achieved.

On traditional vertical green walls, which use hanging planters, the plants cannot grow naturally and effectively, and the use of growth bases and wall irrigation systems does not adhere to the concept of environmental sustainability. Therefore, the purpose of this study was to propose an innovative vertical green wall planting (VGWP) system that integrates sustainability elements such as rainwater harvesting, symbiosis, and environmental protection; incorporates aesthetic planning for exterior walls; and shows potential for development.

## 10.2 Research Design

This study was aimed at designing and modifying visualized VGWP systems and analyzed actual cases for the interaction between VGWP systems and the environment. Brown's theory (2012) was used to examine how planting systems can be implemented in practice, and how the concept of patent implementation can be used. It is an expectation of this study that enterprise resources will be introduced into actual VGWP systems for greening because dense green spaces are scarce in cities (Dobelli 2013). Accordingly, new techniques are being employed to design vertical green wall systems, develop greening methods for exterior walls in various domains, and add new diverse greening targets.

This study involved field trips to construction sites and show houses, and observation of novel construction techniques. Subsequently, qualitative analysis of patented techniques was used to describe the novel system proposed in this study. This study of VGWP systems aimed to develop a new system and increase urban green spaces. By applying the new system to urban buildings, the VGWP system can enhance the greening of urban spaces and create rural characteristics in cities. From this, an innovative concept of promoting green buildings as landscape goals, and improving residents' quality of life and health needs, is derived.

The field trip observations revealed that while some cities have a few buildings that use green walls, green wall applications integrating exterior building walls with aesthetic greening and environmental protection remain scarce. The design is based on the theoretical research framework of Han (2012). This chapter is structured into several sections, which include the study motivation and purpose in the introduction, the design of the plan and structure in order to obtain the study answer in the research design, a literature review adopted to search and evaluate the available literature with respect to the topic, and, finally, the study outcome, illustrating the results, conclusions, and recommendations. This study anticipates that future construction techniques will provide a symbiotic model for long-term coexistence of humans and buildings.

The comparative content of the tables is derived from literature reviews and field research. By analyzing innovative techniques and using a prototype theory derived from cross-disciplinary patents, exterior walls of buildings suitable for living can be designed. This technique's innovation was derived from constructing exterior walls of green buildings based on initial plans, suitably modifying the architecture, and finally applying a "wall within wall" design. The proposed system is feasible for making partial modifications to the appearance and structure of exterior building walls in practice, and part of the designs can be presented as a practical new technological patent. By modifying exterior building walls through greening, art design, and environmental protection concepts, future buildings may feature new looks. Through broader applications worldwide, which may yield further techniques and effects, innovative green buildings with VGWP systems may become possible.

## 10.3 Literature Review

Figure 10.1 illustrates the components, and an example, of a traditional vertical green wall. The leaves of plants on green walls are prone to wilting if the walls lack an irrigation system, thereby reducing the energy conservation effect resulting from reduced ambient temperature (Li 2009). In addition, the traditional landscaping effect shown in Fig. 10.1d fails to integrate with the surrounding environment to form a green building or space.

Technologies for green walls are constantly being improved. The use of VGWP systems as a landscaping approach for exterior walls is feasible in future-oriented construction projects, as the systems can attain environmental sustainability and yield aesthetic effects in green cities. Furthermore, the materials and techniques involved can be adapted to suit the local environment, demonstrating the applicability of this method to urban planning and innovative facilities. The following sections show the traditional case and the innovation case proposed in this study, and compare their systems, techniques, advantages, and disadvantages.

### *10.3.1 Sprinkler System for "Rotating Buildings"*

A field study on a construction project named "Tianyi" in the Danshui district of New Taipei City, which realtors have advertised as a "rotating building," revealed that the building will not literally rotate; instead, vertical green walls will be installed on the corners of the building's balconies to achieve a spatial illusion. In addition, a sprinkler system that will directly irrigate the plants will be adopted. An interview

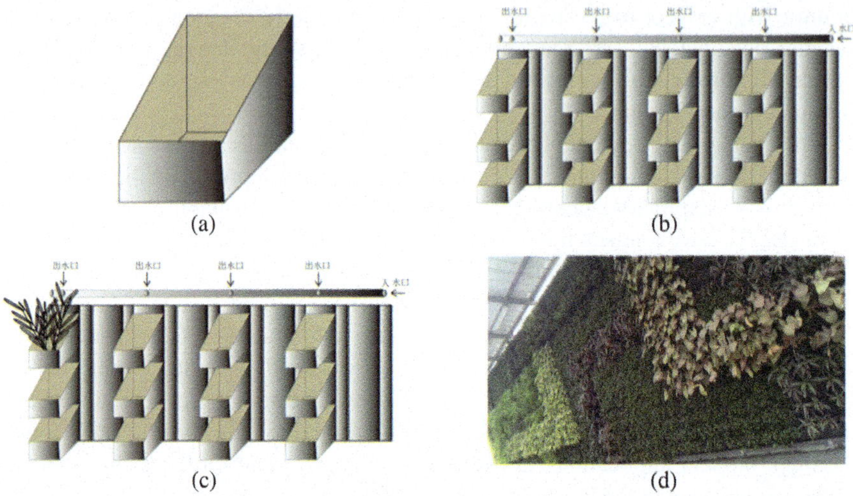


**Fig. 10.1** Components and example of a traditional vertical green wall. (**a**) A regular outdoor wall hanging planter. (**b**) A cluster of outdoor wall hanging planters. (**c**) Irrigation system formed from the planting method. (**d**) A construction project on the Zhongxiao E. Road in Taipei City

revealed that this sprinkler system approach (Fig. 10.2) replaces other traditional approaches, and its use still has the limitations of utilization of plastic materials for an extensive period of time, and expensive replacement.

## *10.3.2 An Existing VGWP System: A 4-Layered Substrate Panel*

Figure 10.3 shows the implementation of another new system. This system has been adopted for a building in the City Branch of National Chengchi University and was examined in a field research interview.

The implementation process is as follows. First, the exterior wall most suitable for installing the system is found. Next, the panel and the water outlet are installed. The water level measurement and control mechanism is installed at the water inlet, and a dripper is installed in the pipe approximately 10 cm above the control mechanism. The four substrate layers are then installed sequentially into the panel, as follows: Insu™ board, planter cell water retention mat, H2O-R foam, and Insu™ felt. The vegetation is planted in the felt layer—the outermost layer—and is affixed using an air pneumatic stapler. This system features a dripper irrigation system above the substrate panel and a water recycling system at the bottom, thus enabling the water circulation loop to be closed.

Fig. 10.2 Green wall "rotating building" design adopted in the Tianyi construction project. (a) Use of a spatial illusion to create "rotating walls." (b) Irrigation system

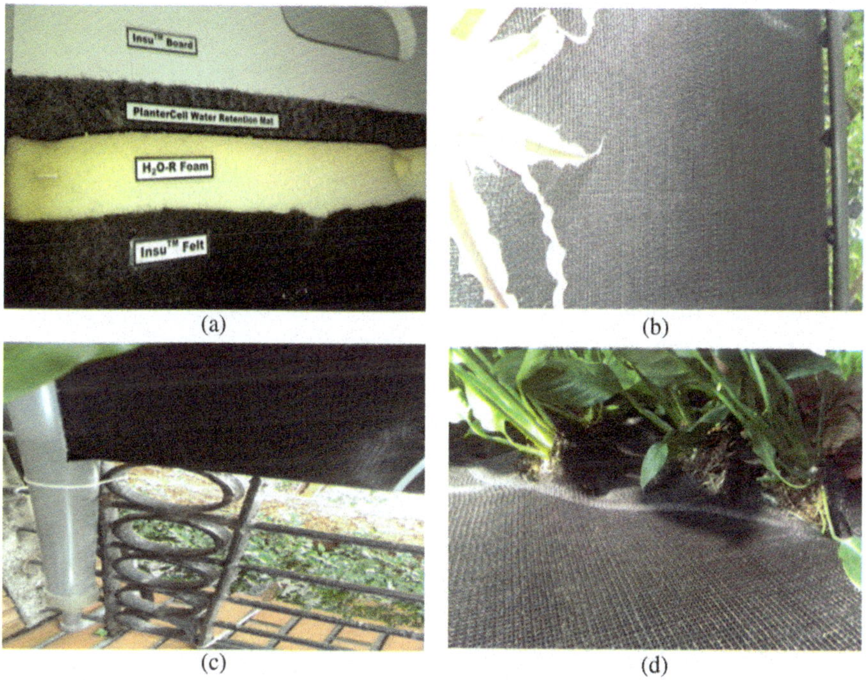


Fig. 10.3 Four-layered substrate panel system adopted in a construction project in the City Branch of National Chengchi University. (a) Four foam-type substrate layers. (b) Water outlet above the substrate panel. (c) Water outlet and recirculation pipe at the bottom. (d) Four-layered substrate panel

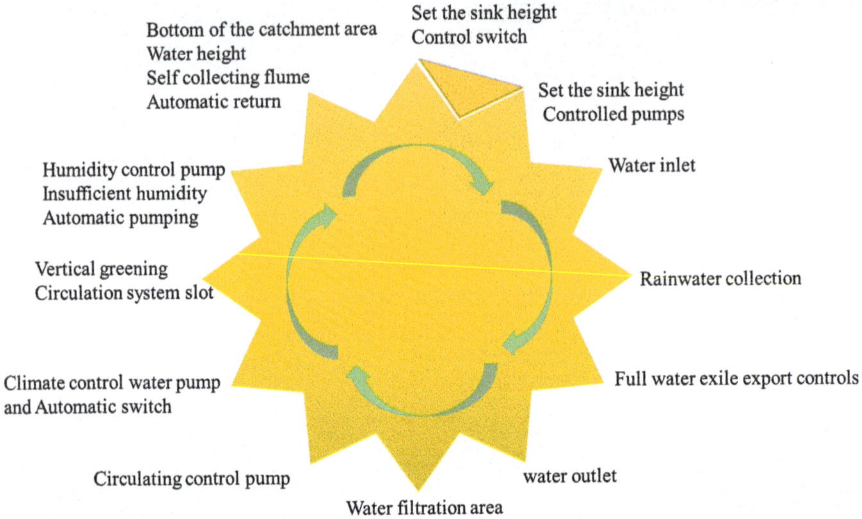


**Fig. 10.4** Process chart of a vertical green wall system

Using field research, the installations of this VGWP systems were analyzed, and the service life of this system was determined to be approximately 5–10 years. However, the recyclability of the system is limited by material deterioration from long-term use, resulting in material toxicity.

## *10.3.3 An Innovative VGWP System*

Figure 10.4 shows the process chart of a novel VGWP system. By setting the sink height control switch, the system enables automatic control of the pumps when the water level reaches the specified height. After the plants are watered, the medium filters the water, completing the circular flow. A humidity control pump is also installed in the planting area of the wall. This pump automatically irrigates the plants when a sensor detects humidity.

The VGWP system proposed in this study is a permanent structure. The system enables rainwater collection from rooftops; when the collected rainwater is insufficient, domestic water is used to replenish the water supply. In addition, temperature and humidity are automatically regulated. The proposed system is a breakthrough from the original system, enabling sustainability of exterior walls of green buildings and establishing a green space. By devising an automated water circulation mechanism with temperature and humidity control, the proposed system is suitable for various types of environment and can be installed on both interior and exterior walls. This system enables the surfaces of exterior walls to be furnished with local green plants and showcases the synergy characteristic of being suitable for various types of plant.

### 10.3.4 *Comparison of Different Systems and Techniques*

A comparison of the three aforementioned systems is summarized in Table 10.1, which lists the traditional and innovative green technique systems adopted in green buildings, and their advantages and disadvantages.

Table 10.2 compares existing study cases (Yin and Tsai 2009; Peng 2013) with the innovative system proposed in this study, showing some of its characteristics, e.g., the exterior walls have waterproof hollow pockets added and a water circulation tank. Because the wall surfaces are not waterproof, the plants can absorb moisture from the wet surface. The hollow pockets feature planting spaces, suitable for planting local plants, and can feature pockets for soil-free cultivation.

This study found that in the Tianyi construction project using "rotating" vertical green walls, the green walls were installed only on the exterior walls of the left and right corners of the balconies. The existing VGWP system has four layers of foam substrates that are embedded within panels to form a four-layered substrate panel, and it is irrigated using domestic water. However, high temperatures in the summer cause these foam materials to produce toxic dioxins.

Table 10.1 compares traditional green buildings and vertical green wall systems. The disadvantage of the Tianyi construction project lies in its use of plastic materials, which are not suitable for long-term use. Considerable costs are incurred to recycle, regenerate, and replace these materials when they deteriorate. Similarly, the disadvantage of the four-layered VGWP system is that the materials become toxic when they deteriorate due to long-term use, rendering them unsuitable for recycling, and their replacement is expensive. Compared with these two systems, the proposed innovative system exhibits the following advantages: use of domestic water for supplementing rainwater collected from rooftops when rainwater is insufficient, automated regulation of temperature and humidity levels, and high weather resistance. Table 10.2 summarizes different vertical greening techniques used in various countries. The coverage of these techniques for walls on lower floors ranges between 50% and 90%. By contrast, the proposed system can provide coverage of 70–95%.

In short, use of green buildings in existing urban spaces requires consideration of five dimensions: land utilization, environmental protection, innovation in construction materials, seismic resistance, and disaster prevention (Landry 2012). In the innovative use of green building structures, existing green wall planting systems can be modified into novel systems using inductive analysis. Figure 10.5 demonstrates the analytical process.

## 10.4 Study Achievement

This study concludes that the relationship between VGWP systems and future construction buildings forms five symbiosis dimensions (Fig. 10.6): green planning, regeneration, symbiosis, architectural scenery, and environmental protection. These

**Table 10.1** Comparison of various vertical green wall planting (VGWP) systems and techniques

| | Sprinkler system (Fig. 10.2) | Existing VGWP system (Fig. 10.3) | Innovative VGWP system (Fig. 10.4) |
|---|---|---|---|
| System | Traditional exterior walls adopting hanging planters and use of the building's balcony corners to create an optical illusion of rotating walls | Use of four foam-type substrate layers, plastic pipes affixed to exterior walls, a water level measurement and control mechanism, and drippers spaced 10–20 cm apart on the substrate panel | Automatic circulation, suitable for various environments, can be installed on interior and exterior walls with various types of plant |
| Innovative technique | Techniques, energy targets, green effects, and innovative technological system chosen according to their use for general buildings | Innovativeness in energy targets, and water collection tanks featuring automated water circulation and resupply cycles | Rainwater harvesting and reuse, environment symbiosis, and greening |
| Disadvantages | An old system utilizing plastic materials that cannot be used for an extensive period of time; replacement is expensive | Long-term material deterioration results in material toxicity or inappropriateness for recycling; replacement is expensive | Requires additional installations |
| Advantages | Hanging planters are suitable for 2 years of use | Feasible simulation principles and concept implementation; suitable for various types of building | Domestic water is used to supplement the collected rainwater when inadequate; use of an automated water circulation mechanism with temperature and humidity control enables the system to be a permanent structure |

dimensions form the research outcome of the VGWP system. The following sections introduce the concept of the proposed innovative VGWP system and present the technical outcome.

### *10.4.1 Comprehensiveness of the Greening Concept*

Cities can benefit from innovation using technology, and new symbiotic relationships can be developed from the spirit of environmental management. Regarding future green urban buildings, the innovative VGWP system patent will contribute to

**Table 10.2** Comparison of existing research cases with the innovative vertical green wall planting (VGWP) system proposed in the present feasibility study

| | Research case location | | | | | |
|---|---|---|---|---|---|---|
| | Taipei, Taiwan | Taipei, Taiwan | Miaoli, Taiwan | Paris, France | Aichi Prefecture, Japan | Innovative VGWP system |
| Building type | School campus | Concert hall | Hot spring resort | Museum | Museum | |
| Technique type | Climbing plants on walls | Soil-free cultivation | Climbing plants on trellis | Soil-free cultivation | Planting pockets on walls | Exterior walls have waterproof hollow pockets added and a water circulation tank. Because the wall surfaces are not waterproof, the plants can absorb moisture from the wet surface. The hollow pockets feature planting spaces |
| Vegetation type | Climbers along walls | Soil-free cultivation—planting on felt layers | A trellis is installed in front of the wall to serve as support for climbers | Soil-free cultivation—planting on felt layers | Hanging plants hooked up to the panel | Suitable for planting local plants and can feature pockets for soil-free cultivation |
| Holder material | Not applicable | Nonwoven felt fabric | Scaffolds | Nonwoven felt fabric | Embedded into walls | Retention of planting pockets enables landscape changes |
| Scale ($m^2$) | 1200 | 160 | 320 | 450 | 300 | Limits may or may not be set |
| Irrigation | Rainwater | Drip pipes | Rainwater | Drip pipes | Drip pipes | Rainwater harvesting or recycling, automated irrigation by the water flow tank |
| Plant types | Japanese ivy, climbing fig | Common Taiwanese plants such as goosefoot plant | Chinese wisteria, climbing fig | Ferns | Ground cover plants, ornamental plants, shrubs | Local plants, plants suitable for soil or soil-free planting pockets, ornamental plants, vegetables |
| Maintenance and management | No special maintenance required | Regular inspection and maintenance | No special maintenance required | Regular inspection and repair | Regular replacement and repair | The elevator on the external wall requires regular maintenance |
| Coverage rate | 90% | 90% | 50% | 90% | 90% | 70–95% |

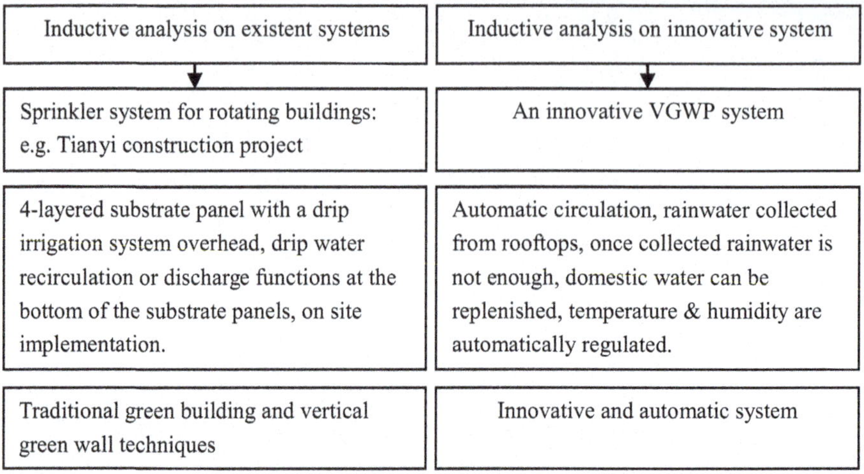


**Fig. 10.5** Application of inductive analysis to vertical green wall planting (VGWP) systems

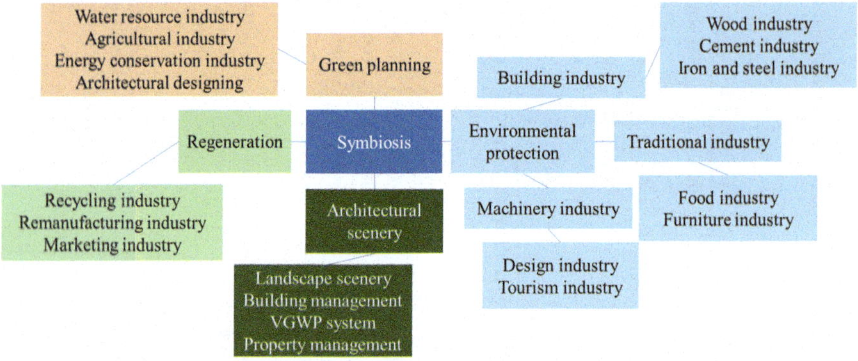


**Fig. 10.6** Mind map of the five symbiosis dimensions

a long-term increase in green spaces on exterior building walls. Through long-term contributions of applying greening techniques to building walls in light of limited land availability, new greening goals, indicators, and patented technological designs can be obtained using the proposed system design. With innovative systems that reflect these green features through exterior building walls and application development, symbiotic and synergistic effects of urban greening on future architectural greening can be attained.

A comparison of research results revealed that vertical green wall techniques and innovative urban green application techniques, as observed in the case study, are currently extremely few in number; no new techniques were identified. Review of techniques in the literature and the concept of innovative greening showed that the addition and application of green vegetation to exterior walls made related building systems more complete, enhancing energy use efficiency and reducing energy consumption. Most techniques in the literature propose a short-term symbiotic relationship between green walls and humans and lack enduring innovative applications over a long period. Such a technological gap has prompted the patent design in this study, which sustains the long-term effects of these greening structures, showcasing synergy in the patent invention scope of VGWP systems.

### 10.4.2 Formation of the Regeneration Concept

Following greening, water resources can be retained within the walls and circulated or recycled, thereby providing water or moisture for regrowth of the plants on the exterior walls. This regrowth extends from the concept of regeneration. Innovation in urban buildings is thus attained from innovating the regrowth concept of plants and applying it to exterior building walls.

### 10.4.3 Symbiotic Water Circulation Technology

The environmental protection mechanism is within the walls, where circulated water cools the ambient temperature or regulates the temperature within a specific range, reducing the use of heating and cooling systems. This saves energy and reduces carbon emissions, fulfilling the basic requirements of symbiotic environmental protection.

### 10.4.4 External Appearance Changes on Buildings

Architectural scenery suggests unconstrained changes to building appearances according to actual needs. Changes may involve changes in size, shape, structure, or facility, which together create artistic scenery. As a principle, local plants suitable for the environment are used for this purpose.

### 10.4.5 An Innovative VGWP System Comprising Five Symbiosis Dimensions

The VGWP system comprises five symbiosis dimensions: green planning, regeneration, symbiosis, architectural scenery, and environmental protection. The implication of symbiosis lie in the first four dimensions; environmental protection implies the coexistence of humans and their environment. These dimensions drive humans to exert interactive and positive effects on green urban spaces, aiming at a harmonious coexistence of air, water, buildings, and humans; improving the environmental quality of cities; and motivating the goal of coexistence between humans and green buildings. Consequently, a synergetic thinking map comprising the five symbiosis dimensions contributing to innovative urban greening is formed.

### *10.4.6 Technical Outcome of the Innovative VGWP System*

The following illustrates the technical outcome of the VGWP system, which has been examined and approved for a patent in Taiwan:

- VGWP system patent no.: 102148601.
- Applicants: Shang-Jen Chen and KuangHui Peng.
- The following sections provide some information pertaining to the patent; for more details, refer to the data filed with the Taiwan Patent Search System under the application number 102148601.

#### 10.4.6.1 Design Implementation and Brief Explanation of the Techniques Involved

Nomenclature

The numbers shown in Figs. 10.7, 10.8 and 10.9 represent different components of the system.

Technological Scope

The VGWP system comprises the following: a planting wall made of water-permeable materials featuring at least one cell for the substrate on the front side; a water container wall designed on the back of the planting wall with several water containers—each with at least one water outlet—distributed among various levels; an impermeable wall designed on the opposite side of the planting wall and with at least one plate capable of waterproofing; a water circulation module that connects the water containers at the bottommost level through their discharge outlet with the topmost-level containers through a water return line connected with a water pump; at least one water storage barrel for collecting rainwater, the discharge outlet of which is connected to the topmost-level containers and features a valve; several case frames with numerous cells containing substrates; and a shield over the top of the topmost-level containers. The container outlets are alternately paired with those of the upper and lower containers.

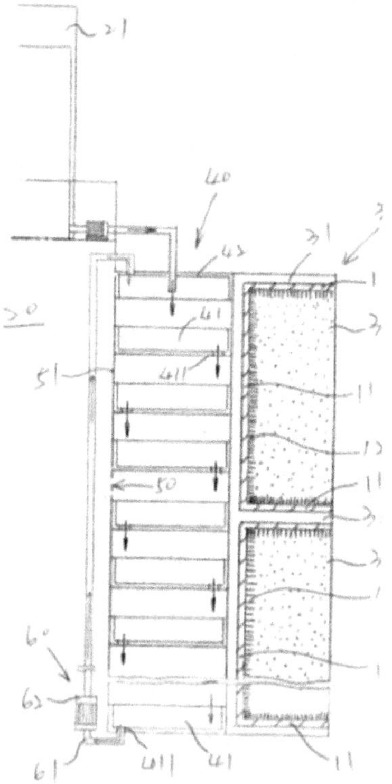

**Fig. 10.7** Side view of the vertical green wall planting (VGWP) system structure. The numbers shown in the figure represent different components of the system: *11* represents the substrate, *12* represents the plant, *20* represents the building, *21* represents the water tower, *30* represents the planting wall, *31* represents the case frame, *311* represents the cells, *40* represents the water storage wall, *41* represents the water container, *411* represents the water discharge point, *42* represents the shield, *50* represents the impermeable wall, *51* represents the plate, *60* represents the circulation module, *61* represents the water return line, *62* represents the water pump

### 10.4.6.2 Technical Outcomes and Conclusion

The proposed VGWP system—which exhibits the characteristics of green planning, regeneration, environmental protection, architectural scenery, and symbiosis—stores water within the walls, irrigates plants on the exterior walls, and provides nutrients to the substrate. Actual applications of patented technology related to VGWP systems have not been identified anywhere in the world. Therefore, use of favorable building materials and biodiversity in building environments, transformation of these techniques into patented technologies, and application of them to improve the sustainability of exterior building walls can contribute to research on VGWP systems.

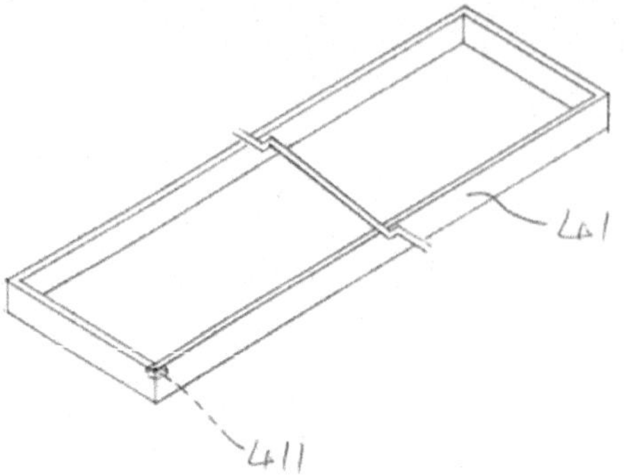


**Fig. 10.8** Structural diagram depicting the appearance of the water container inside the "wall within a wall." The numbers shown in the figure represent different components of the system: *41* represents the water container, *411* represents the water discharge point

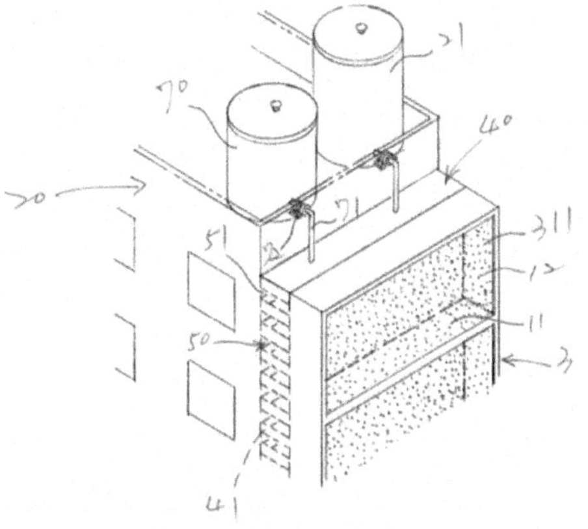

**Fig. 10.9** Reference figure for configuring the vertical green wall planting (VGWP) system. The numbers shown in the figure represent different components of the system: *11* represents the substrate, *12* represents the plant, *20* represents the building, *21* represents the water tower, *30* represents the planting wall, *311* represents the cells, *40* represents the water storage wall, *41* represents the water container, *50* represents the impermeable wall, *51* represents the plate, *70* represents the water storage barrel, *71* represents the water outlet, *72* represents the valve

## *10.4.7 Conclusion and Recommendations*

Green walls composed of hanging planters are not innovative; the field research we conducted did not reveal any innovative or advanced green wall techniques or systems being applied to existing building construction projects. In reviewing and comparing VGWP techniques and systems, the qualitative analysis revealed that most systems use polypropylene materials, which has a service life of 10–20 years. However, when these materials deteriorate, they cause secondary pollution. In the proposed system, these materials have been replaced to mitigate pollution.

VGWP addresses national, social, and human needs for new building structures by replacing structures with live vegetation, thereby attaining feasibility through innovation, activation, and functional decoration. Urban greening can be realized through the practice of system innovations for green building walls. The innovative research outcome of this study is based on technology. By applying technology to innovation of greening methods, sustainability is achieved through five symbiosis dimensions. In addition, regeneration and environmental protection approaches are used to create architectural scenery, implying the use of materials for active symbiosis and creating future green cities that exhibit symbiotic relationships between objects and biological life, and between humans and objects. This study has reviewed VGWP systems through field research, with innovation and modification of existing systems into a feasible and practical system.

The following technical suggestions are limited to applying greening techniques to the exterior walls of existent buildings. Because no advanced technology in theory or practice was identified in the literature, the proposed system is innovative. The proposed technique is a feasible and innovative technique, which can be applied to exterior building walls or surfaces made of concrete cement, glass, or marble.

Breakthroughs in existing technologies for use of solar energy to harvest rainwater for cleaning purposes are necessary. Exterior walls can be designed with previous designs, colors, and materials, or with solar panels. Most old buildings have not undergone greening; this study will continue to work toward increasing greening areas on future new buildings.

Building size is not a problem in green system planning in the building preplanning stage. Exterior wall components and technological content only need to be added to original building designs to attain greening effects and reorient them toward the development direction of green urban innovation. Regarding green planning, regulations and designs of future buildings can contribute to creating a beautiful city that considers environmental protection, air quality, and green scenery. Each green building could eventually become an international tourist hotspot or develop opportunities related to green wall cultural heritage.

The findings of this study may serve as a reference for future research on VGWP. Scholars and business enterprises can consider scale-up studies on these findings or improve on the limitations.

**Acknowledgments** We thank the "Low Carbon Quality of Life: Applying Greening Technology to Sustainable Green Buildings" research team in the Department of Architecture, National Taipei University of Technology, and KuangHui Peng, Chih-Hong Huang, Shu-Ying Tsai, Yeou-Feng Li, Ying-Chun Hung, and Chao-Shun Yang for their inspiration and participation in this study.

## References

Brown T (2012) Change by design: how design thinking transforms organizations and inspires innovation. HarperCollins, New York

Dobelli R (2013) The art of thinking clearly. HarperCollins, New York

Han P-T (2012) Innovation view. World Magazine, Taipei (in Chinese)

Landry C (2012) The creative city: a toolkit for urban innovators. Taylor & Francis, London
Li C (2009) Building Shēng Wài study on the energy-conservation result analysis of graft. Taipei (in Chinese)
Peng K-H (2013) The application of vertical greening to urban rehabilitation and maintenance. Int Rev Spat Plann Sustain Dev 1(3):41–52. Kanazawa, Japan
Yin D-Q, Tsai J-H (2009) Wall greening system of existing buildings, Taiwan Architectural Society of the 21st Session of the Second Architectural Study, published essays, Taipei. Taiwan Architectural Society. (in Chinese)

# Chapter 11
# Creation of Green Culture and Values in the Hanul Madang Rooftop Garden at Seoul National University

**Yong-hoon Son**

**Abstract** This chapter discusses green roof policies and practices in Korea and introduces a case study of a rooftop garden at Seoul National University. Seoul City has greatly grown since the 1970s and is still expanding through urban development projects. This rapidly developing urban environment makes it very difficult to conserve green spaces. Since the 2000s, Seoul City's policy has been increasingly concerned with qualitative growth indicators for green and livable cities. Impervious surfaces cover 48% of the total area of Seoul, i.e., the majority of the urbanized area is covered by impervious surfaces. Therefore, green roofs were included in the 2030 Seoul City Park and Green Space Master Plan, because they are a very effective way of increasing the quantity and quality of urban green spaces. Rooftop greening is a very effective way of contributing to green cities, because it is an economical way to utilize the available space. Previous studies on the value and benefit of green roofs in sustainable cities are summarized, and two types of green roofs are described. The first type is a leisure and recreation space to enrich the green culture in the urban environment, where there is a lack of greenery, while the other type has more ecological effects and aims to enhance the environment through greening of artificial grounds. The type of rooftop garden is decided at the beginning of its installation, which can be distinguished by an intensive or extensive system. In this chapter, the case of the Hanul Madang rooftop garden, built at Seoul National University, is introduced as an example of the use, as well as the ecological and cultural values, of rooftops.

**Keywords** Green roof · Green culture · Seoul City · Seoul National University · Sustainable city

---

Y.-h. Son (✉)
Graduate School of Environmental Studies, Seoul National University, Seoul, South Korea
e-mail: sonyh@snu.ac.kr



Z. Shen et al. (eds.), *Green City Planning and Practices in Asian Cities*,
Strategies for Sustainability, https://doi.org/10.1007/978-3-319-70025-0_11

## 11.1 Benefits of Green Roofs in a Sustainable City

The modern green roof concept originated at the beginning of the twentieth century in Germany, where vegetation was planted on roofs to mitigate the damaging physical effects of solar radiation on roof structures (Mansour 2017). Over the last 40 years, green or vegetated roofs have become widely accepted as a preferred solution for a sustainable urban environment (Mansour 2017).

Green roofs offer many benefits such as slowing water runoff, aesthetic improvement, moderation of the urban heat island effect, improved air quality, fire retardation, improved energy efficiency, noise reduction, and use as an educational tool (Wescott and Sundell 2016). In particular, green roofs have the potential to provide attractive green space in urban areas where it is limited or nonexistent on the ground (Berndtsson 2010). For these reasons, green roofs have become an important tool for planning and design of cities where development has damaged the natural environment.

Green roofs can be classified into two types according to their purpose and environmental characteristics. These are extensive systems with shallow substrates as a growing medium, and intensive systems with a deep soil layer for planting trees and flowers (Tolderlund 2010). Each green roof type requires different levels of plant maintenance, ranging from two or three inspections annually to monitor weeds or damage, to weekly visits for irrigation, pruning, and replanting (Peck and Kuhn 2003).

Green roofs are very valuable regarding their ecological functions and control of environmental conditions. They also act as a place for appreciating scenery, outdoor recreation, and relaxation. While extensive green roofs have ecological functions, intensive green roofs function as a space for people as well as these functions. However, the cost of these spaces is relatively high. An intensive green roof is a green space like an urban park, forest, or garden created on a roof. In this sense, the value of cultural ecosystem services from green roofs is high in urban environments with limited green space. Jafari et al. (2015) studied urban residents' preferences for cultural ecosystem services of green roofs to identify everyday functions of green roofs as spaces for leisure and relaxation, community interaction, environmental aesthetics, observation of landscapes, exercise, and better lifestyles.

## 11.2 Policies and Trends for Green Roof Projects in Seoul City

The urbanized area of Seoul comprises 58% of its total area. The remainder is green space such as forests, rivers, and parks. Impervious surfaces cover 48% of the total area of Seoul. That means that the majority of the urbanized area is impervious (Oh 2011). Excessive urban development causes environmental problems such as increased energy consumption for heating and cooling, rapid storm water runoff, and increased flood risks. Green roofs are an effective approach for creating urban green space without land compensation in Seoul, where land is expensive and there is a shortage of land to plan additional parks (Choi 2011).

The *2030 Seoul City Park and Green Space Master Plan* proposes a plan for allocation and networking of green spaces that require creation and/or conservation in Seoul City. In the master plan, the concept of the green roof is introduced as an effective way to increase the limited green areas in Seoul City, improve the aesthetics, reduce the urban heat island phenomenon, save energy, and solve some ecological problems (Seoul City 2015).

In addition to green roofs, Seoul City implemented the Green Way project, which connects green areas that are disconnected from the pedestrian environment. The Green Way project was applied to 55 sites totaling 29.1 km of Seoul. The project includes the following improvements: lane diets, sidewalk improvement, and riverside utilization. By the year 2015, the riverside greenway had been applied to a distance of 119 km in 10 rivers, and an additional 21 km was planned to be included (Seoul City 2015). Moreover, the Eco School project, in which, by 2010, 1425 school parks (967,140 $m^2$) were created in the schools of Seoul, is another interesting greening project. The project initially proposed the establishment of community gardens, biotope ponds, or green walls in schools (Seoul City 2015).

Yang (2011) specified the principles of green city planning to be self-support, diversity, circulation, and stability. Greening the city through green roofs or other projects directly or indirectly contributes to the above four principles, because the city is an urban environment that is being integrated with a natural ecosystem.

Many studies have recommended that the policy direction of cities should focus on promoting green roofs. Na and Byun (2006) analyzed research trends on rooftop greening in Korea and classified them into three categories. These include technical approaches to developing green roofs, the benefits of green roofs, and the policies supporting the expansion of green roof infrastructure. Park et al. (2012) calculated the available area for rooftop greening in Gyeonggi Province, Korea, using the building management register data and remote-sensing data. They found that 102,454,000 $m^2$ of potential green roof space exists in Gyeonggi Province, which is a significant area, larger than the total area of urban parks (62,018,000 $m^2$) or open green spaces (52,001,000 $m^2$). Considering that it is very difficult to increase the area for green spaces in urbanized areas, rooftop gardens offer a useful approach to expand green space in the city.

In the late 1970s, green roofs were first introduced to Korea with the installation of rooftop gardens in department stores and hotels. However, it was not until 1998 that green roof research became more prevalent through the formation of the Green Roof Research Society by members of academia and related industries. Emeritus Professor Byeong-I Yang of the Graduate School of Environmental Studies, Seoul National University, was the first chairperson of this study group, which is a joint space where various experts and practitioners from Seoul National University and related institutions and industry can conduct research.

In 1999, the Ministry of the Environment published a green roof manual. The Green Roof Research Society was actively involved in researching international case studies in Japan and Germany, as well as presenting local trends for this publication. Since 2000, local governments in Seoul, Bucheon, Suwon, and Seongnam have developed green roof projects (Zhao and Kang 2011). In 2003, the Green Roof Research Society was incorporated and registered as the Korean Green Roof and Infrastructure Association, as a subsidiary of the Ministry of Environment. It continues to conduct academic research, seminars, and educational activities regarding green roofs.

Seoul City is a leading area in the green roof projects of Korea. A green roof project called the Seoul City Hall Seosomun Annex commenced in 2000 with a 298 $m^2$ rooftop garden. Following this experimental project, green roof projects have spread nationwide (Choi 2011). The Seoul City Government has implemented many green roof support projects since 2002, as outlined in Table 11.1.

By 2010, a total of 446 green roofs had been built, with a total area of 202,448 $m^2$. Since 2007, Seoul City has been conducting the 10 Million Green Roofs project with the aim of creating an additional 100,000 $m^2$ of green roofs in living space per autonomous region in Seoul (Lee et al. 2014). The government supports the entire cost of building new green roof projects on government-owned buildings and supports 50–70% of the project cost for other public and privately owned buildings.

Geographically, Seoul City is a basin surrounded by mountains, with the Han River running through the city. Urban forests surrounding the city and its water system create a pleasant urban environment. However, due to rapid urban development, Seoul lacks green spaces in residential and commercial areas. Figure 11.1 displays urban green spaces, water systems, residential and commercial areas, and green roof project sites. These projects play a big role in providing green space for living environments where there is currently insufficient green space in Seoul.

**Table 11.1** Numbers of green roof projects in Seoul in 2002–2010

| Year | Number of projects | Total area ($m^2$) | Project expense (MKW) |
|---|---|---|---|
| 2002 | 11 | 5455 | 640 |
| 2003 | 10 | 3322 | 640 |
| 2004 | 9 | 2670 | 320 |
| 2005 | 6 | 3970 | 430 |
| 2006 | 14 | 5611 | 803 |
| 2007 | 62 | 24,154 | 5588 |
| 2008 | 106 | 59,221 | 12,323 |
| 2009 | 120 | 46,971 | 9662 |
| 2010 | 108 | 51,074 $m^2$ | 11,229 MKW |
| Total | 446 | 202,448 $m^2$ | 41,635 MKW |

Source: Choi (2011)
*MKW* million Korean won

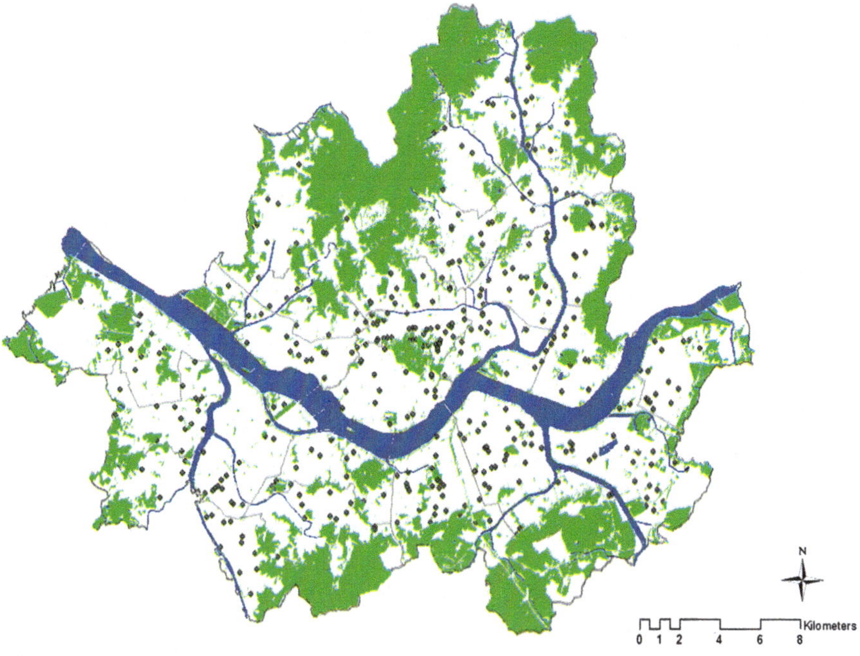


**Fig. 11.1** Green space in Seoul City and green roof project sites (created by the author using the biotope map in Seoul and Open Data Portal (data.or.kr) using Arc GIS 10.2.2). Urban green spaces are shown in green, water systems in blue, residential and commercial areas in white, and green roof project sites as black dots

## 11.3 Hanul Madang Rooftop Garden

### *11.3.1 Rooftop Garden Project of Hanul Madang*

The university campus has great potential and feasibility for developing green roofs. The site can be used for multiple functions after construction such as recreational, experimental, educational, and research purposes (Wescott and Sundell 2016). In Korea, there are already some examples of development of rooftop green spaces for research at universities. At these sites, the vegetation, soil characteristics, environmental conditions, and population dynamics of plant species after the installation of the green roof are studied (Lee et al. 2013a, b).

The development of a rooftop garden at the Graduate School of Environmental Studies (GSES) at Seoul National University (SNU) is a project to promote green roof infrastructure at public institutions in Seoul. As mentioned previously, the GSES promotes the Green Roof Research Society, which is a primary resource for research on green roofs in Korea. In 2010, the Seoul City Government and SNU planted 746 $m^2$ of flowers and trees on the rooftop of the GSES building to create an ecological habitat and a recreational space for people. The total project cost was 200 million Korean won. Emeritus Professor Byeung-rim Yu and the Eco-Engineering company designed the green roof, and Hwa-am Landscape Architecture Co., Ltd. constructed it.

This rooftop garden is named Hanul Madang, which translates as "garden in the sky." Hanul Madang is the first example of an existing roof retrofitted with a green roof at SNU. The Hanul Madang rooftop greening project has attracted significant attention at the university. Since its construction, rooftop gardens have also undergone development on buildings in the College of Engineering, College of Natural Sciences, and College of Agricultural and Life Sciences.

### 11.3.2 Using Green Spaces to Connect a Green Corridor to the Natural Landscape

The SNU campus originally contained abundant natural areas at the foot of Mt. Gwanak. However, since building densities have increased, the number of green spaces has decreased. The campus master plan regulates the development of the site. However, these regulations are often not followed, and existing green areas are also at risk of decline. This means it is necessary not only to be concerned with passive protection of existing green space but to identify spaces that can be developed on the campus to promote new green space.

Hanul Madang is a significant initiative for reconnecting green space that has disappeared due to the development of buildings. Rooftop gardens may reconnect previously disconnected green spaces. The rooftop gardens offer panoramic views that connect with the surrounding natural areas and scenic views. If the building development continues at SNU, with rooftop gardens, these areas can be linked to reconnect disconnected natural areas at the ground level.

Rooftop gardens also offer advantages of accessibility, views of surrounding areas, and convenience compared with the use of other green spaces. Therefore, they have a high cultural value in addition to their ecosystem service values. The rooftop garden provides leisure, recreational, healing, educational, and tourism values. To elaborate, it offers a place for new experiences, a healing space that relieves stress from everyday research and study, and a space for events and welcoming visitors.

It is difficult to ascribe an economic value to these cultural ecosystem services that achieves universal agreement. This value is qualitative, and the landscape and cultural values are very relative to the perspective of different users. Hanul Madang at the GSES offers a comprehensive range of values (Figs. 11.2, 11.3 and 11.4).

**Fig. 11.2** Rooftop gardens connect greenery (Photo taken by Son Yong-hoon, July 15, 2017)

**Fig. 11.3** Roof garden landscape with surrounding nature (Source: Department of Landscape Architecture, Graduate School of Environmental Studies, Seoul National University)

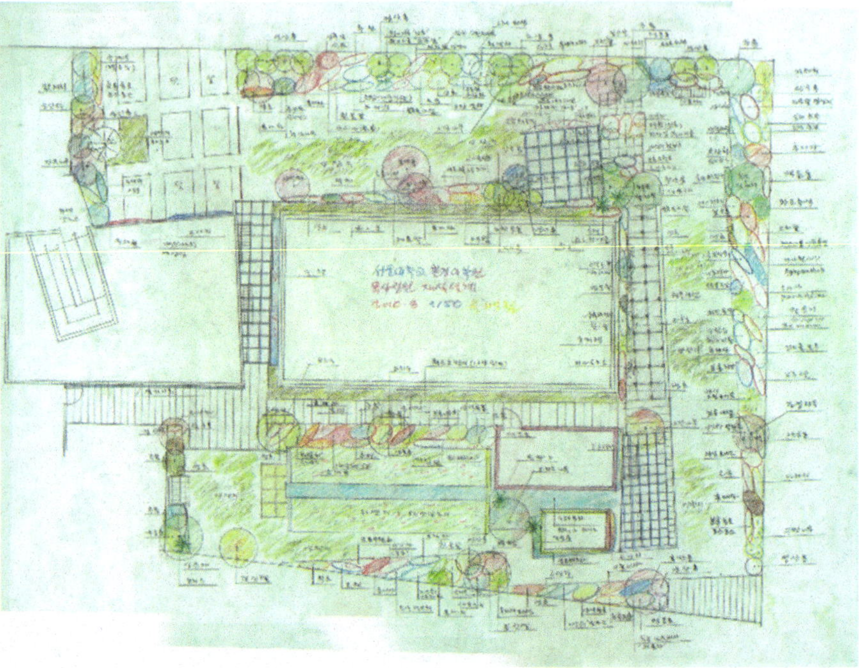

**Fig. 11.4** Rooftop garden design of Hanul Madang by Emeritus Professor Byeung-rim Yu (scale 1:50) (Source: Department of Landscape Architecture, Graduate School of Environmental Studies, Seoul National University)

### *11.3.3 Hanul Madang as a Living Landscape*

Hanul Madang is open to the public, as well as students and staff. The roof garden is becoming a popular place for students and visitors. There are approximately ten varieties of trees and shrubs, and about 120 kinds of herbaceous plants, attracting birds and butterflies. Hanul Madang is both a pleasant resting place for people and an ecological space.

Hanul Madang is managed by landscape architecture students from the GSES. The rooftop garden is a dry area, which is less fertile than ordinary soils. The area is subject to high evaporation rates and requires a high level of maintenance to manage weeds and maintain irrigation of the plants. Students who are interested in garden design use this as a learning experience.

Between early summer and early autumn, maintenance focuses on flowers such as *Lythrum*, *Liatris*, *Echinacea*, *Eupatorium japonicum*, and *Iris domestica*, which are fixed by lines. Bulbs of *Muscari*, *Hyacinthus*, and lilies, which provide a rich scent, are stored in the shade and planted during the following spring.

There are some spaces in the garden where students can directly grow tomatoes, peppers, lettuce, cucumbers, pumpkins, and other vegetables, as well as basil, lemon balm, peppermint, bergamot, chamomile, and other herbs.

**Fig. 11.5** Event and garden party at Hanul Madang (Source: Department of Landscape Architecture, Graduate School of Environmental Studies, Seoul National University)

From early 2013, the Student Council of the GSES obtained permission to commence beekeeping activities. These students now raise bees and collect honey. The bees move between the rooftop and the nearby Mt. Gwanak. In early May, when the acacias are flowering, chestnut honey is collected, and the bees gather pollen from various flowers to enrich the beehive. The students have designed labels for bottles of honey, which are used to sell harvested honey as a souvenir to visitors to the rooftop garden. In areas of the garden where the honey bees are more active, the pollination of flowers noticeably increases (Figs. 11.5, 11.6, 11.7, and 11.8).

## 11.4 Discussion on Green Culture Experienced Through Green Roofs

Hanul Madang is managed by the students themselves, and the landscape of the garden continues to develop through the participation of the users, not as a fixed form. It is not only a beautiful garden to view but also a space for living things to coexist and operate within this ecosystem. As such, Hanul Madang is a place where a variety of activities take place and where direct experience in garden design and management is achieved.

In a city lacking green spaces, green roofs are very valuable ecologically functioning areas. They can be impressive garden spaces with natural landscapes and also serve as places to provide users with a richer experience of daily life.

**Fig. 11.6** Beekeeping activity and Graduate School of Environmental Studies honey (Source: Department of Landscape Architecture, Graduate School of Environmental Studies, Seoul National University)

**Fig. 11.7** Participation in garden management activities (Source: Department of Landscape Architecture, Graduate School of Environmental Studies, Seoul National University)

**Fig. 11.8** Participating in garden management activities (Source: Department of Landscape Architecture, Graduate School of Environmental Studies, Seoul National University)

Rooftop gardens have high cultural values among all four types of ecosystem services, which increases their potential value through various uses. The rooftop garden is a good place to connect with nature and develop relationships with other people in situations where inadequate green space exists. It is a superior educational area for students studying garden design. The rooftop is an attractive place to meet and interact with nature. It also provides various benefits in promoting green culture to create opportunities for experiences with nature in everyday life. These activities could be transformed into values, which enable them to have diverse effects that would possibly increase the qualitative value of green roofs.

Finally, it should be noted that rooftop gardens are installed on artificial surfaces, whose management requires a lot of effort. In addition, the lifetime of an ordinary rooftop garden is about 10 years. Therefore, it is more expensive than an open green space or a park. However, rooftop gardens create additional opportunities to experience various activities. Thus, our societies need to learn how to manage and enjoy them to get their maximum value and to increase their number (Fig. 11.9).

**Fig. 11.9** Vegetable gardens that student cultivate at Hanul Madang (Photo taken by Son Yong-hoon, May 31, 2012)

## References


Berndtsson JC (2010) Green roof performance towards management of runoff water quantity and quality: a review. Ecol Eng 36(4):351–360

Choi GB (2011) Performance report of rooftop greening in Seoul. In: Proceeding of re-discovering Roof Gardens symposium 2011. Environmental Planning Institute in SNU, Seoul, pp 37–44

Jafari N, Yunos MYM et al (2015) Residents' preference towards rooftop garden: a case of the heritage condominium, Selangor. Adv Environ Biol 9(4):79–81

Lee SJ, Park GS et al (2013a) Change of vegetation and soil characteristics of green roofs in Dongguk University. J Korea Soc Environ Restor Technol 16(1):193–206

Lee SJ, Park GS et al (2013b) A study on the changes of plant species and soil environmental characteristics on green roofs at Seoul Women's University. J Korea Soc Environ Restor Technol 16(6):109–117

Lee B, Ahn GY et al (2014) A study on green roof planting of Seoul. Proc Korean Soc Environ Ecol Conf 24(2):117–118

Mansour O (2017) Water, energy, and rooftops: integrating green roof systems into building policies in the Arab region. Environ Nat Resour Res 7(2):11–36

Na HY, Byun BS (2006) Research trend of rooftop afforestation. Geogr J Korea 40(1):95–106

Oh CH (2011) Meaning and effect of rooftop garden in the era of climate change. In: Proceeding of re-discovering Roof Gardens symposium. Environmental Planning Institute in SNU, Seoul, pp 28–33

Park EJ, Nam MA et al (2012) Estimation of the potential area for roof greening in Gyeonggi-do and its implications. J Korea Soc Environ Restor Technol 15(1):107–117

Peck SW, Kuhn M (2003) Design guidelines for green roofs. Ontario Association of Architects, Toronto

Seoul City (2015) 2030 Seoul City park and green space master plan. Seoul City
Tolderlund L (2010) Design guidelines and maintenance manual for green roofs in the semi-arid and arid west. Green Roofs for Healthy Cities, Denver
Wescott S, Sundell V (2016) Rooftop garden initiative. Macalester College
Yang BI (2011) Green city. Seoul National University Press, Seoul
Zhao HX, Kang TH (2011) A comparative study on the legal system of building a rooftop gardening between Korea and China. J Korean Inst Landsc Archit 39(4):11–17

# Part III
# Green Building Practice and Innovation

# Chapter 12
# Winter Wind Environmental Comfort in Courtyards in Vernacular Architecture: A Case Study of the Fuyu Building in Yongding, Fujian

**Ying-Ming Su and Hui-Ting Chang**

**Abstract** Global warming is worsening. Sustainability is the main goal for reducing the consumption of energy and carbon dioxide emissions worldwide, and ecological architecture is the optimal means of reaching this goal. Vernacular architecture highlights local characteristics, emphasizing harmony with the natural environment and local conditions. Tulou, which have been listed as World Heritage Sites, maintain a livable internal environment that is warm in winter and cool in summer through the appropriate placement of openings and patios. People feel comfortable even without the use of appliances. This design adheres to the concept of sustainable development through ecological and energy conservation, providing an excellent example for modern architectural planning and design. Through field measurements of wind environment comfort, this study adopted a square tulou—the Fuyu Building in Yongding, Fujian, China—as an example, because its climatic conditions and architectural forms are in line with Taiwan's current environmental status. Through comparisons of temperature, humidity, and wind field, the differences between the internal and external environments of the traditional communal residence were explored. The results showed that ventilation produced positive effects on indoor environmental comfort in the tulou through the placement of windows and patios. This design approach can serve as a reference for future ecological building design.

**Keywords** Wind field · Tulou · Vernacular architecture · Ecological architecture · Sustainable development

Y.-M. Su (✉)
Department of Architecture, National Taipei University of Technology, Taipei, Taiwan
e-mail: ymsu@ntut.edu.tw

H.-T. Chang
Graduate Institute of Architecture and Urban Design, National Taipei University of Technology, Taipei, Taiwan



Z. Shen et al. (eds.), *Green City Planning and Practices in Asian Cities*,
Strategies for Sustainability, https://doi.org/10.1007/978-3-319-70025-0_12

## 12.1 Introduction

Since the industrial revolution of the nineteenth century, International Style architecture has tried to satisfy human needs and functional requirements through design-oriented building models, ignoring the importance of local conditions (Fu et al. 2010). The rapid development of cities for economic reasons has exacerbated climate change, with energy consumption and emissions contributing heavily to the environmental impact of urban growth. To mitigate the environmental damage caused by external effects, the United Nations Conference on Environment and Development, held in Rio de Janeiro in 1992, proposed the concepts of sustainable development and sustainability from an ecological viewpoint (Sustainable Development Knowledge Platform 2014), which have drawn increasing attention to the concept of ecological architecture.

Over the centuries, people have designed and built buildings with engineering technologies adapted to the local climate, materials, topography, and cultural practices. An ecological building is built with local natural materials and is suitable for human habitation without the use of modern energy and electrochemical equipment. Vernacular architecture is local architecture in its original form, such as traditional houses that have not been influenced by modern society. Tulou, which have been designated as World Heritage Sites, use immature soil as the main building material. In addition to enabling residents to defend against outside aggression, this form of construction is able to self-regulate and adapt to modern climate change through capillary action in the soil, creating a pleasant living environment and representing a successful interpretation of the eco-building spirit.

The study subject was the Fuyu Building, built in the wufenglou style, located in Yongding, Fujian, China. From the perspective of improving human comfort through the wind field, the influence of tulou on the environment was surveyed. Through analysis and discussion, we outline the evolution of this architectural form and how it accommodates environmental change. These findings are expected to be useful in future ecological building design.

## 12.2 Background to the Case Study

### *12.2.1 Tulou*

Fujian tulou were first built in the Tang dynasty, and their construction and development processes are closely connected with Hakka history and culture in western Fujian (Xie and Von 2011). Several large-scale migrations after a period of chaos in Yongjia resulted in the formation of clans, the settlement of a new area, and a largely closed-off way of life (Chen 2005). Tulou are mainly distributed in hilly areas, valleys, and basins in subtropical climate zones at lower latitudes (Huang 1994) (Fig. 12.1).

Over the long history of tulou, diverse architectural forms and spatial characteristics have been developed, including strong defensive capabilities and a communal

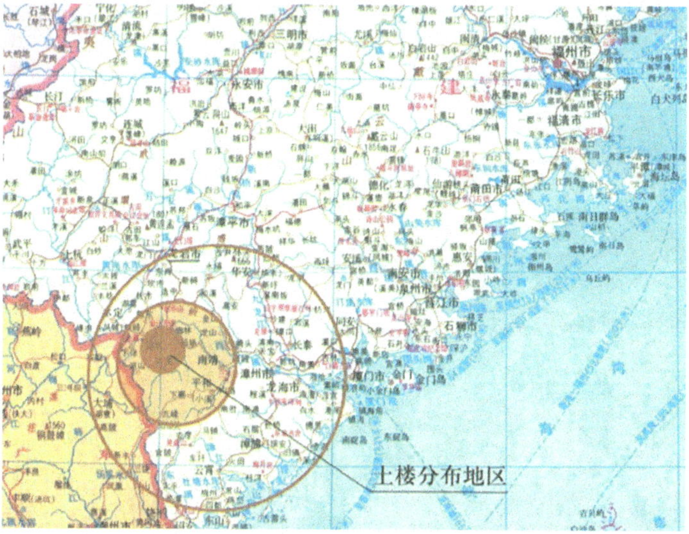


**Fig. 12.1** Location of tulou (From: China Architecture & Building Press 2010)

ethic emphasizing equality. In addition, the use of sedimentary soil as a construction material keeps the interior environment warm in winter and cool in summer (Su and Yang 2012).

This vernacular eco-building form has its origins in the courtyard houses of the Central Plain. These gradually transformed from wufenglou into round forms, affected by geology, climate, and the prevailing social environment in Fujian. In round buildings, square buildings, and wufenglou, the interiority, symmetry, and axis-symmetrical layout of the interior space are associated with specific customs and a particular philosophy of living (Huang 1994) (Fig. 12.2).

Round and square tulou are divided into two styles: the corridor style, mainly found in western Fujian (Minxi), and the unit style, mainly found in southern Fujian (Minnan). These styles have similar external appearances but different internal layouts (Liu and Tang 2009). As tulou are adapted to the local climate and geographical environment, they are considered regional eco-buildings. The advantages of this architectural style, such as rammed earth technology, special structural design, and load-bearing walls up to 1 m thick, are listed in Table 12.1 (Yang 2013).

### *12.2.2 Wind Environment Comfort*

Appropriate ventilation design is a prerequisite for reducing building energy consumption by air conditioners, the most natural means of energy-efficient construction, and the most important green climate control strategy (Lin 2006). Changes in air flow around a building are highly complex and influenced by many factors,

**Fig. 12.2** Tulou (From: China Architecture & Building Press 2010)

**Table 12.1** Design concept of tulou

| | |
|---|---|
| Structure | The earthen walls can effectively disperse the building load, with the combination of masonry foundations and the wooden structure, not only prevent damage of fire and earthquakes but also resist invasion of strong winter the northeast monsoon |
| Comfort | Due to high specific heat, the material of earth has superior heat and insulation to facilitate building warmth in winter and cooling in the summer and the living environment. Immature soil with high porosity can absorb moisture in the air and the walls can adjust heat dissipation, adapting to the humid mountain climate |
| Recycling | Building waste can return to nature and achieve the purpose of recycling and utilization of natural resources and ecological balance |
| Suitability | With resistance to pest erosion, the main structure is robust and can regulate the physical environment, further preserving the natural environment and harmonious coexistence |

From: Tz-Yue Yang (2013)

including architectural geometry, the surface projection of the buildings, wind direction, and nearby buildings (Zhu 2003). Natural ventilation affected by temperature and humidity can adjust air flow distribution, eliminating indoor air pollution. The inflow of air currents from windows and doors in a manner that has a

**Table 12.2** Effects of wind velocity on the human body's operation

| Air velocity (m/s) | Effects |
|---|---|
| 0–0.25 | Imperceptible |
| 0.25–0.5 | Pleasant; does not affect work |
| 0.5–1.0 | Pleasant; need to avoid tissue fly away |
| 1.0–1.5 | Slightly annoying wind force; grass surfaces and paper are blown |
| 1.5–7.0 | Obvious wind force; thin paper drifts and thick paper stock is blown; appropriate action is required to correct air volume and wind control to maintain good and healthy working conditions |

From: Rong-ping Lai et al. (1991)

subtle effect on comfort in areas used for everyday activities can moderate the body's heat and prevent excess heat and moisture from entering the building. The effects of wind velocity on the human body are detailed in Table 12.2.

The skin temperature causes heat transfer and changes according to the external environment. Air velocity can make the skin temperature drop more rapidly and approach the air temperature. For humans, comfort in low-temperature environments is influenced by the temperature difference, which is closely related to air velocity.

To advance the goals of increasing indoor comfort and saving energy, this research promotes natural ventilation effects through wind guidance in courtyards, with the air velocity as the most important factor in environmental comfort.

## 12.3 Methodology

This study measured physical environmental factors, including air velocity, average air temperature, and relative humidity in indoor and outdoor spaces. To evaluate environmental comfort in winter wind, the experiments also investigated factors such as form, structures, and construction materials.

### *12.3.1 Fuyu Building*

The Fuyu Building, an outstanding representative of the mansion style of Yongding tulou, is located in Hong Keng Village, Yongding County, Fujian.

Yongding is in the vicinity of the Tropic of Cancer (24°40′N, 116°58′E), with an average annual temperature of 20–21 °C. There is a small temperature difference between winter and summer, but a large gap between the monthly high and low temperatures, as well as abundant rainfall (Fujian Province Local History Compilation Committee 1994). The annual average wind speed is 13.89 km/h, the prevailing wind is northeasterly (and southwesterly in the summer only), and the relative humidity is 72.52% (Weather Underground 2014) (Figs. 12.3 and 12.4).

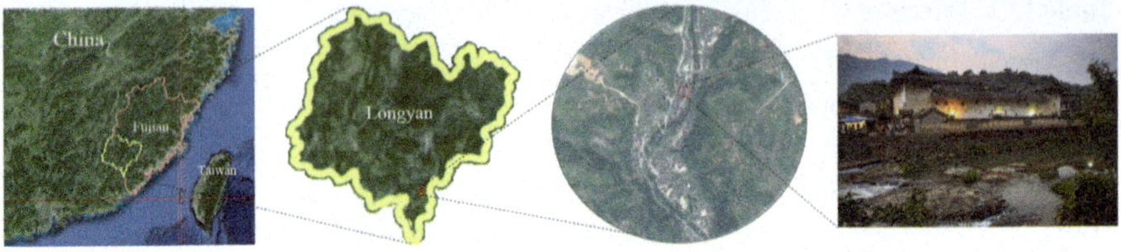


**Fig. 12.3** Location of the Fuyu Building (From: Thematic Database for Human–Earth System & Weather Underground)

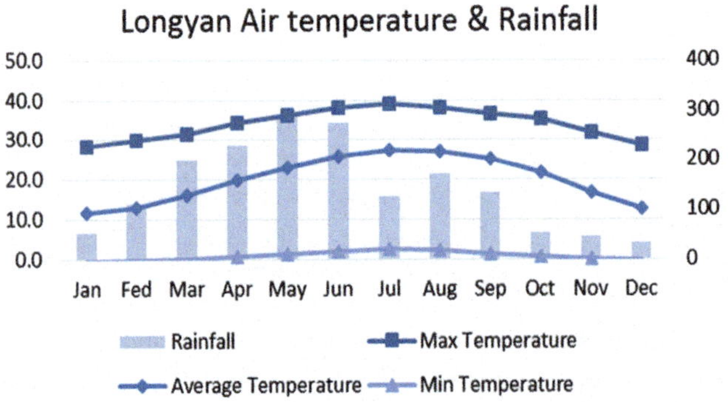


(a)

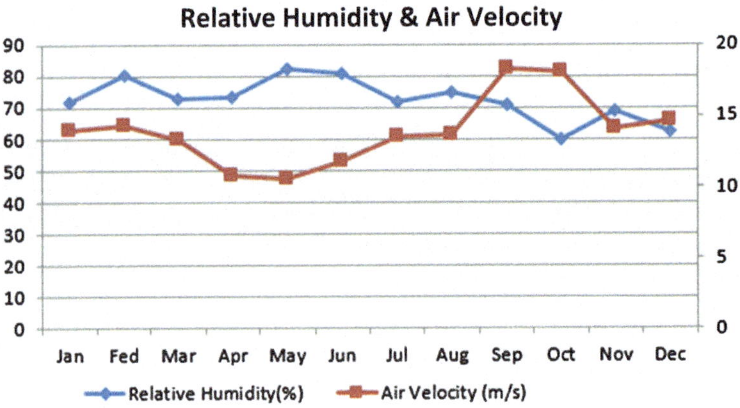


(b)

**Fig. 12.4** Weather information in Yongding (From: Thematic Database for Human–Earth System & Weather Underground)

**Fig. 12.5** Pictures of the Fuyu Building

The Fuyu Building covers more than 7000 m$^2$. It includes an antechamber on the lower two stories that is used as the ancestral hall, with a total of five stories for the main hall. The horizontal houses on both sides of the cross-connected, tall buildings surrounding the enclosure constitute a highly defensible layout. The building's form is characterized by elaborate variation. The roofs of its front, center, and back halls have progressively higher layers, with a steeper roof slope than other types of tulou. The passage and rooms constitute six courtyards, framed by two stories on two sides. There are three wings, and the back hall contains nine units, further subdivided into three groups of three each. All staircases are located in the front of the halls. There are two rooms in the back part of each hall. Each building taller than two stories has the same structure. For example, each has an attic, and there are wells on both the left and right sides of the main hall. The bilateral interior corridor that connects the front and main halls is 6 m tall and is made of brick and wood. Brick and wood bungalows, likewise bilaterally arranged around the halls, are used as restrooms, sties, warehouses, and mills (Huang 1994) (Fig. 12.5).

## *12.3.2 Field Measurement*

Measurements were continuously recorded from 0900 h on November 19th, 2013, to 0900 h on November 20th, 2013. Rapid temperature change and low temperature in key positions at the measured points were recorded over these 24 h. The experimental data, which recorded spatial differences, were measured at a height of 1.5 m, to use the range of pedestrian comfort in the wind field as a benchmark. Spatial differences and the regulation of the physical environment were observed in the tulou in different periods.

Figure 12.6 shows images of the status of the environment in the passage, front yard, and courtyard in the Fuyu Building. Coordinates at 20 monitoring points were recorded, considering that a high number of measuring points would be beneficial to experimental accuracy.

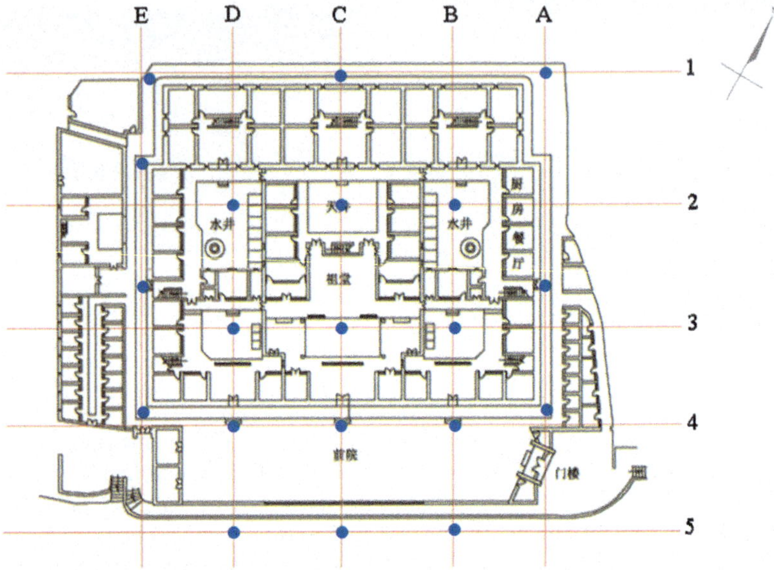


**Fig. 12.6** Measuring points in the Fuyu Building

The air velocity magnitude and air temperature were measured by a thermal flow velocity sensor—irrespective of flow direction—and a relative humidity sensor, respectively. The related accuracy values were within 0.03 m/s and 0.5 °C, respectively, in the ranges of 4–10 m/s and 20–70 °C. To avoid the disturbance of airflow caused by device movement, the measuring period was at least 10 min, with data recorded for 1 s at each point. The mean radiation temperature was recorded by a globe sensor with a diameter of approximately 150 mm and accuracy of 0.5 °C. All instruments were installed on tripods, and the aforementioned probes were connected to a multiparameter indoor climate meter (Testo 480) to ensure accurate measurement of the flow velocity, temperature, and relative humidity (Fig. 12.7). These data were used to evaluate upper-story building arcades in summer and autumn wind environments.

## 12.4 Results and Analysis

### *12.4.1 Air Temperature*

As shown in Fig. 12.8a–c, the air temperature range was 10.6–26.5 °C. The maximum temperature at monitoring point B-05, 24.8 °C, was recorded at 1200 h. Monitoring points B-05, C-05, and D-05 had their minimum temperature, only 10.6

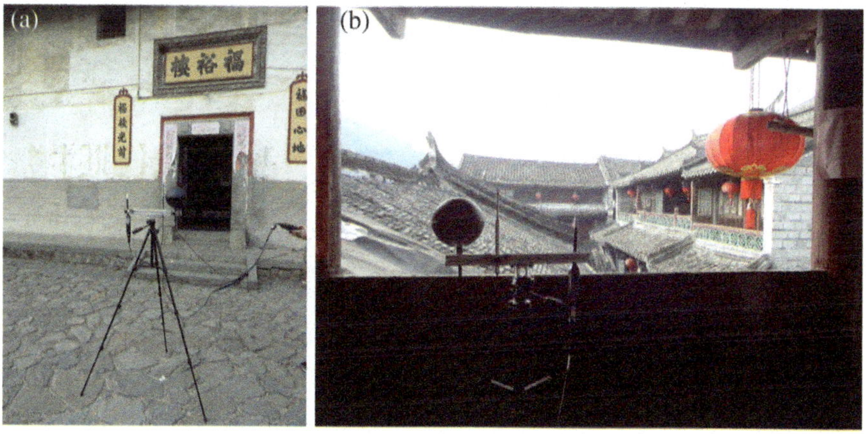


**Fig. 12.7** Measuring photo of the Fuyu Building

°C, at 0300 h. Thus, there was a difference of 14.2 °C between day and night. Due to adjacent waters and the impact of geography, the temperature during the day was moderate and therefore evapotranspiration was also moderate. After nightfall, the dew point was lower, and condensation caused the temperature to plummet. As shown in Fig. 12.8d, the highest temperature, 26.5 °C, was recorded at monitoring point D-04 in the front yard in the afternoon. The lowest temperature, 10.6 °C, was recorded at monitoring point D-04 before dawn. This approximately 15.9 °C temperature difference was the greatest difference between day and night.

According to Fig. 12.8e, the temperature around the building part of the corridor and front yard underwent marked changes. In comparison, the ambient temperature was more stable in the courtyard inside the building. Monitoring point C-02 near the windward side of the courtyard had the highest temperature, 22.1 °C. The lowest temperature, 11.0 °C, was recorded in the early morning. Because the windward side was blocked by the architectural volume, monitoring points B-03, C-03, and D-03 had smaller temperature changes. In contrast, monitoring point C-03 had the most moderate temperature changes in 6 months, due to its location in the central axis of the building preventing influence from the building's height on both sides.

The overall highest temperature was measured in the afternoon. There was a rapid temperature drop later, after the sunlight was blocked by the mountain and prevented from providing heat. The dissipation of thermal radiation in the afternoon caused the outdoor temperature to gradually drop until it was lower than the indoor temperature, with the rate decreasing in the evening and gradually leveling off until sunrise, when the temperature began to rise again.

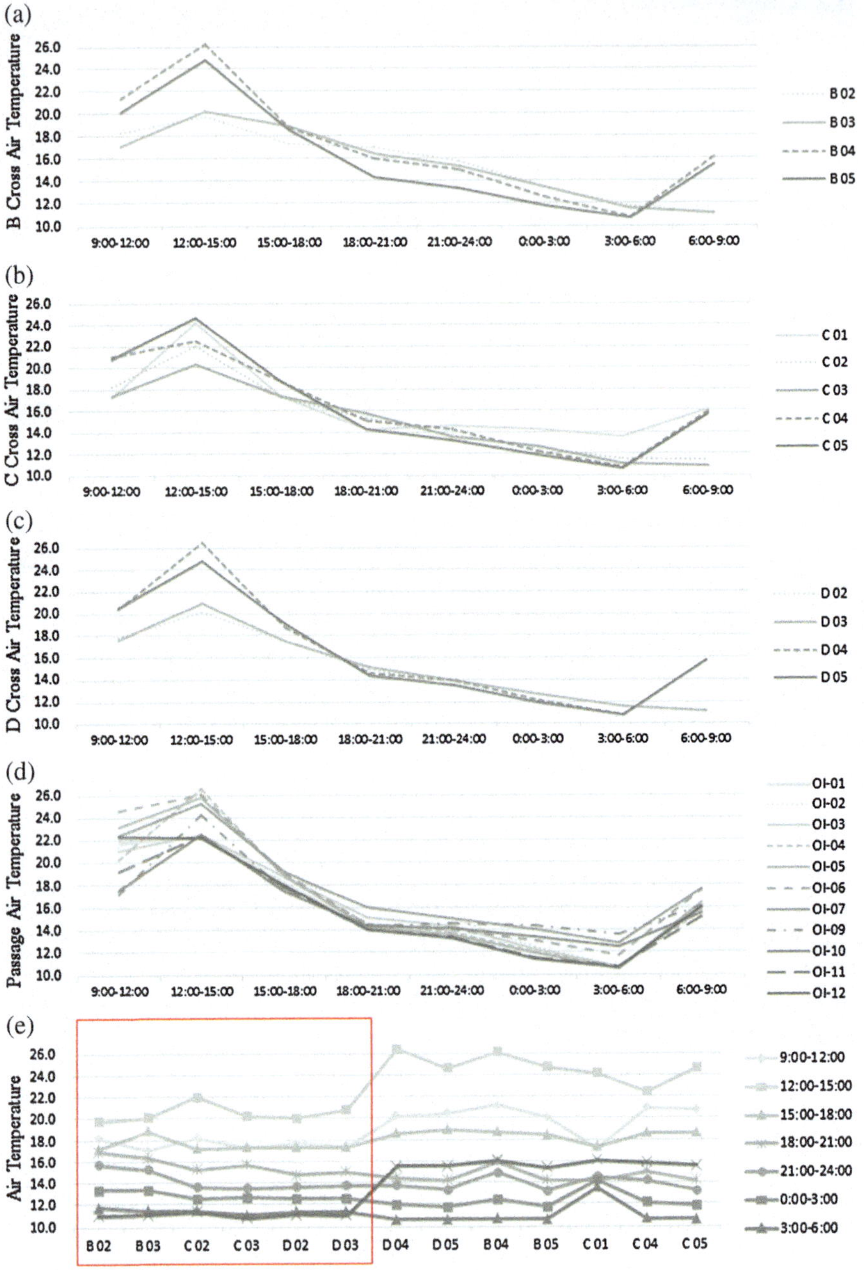


**Fig. 12.8** Air temperature measurement results for the Fuyu Building. (**a**) Air temperature of B cross section measuring points; (**b**) Air temperature of C cross section measuring points; (**c**) Air temperature of D cross section measuring points; (**d**) Air temperature of passage measuring points; (**e**) Air temperature of measuring points

### 12.4.2 Air Velocity

The air velocity measurements are shown in Fig. 12.9. The average air velocity was between 0.11 and 2.09 m/s. Figure 12.9a–c demonstrates that the air velocity was higher outdoors, and that day and night had markedly different wind speeds, with a range of 0.20–2.09 m/s. According to Fig. 12.9d, air velocity increased because the opening at monitoring point E-04 was unblocked, and the temperature difference between land and water, as well as the mountain breeze, caused the maximum air velocity of 2.09 m/s to occur at 1200 h.

As Fig. 12.9e shows, the front yard monitoring point D-04 adjacent to the water was affected by the temperature difference between the land and water that formation windward side vortex make the wind speed increase. The air velocity at monitoring point D-04 was 1.18 m/s in the morning, higher than the velocity at the B-04 and C-04 points. Monitoring point B-02 had a maximum air velocity of 0.49 m/s, which appeared at 0300 h, and a minimum of 0.11 m/s, which appeared at 1500 h. The drop in temperature, the building size, and the courtyard effect meant that the courtyard air velocity exceeded 0.2 m/s, slightly higher than the velocity at the other measurement locations. Because the impact factor was lower at monitoring point C-02, the wind field was more stable due to obstruction by the front living space and the ancestral hall.

In winter, the prevailing wind was northeasterly, but the mountains and buildings at the northeast created a barrier that lowered the air velocity on the windward side during the occasional monsoons. The main reason for the change in air velocity was the temperature differences between the valley, land, and water.

### 12.4.3 Relative Humidity

As shown in Fig. 12.10, the average relative humidity was between 20% and 86%. The measured results in Fig. 12.10a–c show that the lowest relative humidity was in the afternoon in the outdoor space.

At monitoring point D-05, under the influence of E-04, the windward side moisture in the air was blown by the faster wind. The evapotranspiration was intensified by sunshine; consequently, the minimum humidity of 20.4% occurred at 1200 h. As shown in Fig. 12.10d, there was a considerable difference (12.7%) between the windward and leeward corridor humidity at the same times. This was especially true at 0300 h, with maximum humidity of 84.9% at monitoring point A-04 and minimum humidity of 72.2%. The high moisture in the wind caused the humidity to rise. In addition to how measuring point A-04 was located outside the windward side, its position in a corner of the building interior resulted in relatively high humidity. As shown in Fig. 12.10e, point B-04 had only 23.6% humidity at 1200 h. The adjacent opening resulted in the front yard having lower humidity than the other outdoor areas. The measurement began to rise when the afternoon sunshine was blocked by the mountains and declined after sunrise the next morning. In the courtyard, humid-

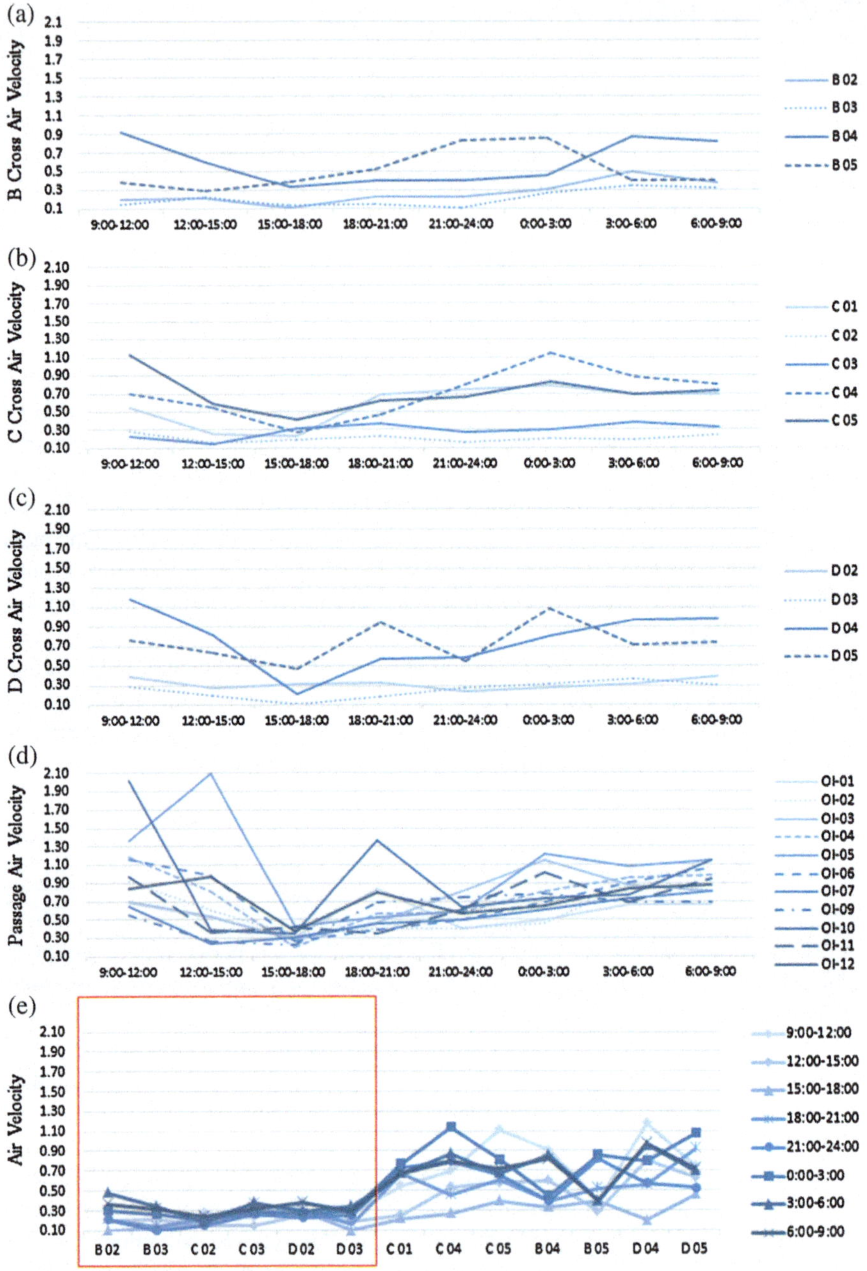


**Fig. 12.9** Air velocity measurement results for the Fuyu Building. (**a**) Air velocity of B cross section measuring points; (**b**) Air velocity of C cross section measuring points; (**c**) Air velocity of D cross section measuring points; (**d**) Air velocity of Passage measuring points; (**e**) Air velocity of measuring points

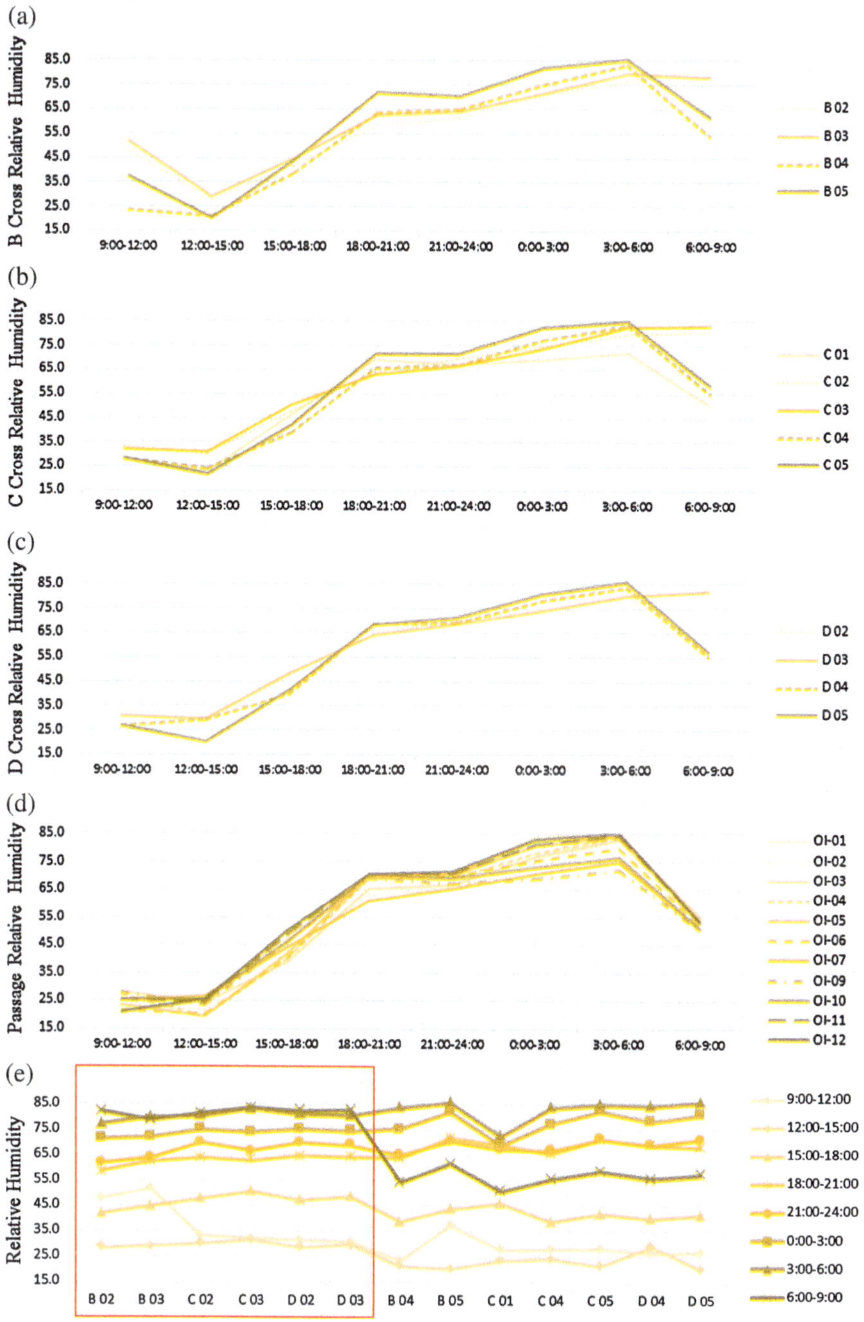


**Fig. 12.10** Relative humidity measurements results for the Fuyu Building. (**a**) Relative humidity of B cross section measuring points; (**b**) Relative humidity of C cross section measuring points; (**c**) Relative humidity of D cross section measuring points; (**d**) Relative humidity of passage measuring points; (**e**) Relative humidity of measuring points

ity fluctuated frequently. In the afternoon, courtyard monitoring points B-02 and B-03 had the highest humidity due to the steep terrain, which impeded the mountain breeze from entering.

As demonstrated in Fig. 12.10, at 1200 h the temperature began to decline and the humidity began to increase. At this time, the outdoor humidity exceeded the ambient humidity inside the building. At monitoring point C-02, the humidity was relatively stable; the environment was more comfortable in the courtyard because it did not face the wind.

## 12.5 Conclusion

Natural ventilation is closely related to building type. This study examined the wind environment of a form of vernacular architecture. The Fuyu Building was observed to be located in a valley, restricted on the east and west sides by mountains and a block of buildings. It has a front yard close to the water, lessening the effect of the winter northeast monsoon on the architectural volume. The main factors influencing the wind field were the wind from the valley and the convection due to the difference in the surface temperature between land and water. Because the antechamber, nave, and purlin height (in increasing stories) are parallel with the water, the wind can be effectively introduced into the atrium space, although it has a closed architectural form. The self-regulating microclimate inside the building can improve the internal cycling of the natural ventilation provided by the wind field, avoiding air stagnation in the interior environment and thereby mitigating discomfort. The wall surrounding the atrium can provide a sufficient barrier to prevent external wind from entering, reducing the temperature difference between day and night and trapping warmth in the colder winter months, potentially reducing the incidence of cardiovascular disease among residents. As demonstrated by the adjustable microclimate of the indoor environment, the atrium is not vulnerable to the effects of changes in the external environment. The natural ventilation of the tulou can provide a healthy and comfortable living environment. The results indicate that the tulou has superior natural ventilation performance. Its form is suited to the topography and landscape and reflects knowledge of how to live in accordance with the environment. The design of the main entrance relative to the river direction, as well as the design of the courtyards along the building axis and the continuous corridors, is intended to ensure satisfactory cross-ventilation and heat dissipation.

Tulou, as World Heritage Sites for vernacular architecture, make use of geography to ensure a level of interior environmental quality that remains stable and appropriate to local conditions and has a minimal impact on the environment through its architectural forms and materials. The use of local materials reduces the amount of energy consumed by transportation and construction. This is especially true considering the large amount of materials used in the construction of tulou. Tulou is thus a prominent example of ecological building that can be used as a reference in the design of future ecological architecture.

**Acknowledgments** This chapter represents part of the results obtained under the support of the National Science Council, Taiwan (Contract No. NSC101-2627-E-027-001-MY3).

## References

Architecture and Building Research Institute, Ministry of The Interior. http://formosa.heart.net.tw/green/eco.htm

Chen L-D (2005) The culture and Hakkas' buildings in West Fujian. J Wenzhou University 18(6):48–52 (in Chinese)

Fu Y-S, Ni J-W, Jiang Z-Y (2010) From modern architecture to ecological architecture. Comments. Forum:21–22

Fujian Province Local History Compilation Committee (1994) http://www.fjsq.gov.cn/ShowText.asp?ToBook=3158&index=30&

Huang H-M (1994) Fujian Tulou. Han Sheng Culture Ltd, Taipei (in Chinese)

Lin H-T (2006) Green architecture. Chan's Arch-Publishing Co., Ltd, Taipei (in Chinese)

Liu Dong, Tang Ji (2009), Interpreting the history and architecture of the earthen building in Fujian. Contemporary China Press (In Chinese)

Rong-ping Lai, Xian-de Lin, Zhou Gu-peng (1991) Physical Environment of Buildings. Liu Huo Press (in Chinese)

Singh MK, Sadhan M, Atreya SK (2009) Building and environment, pp 878–888

Su Y-M, Yang T-Y (2012) A study of Hakka Earthen in the viewpoint of ecological planning. In: Conference on science, technology, and society, Hsinchu, Taiwan (in Chinese)

Sustainable Development Knowledge Platform (2014) https://sustainabledevelopment.un.org/sdgsproposal

Weather Underground (2014) http://www.wunderground.com/

Xie L, Von Y-F (2011) Historical and geographical environment of the formation and development of Fujian Tulou. Hunan Soc Sci 1:212–215 (in Chinese)

Yang T-Y (2013), The research of wind environmental impact in Hakka Tulou as a view of ecological architecture—a case study of Eryi Building in Fujian Hua An (in Chinese)

Zhu J-R (2003) An introduction of wind engineering. Sci-Tech Books Co., Ltd (in Chinese)

# Chapter 13
# Energy-Saving Design, Based on a Climate Adaptation Strategy, of the Dinosaur Egg Ruins Protecting Museum in Hubei

**Baofeng Li and Jia Zhu**

**Abstract** With rapid urbanization, there are more and more construction projects of public facilities in China. The concept of green building has been introduced into the design process for museum buildings, which is one type of public facility pursuing energy savings. At present, artificial ventilation and heat preservation devices are utilized in museum buildings to maintain a stable temperature and humidity range for indoor exhibits. After equipment installation, energy consumption becomes a big cost of operating museum buildings. With the concept of green building, this work explores passive energy-saving strategies from a practical view of museum design. Lighting and heat preservation are pertinently considered under some special condition requirements, including processing of sloping field topography, adjustment of lighting and room temperature, combination with local materials and local culture, etc. As an example, the authors present a case study of implementation of the green building concept for energy saving in China.

**Keywords** Energy-saving · Passive strategy · Site protection · Climate adaptation · Ventilation-friendly and light-resistant · Double-layered roof

## 13.1 Introduction and Background

Energy saving is the most important characteristic of the green building concept. As the construction industry is one of the largest energy-consuming industries, China's construction industry still has a great need for energy-saving solutions (Qian and

B. Li · J. Zhu (✉)
School of Architecture and Urban Planning, Huazhong University of Science and Technology, Wuhan, China
e-mail: zj_alice@163.com



Z. Shen et al. (eds.), *Green City Planning and Practices in Asian Cities*, Strategies for Sustainability, https://doi.org/10.1007/978-3-319-70025-0_13

Yin 2009). For energy-saving building design, there are a lot of methodologies and solutions discussed in the world literature. For example, insulation system structures in external walls and inorganic thermal insulation materials for walls, doors and windows, roofs, and other building parts (Zhong 2015) can be used to improve the energy saving of the envelope and building structure (Bao 2013), etc. It is worth noting that combined lighting and ventilation devices (Deszberg 1983) are becoming one of the most effective methodologies for improving energy saving in building structures.

Generally, sites with conditions of a dry field, smooth drainage, and good ventilation will be chosen for locations of public buildings, which also require adjustment measures corresponding to local conditions in comprehensive planning. There is an increasing interest in ecological design of buildings in warm regions, where natural ventilation has been proved to be the most efficient low-energy cooling technique (Ghiaus et al. 2005).

As the background to this project, geological experts undertook excavations and found cretaceous dinosaur egg fossils at Qinglong Mountain in the Yun County of Hubei Province. Therefore, a museum building was planned as a science education base to explore dinosaur reproduction and extinction, and the local government built a greenhouse with sunlight glass to protect the dinosaur egg fossils. However, greenhouse effects caused severe damage to the eggs, and the colored light of the sunlight glass was not helpful to display real dinosaur egg fossils.

Regarding architectural models of theme museums, firstly it is possible to use pictographic language, such as images of dinosaurs or eggs. This form of imitation is too figurative, lacking imagination and creativity. Secondly, they can follow the archaeological protection and excavation process—for example, starting with a temporary protection canopy in the excavation area when digging starts. Lighting can then be set up after completing the dig. Because this meets both requirements of showing the excavation and protection, we choose the second model.

The discovered dinosaur egg fossil field (Fig. 13.1) is situated on a mountain slope, where site protection is necessary for museum building construction. Usually, there are enclosed spaces in a museum building, therefore it is a challenge to deal with heat prevention during summer. The climate of Hubei Province is very hot in summer and cold in winter. In addition, the temperature changes over the course of a day are very large because the site is located on a mountain slope. The specified range of the museum construction site on Qinglong Mountain is more than 1000 $m^2$, and it would be improper to move the dinosaur egg fossils, according to the geologists' consideration, from the viewpoint of heritage protection. Therefore, there is a conflict between the extreme climate at the local site and the need to maintain a constant temperature. We attempted to find solutions to this issue using green concepts in order to solve the problem of indoor overheating in summer and the need to protect the egg fossils from direct light.

The authors expand on the architectural model of this geological museum, combined with meteorological data, from the whole building form to its detailed structure, as a case study of application of green concepts to the Qinglong Mountain

**Fig. 13.1** Dinosaur egg fossil excavation site

Dinosaur Egg Fossil Ruins Protecting Museum. This is followed by an empirical discussion using actual measurement data, with conclusions and suggestions for further work.

## 13.2 Literature Review

In this design work, the authors recreated the original distribution of the dinosaur egg fossils, respecting historical and cultural relics in the museum building, in which the exhibition floor plan was designed for a one-way tour to present views of the dinosaur egg fossils and provide a spatial–temporal correspondence with the excavation area. In the design process for this museum, light, ventilation, and heat insulation were considered as important factors in low energy consumption for energy saving.

### *13.2.1 Blocking Light*

Natural light is one of the important factors in constructing space (Huang 2007), which can provide a variety of spatial experiences for museum visitors (Jang and Park 2016). How to meet the needs of basic lighting in a room, while avoiding heat radiation caused by a large amount of light entering the room, is a challenge for us. Light has been considered as one of the most important elements in architectural design. Light provides a lot of architectural experiences for visitors to buildings by interrelating the space, shape, and other design elements. In particular, natural light is a valuable source to create better indoor space, in comparison with artificial light. It is a sustainable energy source and offers a more natural environment. It also enables visitors to perceive the form and depth of the space (Kim 2014).

The lighting elements of skylights and high side windows are given priority in this design as one element of the architectural form, and we use light-taking windows in parts of the exhibition space in this museum. Bright light is used to highlight exhibits in key positions in an exhibition area. Other spaces are dark, and the use of brightness and darkness clearly defines the boundary between the exhibition area and other areas in the museum. The visitors' impressions of the details depend on the setting of the light source and the walking path (Lan and Fei Hu 2009).

### *13.2.2 Ventilation*

Green buildings should be designed with controllable natural ventilation. Effective natural ventilation is necessary. It has positive significance for temperature control, discharge of air pollution, and air flow (Fordham 2000). Reasonable utilization of natural ventilation has important significance in improving indoor air quality (IAQ) and saving building energy (Guangcai 2003). Natural ventilation efficiency can be affected by facades, the roof, the outline of the building, and the layout and organization of the interior spaces (Kleiven 2003).

### *13.2.3 Heat Insulation*

There are many ways of incorporating heat insulation into a building. Heat insulation materials for roofs, walls, doors, and windows are a popular research field. Especially appropriate skin materials should be selected for compatibility with the climatic environment where a building is located (Hou et al. 2012).

With regard to the main envelope of the building, the roof is one of the most important parts to be considered in an energy efficiency project. Roof insulation panels with composite materials can be included in the roof structure, whose insulation properties, mechanical properties, and durability can thereby be raised (Ming-Hai et al. 2008). There is a contradiction between hot sunshine and energy conservation in roof design, which can be solved by a double-skin system. That has great energy-saving potential. At the same time, it is possible for designers to create alternatives of interesting facade designs, which are recognized as one of the methods for ecological construction (Baofeng Li 2001). Furthermore, heat-insulating solar glass (HISG) is a recently developed smart building material used to minimize energy consumption in the building sector (Cuce and Riffat 2016). Use of energy-saving doors and windows is important for energy saving in buildings. There are many types of energy-saving door and window products available to choose from (Chai and Zhang 2009).

Low-energy-consumption building technology has attracted more and more attention from government and society, and its influence on the property market in China has been discussed previously (Cheng 2006). Passive building design technology is discussed in this chapter for considering suitability for regional climate, with ultra-low power consumption for energy saving, which is recognized as a fundamental way to realize sustainable development of human settlements (Liu et al. 2015).

## 13.3 Methods

### *13.3.1 Study Area*

Energy-saving buildings are usually dependent on temperature control and lighting control. In order to keep indoor temperatures and lighting control within a reasonable range, we can also use the method of changing some of the building layout or structure to achieve the same effect. The architectural model, building materials, natural lighting, ventilation, heat resistance, and so on, can influence the whole building's energy consumption. These were all factors needed for the study of this museum in the design process. What we suggested for this museum design were two passive strategies for achieving this purpose: (1) we could incorporate energy-saving ideas into the structural design itself instead of relying on air conditioning equipment; and (2) we could choose to use some energy-saving materials to adjust the indoor thermal environment and lighting environment.

### *13.3.2 Design Targets*

1. To integrate buildings into the natural environment without damaging the surrounding natural environment, by reducing the quantity of earthworks
2. To create a dark display environment with moderate skylighting to highlight exhibits
3. To recycle old materials at low cost and with low technology for adaptation to regional climate and energy saving

### 13.3.3 Design Studies Approach

The design studies approaches in this work were divided into two parts: one was how to complete the design based on thermal insulation roofing and ventilation windows for energy saving, and the other was experimenting via field measurements.

Based on design theory such as phenomenology (Moran 2000), the authors studied the surviving arrangement of the dinosaur egg fossils to determine the exhibition path. Thereby the layout of lighting shafts and devices that were light resistant but allowed natural ventilation were determined for optimizing the exhibition and energy-saving efficiency.

Based on the principle of diffuse light propagation and airflow, relative analysis of the museum building system was conducted in order to choose the most suitable building structures, and a field experiment was carried out to choose the precise locations for the building elements.

In this museum design, all building elements, such as the size and orientation of the building, the relative positions of doors and windows, ventilation, and lighting system were studied. Due to the complicated terrain shape, the most difficult problems were how to locate a large building 70 m long on a construction site with a 15 m difference in elevation, how to achieve the goal of allowing ventilation but with light resistance, how to ensure energy saving under the premise of good display effects, and what kind of low-technology strategy would be feasible and effective for this project.

In order to solve those problems, the following strategies were chosen for the design principles of this museum building:

1. To divide a large construction of a 70-meter-long building into several small blocks following the terrain shape naturally, minimizing the required quantity of earthworks (Fig. 13.2)
2. To create a light-resistant structure that would allow natural ventilation, using passive methods to maintain a steady and relatively comfortable indoor temperature
3. To learn from the local traditional double-roof strategy, recycling old tiles for use on the second roof to reduce the massive entry of solar radiation and thereby reduce the indoor temperature
4. To use bamboo springboards for the concrete wall template, adjusting the temperature and creating a rich skin texture
5. To use a limited skylight focus on exhibits and use only a small number of LED lights to provide weak illumination of footpaths, which would not only guarantee the best presentation but would also save energy

**Fig. 13.2** Building blocks

## 13.4 Result and Analysis

### *13.4.1 Integral Configuration*

As is customary in the archaeological protection and excavation process, the scale of the construction site needed to be roughly investigated for construction of the museum, followed by implementation of excavation of the egg fossils in the museum. Finally, lighting could be set up after completion of the excavation process. However, there was no instrument that could accurately detect egg fossils under the surface of the construction site, thus it was too difficult to clarify the form of the museum based on the distribution of the egg fossils. Our approach was to conduct detailed mapping of the distribution of excavated egg fossils after the geologists had excavated under the shelter. On that basis, we set up wooden plank paths among the egg fossils and also provided observation platforms for large groups of visitors. Accordingly, the outline of the museum envelope was based on the layout of the egg fossils, plank paths, and platforms, with slight adjustments according to aesthetic sense.

As to the site planning, it was decided that the basic shape of the museum would follow that laid out by the dinosaurs 70 million years ago; for this the geologists had been excavating for one year with our design team. This design approach was inspired by the gradient-varying form, which adapted the irregular changes in the

**Fig. 13.3** Irregular concave side windows

topography, breaking up the building into several blocks connected to each other with irregular concave side windows (Fig. 13.3).

## *13.4.2 Energy-Saving Structure Analysis*

### 13.4.2.1 "Ventilation-Friendly and Light-Resistant" Device

The first energy-saving structure came from the principle of natural ventilation (Brager 2000). The museum project discussed in this chapter is located in the mountains of western Hubei, and the weather in this region is cold and damp in winter, while in summer the day-and-night temperature range is very large. We studied a scenario adopting passive methods used in traditional local dwellings, for which we designed a "ventilation-friendly and light-resistant" device (Fig. 13.4). In this scenario, there are adjustable side windows inside and double-deck louvers outside, and the leaves of the louver are set up at 90° to each other. Such a structure ensures ventilation while blocking direct sunshine. The natural wind helps with indoor cooling on summer nights. We installed this device into the irregular concave areas on

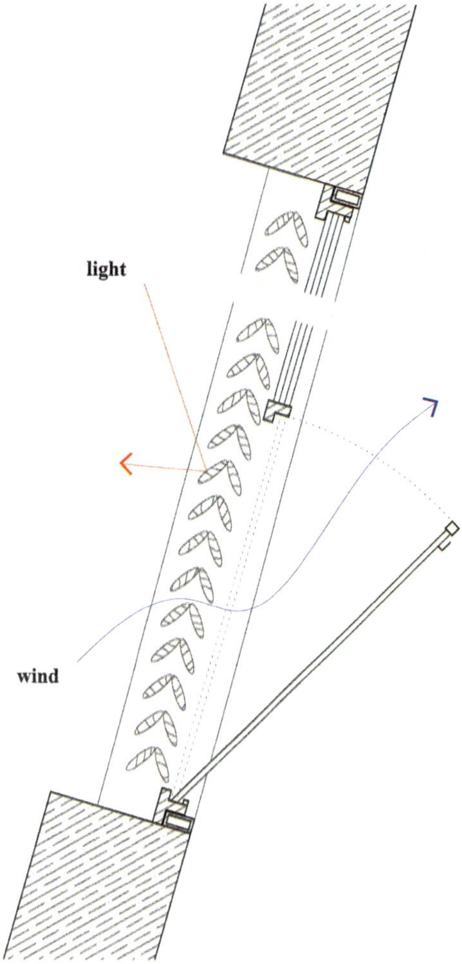


**Fig. 13.4** Ventilation-friendly and light-resistant structure

the exterior wall. After the building construction was finished, we measured the actual temperature in the museum for some days in summer 2016, and the results are shown in Fig. 13.5.

#### 13.4.2.2 Skylight Cylinder Device

The second creative structure we used was skylight cylinders (Fig. 13.6), which were designed like chimneys in form so that the skylights could be shut with glass. In the darkness indoors, those devices can provide basic lighting for each set of dinosaur egg fossils. The size of each cylinder is proportional to the size of the corresponding egg fossil, highlighting the displays. This kind of design both highlights the exhibits and saves energy. We beveled the top of the cylinder to let the glass face

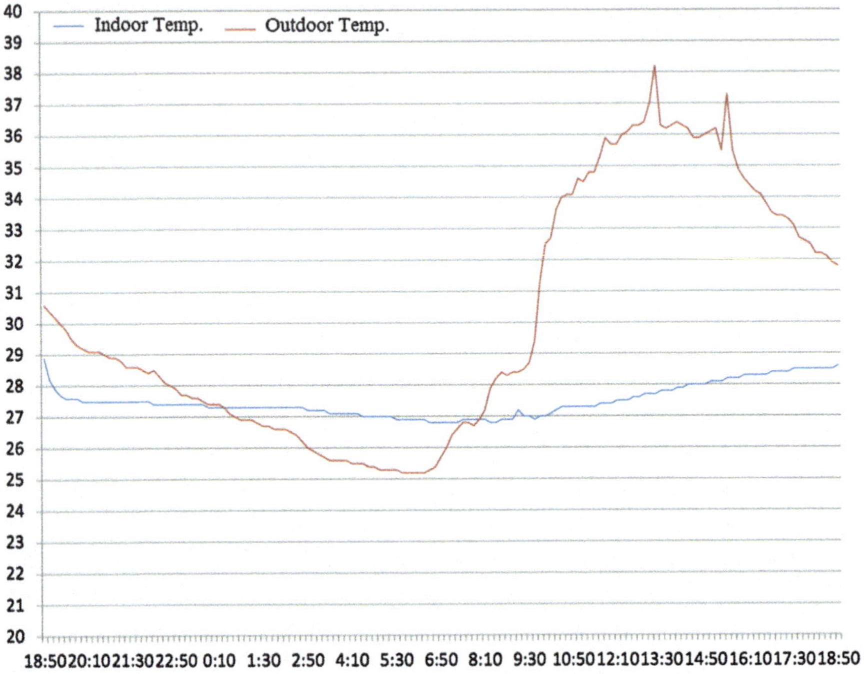


**Fig. 13.5** Indoor and outdoor temperature variations between day and night during summer

north, avoiding the direct sunlight coming in. This kind of structure can significantly reduce heat radiation from the outside.

#### 13.4.2.3 Double-Layered Roofing Strategy

Owing to historical conditions, most local residential buildings use a double-layered roof structure (Pröckl 1995) (Fig. 13.7).

The double-layered roof structure with moving air between the two layers has excellent effects in summer for reducing the heat indoors. This traditional architectural strategy is still enlightening for architectural design of the present day.

We found that the site was surrounded by some old adobe houses (Fig. 13.8). These buildings, with their poor durability, were on the verge of collapse. For safety reasons, the residents had all moved out and the buildings had been abandoned.

We collected the old tiles from the abandoned local buildings, and we used them on the second roof of the museum. There is a 20 cm air layer between the old tiles and the real roof of the museum, and the flow of air within the layer helps to reduce heat transfer. The old tiles also continue the local cultural context and increase the historical sense of the new building, as shown in Fig. 13.9.

For the purpose of cost control, we use locally grown bamboo as a facade template for cast-in-place concrete. As a crude-surface building, it has a high error

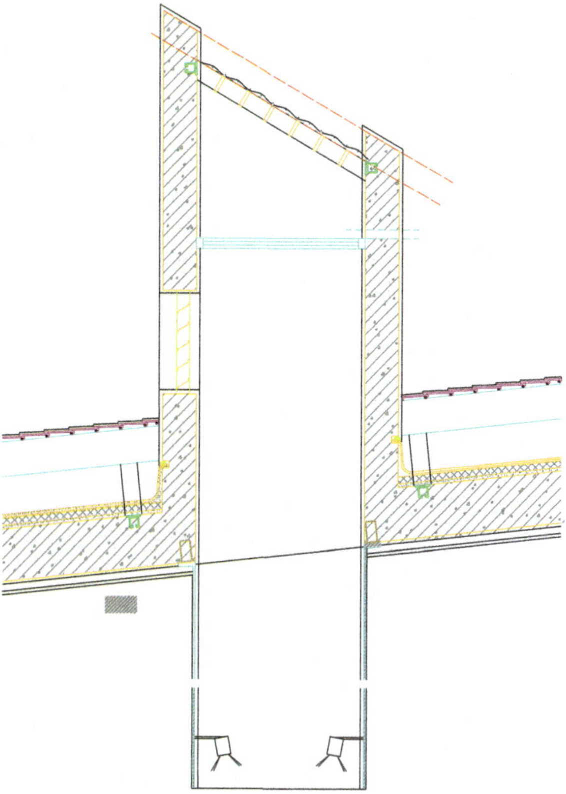

Fig. 13.6 Skylight cylinder

Fig. 13.7 Double-layer roof structure

**Fig. 13.8** Local residential buildings

**Fig. 13.9** New building with an old roof

**Fig. 13.10** Bamboo template wall

tolerance, so the construction of this rough surface was not very difficult. This concept should be suitable for a building in the mountains. Bamboo is reproducible and easily degradable, thus the use of bamboo increased the building's ecological benefits. Moreover, the crude skin formed by the bamboo provided a sense of vicissitude to the building (Fig. 13.10), which is consistent with the characteristic of the ruins museum.

### *13.4.3 Interior Display Effect Analysis*

For creating a good interior atmosphere, the most reliable and elegant way is to create a dark environment with a focused skylight, which is useful for displaying the exhibits and also can help to reduce energy consumption. We used light cylinders for displaying the dinosaur egg fossils. With only a small amount of LED lighting to provide weak illumination for footpaths, the weak LED lights can guide tourists through the walking areas in the dark indoor environment. When tourists enter the exhibition space, their gaze is caught first by dinosaur egg fossils illuminated by a shining tube, and each group of dinosaur egg fossils gradually appears clearly as the visitors move forward through the space, as shown in Figs. 13.11 and 13.12. The primary structural decomposition is shown in Fig. 13.13.

Fig. 13.11 Interior display effect

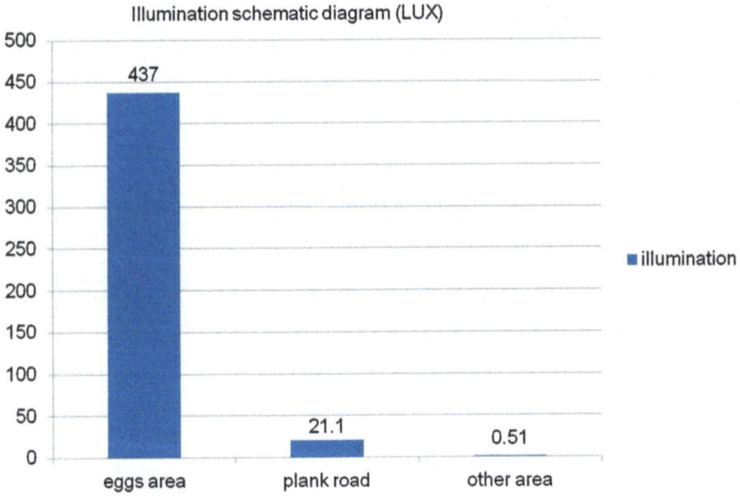


Fig. 13.12 Illumination schematic diagram

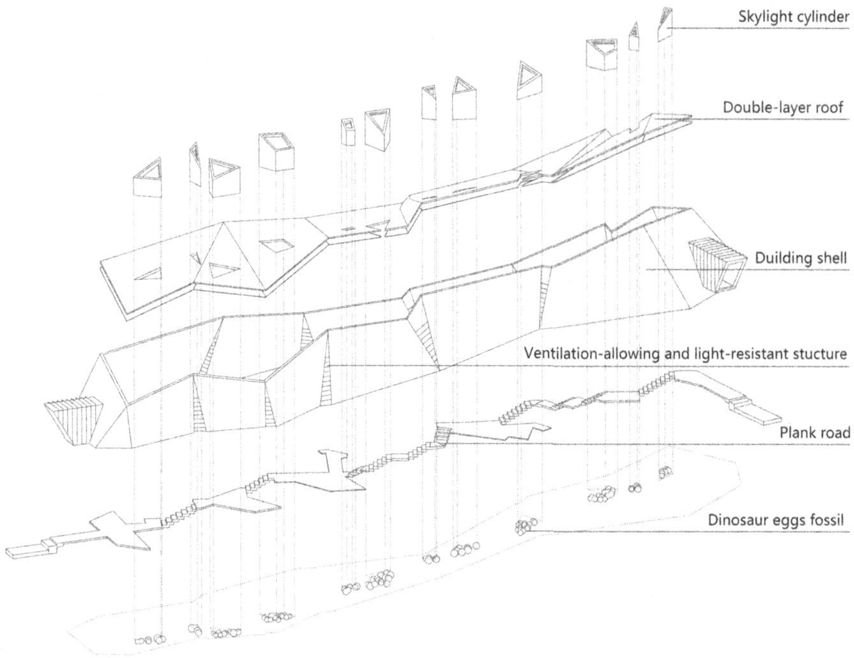


**Fig. 13.13** Structural decomposition

## 13.5 Conclusion and Discussion

As green building becomes increasingly popular in the future, energy saving will be more important in architecture design study. This chapter focuses on an energy-saving strategy for the design of a dinosaur egg fossil museum in Hubei, China. The form of the museum building is derived from the locations of the dinosaur egg fossils, plank paths, and observation platforms. The broken form of the museum building is also designed based on the ground surface for adaptation to its natural environment.

In the design process, lighting cylinders, a double-layered roof that is ventilation-friendly, and light-resistant structures are integrated for display of exhibits, indoor comfort, and energy saving. In short, the design approach is to consider how to display exhibition areas of dinosaur egg fossils in a museum building from the perspectives of ecology, place, and cultural context. As result, energy savings based on the climate adaptation strategy that we adopted for the design have been achieved in the Dinosaur Egg Fossil Ruins Protection Museum. The concept of green building design depends on many other factors, such as outdoor temperature variations and wind direction, rainwater collection, sponge city, combined active and passive strategies, etc. We can improve and refine this research in further studies.

**Acknowledgments** Design of the Dinosaur Egg Ruins Protection Museum started in 2010. Construction was started in 2011, and it was completed in 2012 and opened to the public in 2013. *Dezeen*, the famous architecture and design magazine, appraised and awarded this design as one of the "2016 Top Ten Global Outstanding Public Buildings." It is the only Chinese building to have received this honor. Thanks to the team members for their contribution to this project. They are master candidates Zeng Zhongzhong, Guo Fan, and Qu Tianming.

## References

Bao Y (2013) Discussion on the application of energy-saving technology in building external envelope structure design. Urbanism & Architecture

Baofeng Li (2001) Analysis of types of "double layer" curtain wall and prospects of their application. Architect J

Brager GS (2000) A standard for natural ventilation. ASHRAE J 42(10):21–23+25

Chai W, Zhang H (2009) Study of heat insulation performance of building door and window. Construction Technology

Cheng HT (2006) Application of low energy consumption architecture technology in Nanjing Landsea International Block. Building Science

Cuce E, Riffat SB (2016) A smart building material for low/zero carbon applications: heat insulation solar glass—characteristic results from laboratory and in situ tests. Int J Low-Carbon Technol 12(2):126–135

Deszberg TM (1983) Combined lighting and ventilation device. EP, EP0088567

Fordham M (2000) Natural ventilation. Renew Energy 19(1–2):17–37

Ghiaus C, Allard F, Wilson M (2005) Natural ventilation in the urban environment assessment and design

Guangcai BG (2003) Situation and development of natural ventilation. Building Energy & Environment

Hou WS, Chuang TF, Hsieh EC et al (2012) An analysis of heat insulation efficiency of building outer skins used for green building. Build Serv Eng 33(33):407–422

Huang C (2007) Light in architecture. Fujian Architecture & Construction

Jang HJ, Park CI (2016) A study on the phenomenal expression methods and the characteristics of spatial experience in the Steven Holl's museum architecture. Korean Inst Inter Des J 25(2):151–162

Kim CS (2014) Architectural expression of light appeared in museums designed by Henri E. Ciriani. Korean Inst Inter Des J 23(2):3–11

Kleiven T (2003) Natural ventilation in buildings architectural concepts, consequences and possibilities 38(5):491–501

Lan L, Fei Hu C (2009) The understanding of the optical element using in architectural design practice. Chin Overseas Archit 7:81–83

Liu Y, JingJing Y, Bing S, XinRong Z (2015) Fundamental research and practice of passive and ultra-low energy consumption buildings. Chin Sci Bull 60(18):1698–1710

Ming-Hai LI, Wang WW, Wang Z et al (2008) Study on the technology of new heat insulation board for building roof. Construction Conserves Energy

Moran D (2000) Introduction to phenomenology. Routledge

Pröckl GI (1995) Furring construction for double layer roof systems. EP, EP 0685612 A1

Qian K, Yin XS (2009) Analysis of energy-saving structure of green building in cold region. Construction Energy Conservation

Zhong A (2015) Application of thermal insulation mortar in energy-saving retrofitting on the external wall of an existing structure. ICCREM 2014: Smart Construction and Management in the Context of New Technology. ASCE, pp 140–146

# Chapter 14
# Development of a Simplified Green Building Model in Taiwan: The Case of the AGS1 Experimental House

**Chih-Peng Liu and Ding-chin Chou**

**Abstract** This chapter firstly describes the local climate characteristics of Taiwan, green buildings, and the manner of using green materials in Taiwan by using a literature review analytically, then explains the development model of green buildings for a green environment, green structure, green materials, green living, and green decoration and finishing under a "simplified" concept. Secondly it describes a practical and appropriate residential environment model encompassing a combination of correct temperature and humidity values indoors, reduced building weight of envelope structures and heat gain through windows, avoidance of excessive interior decoration, choice of an envelope material with high resistance to moisture permeability, an air exchange mechanism with a proper intermediary space, and a suitable fitted air conditioning model. Last, but not least, this chapter describes the concept of the "Development of A Simplified Green Building Model in Taiwan" by aggregating case study data and presents a description of the development of the AGS1 experimental house on the basis of the effectiveness evidence of some of the building functions of the AGS1 experimental house, and also presents suggestions for the direction of short thesis research in the future, hoping to facilitate positive development of the green building sector in Taiwan.

**Keywords** Simplified · Green buildings · AGS1 · Heat gain · Building structure

C.-P. Liu (✉)
Doctoral Program in Design, College of Design, National Taipei University of Technology, Taipei, Taiwan
e-mail: aghouse.liu@gmail.com

D.-c. Chou (✉)
Graduate Institute of Architecture and Urban Design, Architecture Department, National Taipei University of Technology, Taipei, Taiwan
e-mail: dcchou@ntut.edu.tw

 
Z. Shen et al. (eds.), *Green City Planning and Practices in Asian Cities*, Strategies for Sustainability, https://doi.org/10.1007/978-3-319-70025-0_14

## 14.1 Foreword

Global warming and the energy crisis constitute twin threats to Taiwan, so we are faced with the task of increasing both residential comfort and energy efficiency. In tackling these threats, the development of green buildings needs to focus on weight reduction, carbon reduction, and energy conservation. Taiwan faces the problems of excessive cement use and illegal gravel mining. It ranks second in the world for cement consumption per capita thanks to reinforced concrete (RC) building structures accounting for over 95% of all building structures. Furthermore, environmental illness sources such as high thermal transmittance and heat and cold cumulative effects in reinforced concrete structures (Fig. 14.1), and excessive decoration and finishing indoors, lead to sultry summers and cold wet winters, resulting in dew formation and fungal contamination of walls, which seriously affects the health of residents in Taiwan. Furthermore, in view of the issues and concerns about high carbon emissions, high energy wastage, and the threat to comfort indoors during the life cycle of a reinforced concrete building structure, it is essential to make changes in the structure of reinforced concrete buildings (Figs. 14.2 and 14.3).

The underlying concept of simplified green buildings is environmental conservation. As such, green buildings should be developed through introspective thought rather than by mindlessly adding materials and facilities and wasting the earth's resources. To "simplify" means to minimize complexity, to reduce, and to improve from within. This concept was proposed by Liu Chih-Peng in June 2015. The AGS1 is a disaster-proof green building structure developed by the Taiwanese architectural firm AG-House. Following the successful integration of geothermal green energy structures with health-conscious, non-toxic wood interior decoration, AG-House accomplished the construction of its first green building in 2016 using a seventh-generation construction method consisting of the aforementioned features. The building, named AGS1, takes its name from the company, meaning a good house or a

**Fig. 14.1** Wall in a conventional reinforced concrete building structure or a brick masonry structure

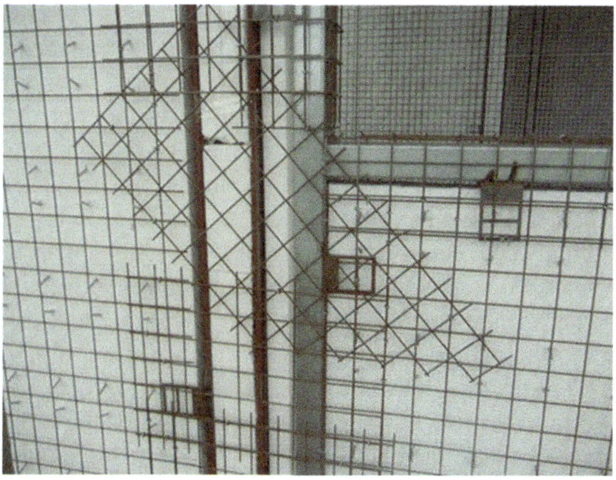

**Fig. 14.2** Wall in the AGS1 composite building structure

**Fig. 14.3** AGS1 House

global house. “S” stands for “snail” since in Taiwan the homeless are commonly referred to as shell-less snails. AG-House was founded to realize the concept of enabling inhabitants to own their homes. AGS1 perfectly embodies the concept of simplified green building by being disaster proof, safe, health conscious, energy efficient, and comfortable. The fully developed structural system, appropriate materials, and economical facilities completely resolve issues typically incurred by Taiwan’s humid climate and provide the inhabitants with access to healthy air quality.

Among the various green building techniques applied in mainstream green buildings in Taiwan, quite a few were introduced into Taiwan from experiences accumulated in countries in high-latitude regions. Numerous certified green material products have been introduced into Taiwan from overseas, and excessively promoted green building designs mislead us through predominant product promotion and marketing calls despite our concerns about real benefits and effectiveness. This chapter firstly interprets the development model of green buildings under the "simplified" concept, which encompasses a green environment, green structure, green materials, green life, and green decoration and finish. Secondly, by taking the AGS1 experimental house as the object in question, it assesses and calculates the carbon footprint and overall carbon emissions of the composite building structure as well as a reinforced concrete building structure, and also the thermal transmittance (Ui) of structural elements; examines the ratio of the influence of the heat gain of the envelope and that of the window; conducts tests and verifies, in terms of building function, the effectiveness and benefits of the AGS1 experimental house; and finally presents the research results and conclusions and also some suggestions.

## 14.2 Introduction

### *14.2.1 Motivation and Purpose*

The motivation of this research is to establish a development model of green buildings centering on safety, healthiness, comfort, and energy efficiency, using the "simplified green buildings" concept and particularly taking the regional and natural environment and climate characteristics of Taiwan into consideration. The purpose of this research is to present "The Development Model of Simplified Green Buildings in Taiwan," which encompasses topics such as the argument over the efficiency of energy conservation windows; the "Simplified Green Building Model for a Detached House on the Plains of Western in Taiwan"; the "Construction Method of the AG Innovative Green Structure"; and the Development Model of the AGS1 experimental house. Furthermore, it presents both the research results of "Tests and Verification of Some Architectural Benefits of the AGS1 Experimental House" and suggestions for the direction of future development in the hope of facilitating positive development of the green building industry sector in Taiwan.

### *14.2.2 Study Method and Process Flow*

This research makes use of research methods including a literature review and case study by compiling "The Development Model of Simplified Green Buildings in Taiwan," comparing the respective contrasting data on carbon emissions and

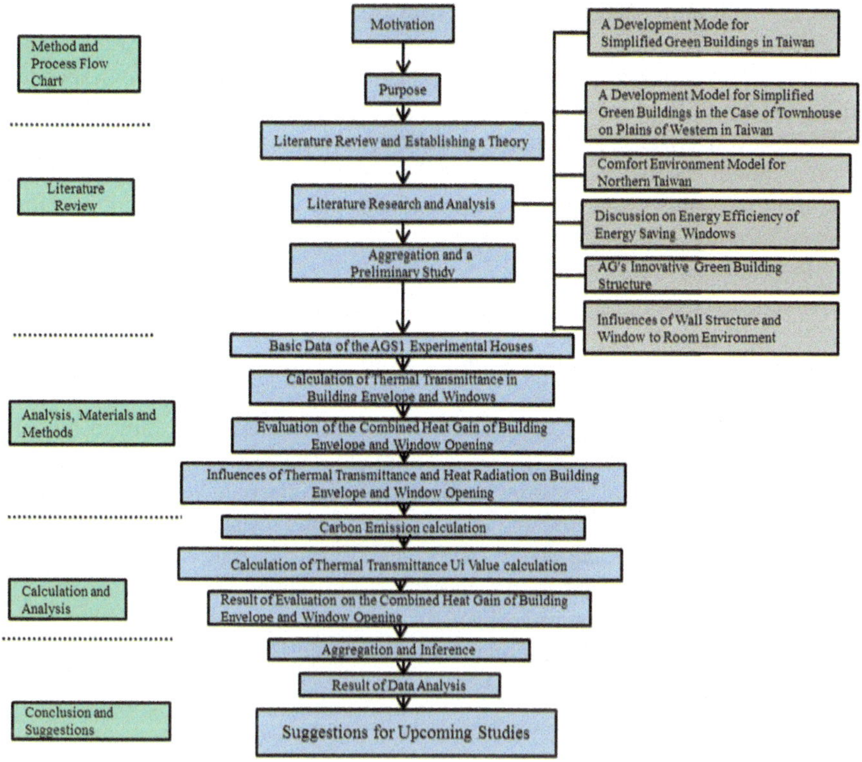


**Fig. 14.4** Study method process flow

thermal transmittance (Ui) of the building envelope and window combinations in the AGS1 experimental house building structure and a conventional reinforced concrete building structure, and, finally, further exploring and concluding what are the optimal and most appropriate combinations (Fig. 14.4).

## 14.3 Literature Review

### *14.3.1 Development of the Simplified Green Building Model in Taiwan*

This section is a summary of "The Simplified Green Buildings" (Chih-Peng Liu 2015) and gives an overview of each article as follows.

#### 14.3.1.1 Climate Characteristics of Taiwan

In reply to climate change impacts on the natural environment, Taiwan desperately demands the development of building structures that are best suited to her regional climate characteristics. This article discusses problems such as earthquakes, wind storms, floods, landslides, acid rains, the ultraviolet index (UV), moisture, mites, and termites; sun-shading building conventions; and the results of a pilot exploration on forming a green building–supportive environment under the climate conditions in north Taiwan (Fig. 14.5).

#### 14.3.1.2 Discussion on Green Buildings and Green Materials in Taiwan

The practical benefits of green buildings and green materials in Taiwan must be clarified with further practical studies and thinking. This article explores issues such as industrialized housing, energy conservation, carbon reduction, Leadership in Energy and Environmental Design (LEED) myths, and the significance of devising on-site water retention and water saving, the benefits of diatomaceous earth and photo tiles, indoor daylight versus energy-saving lighting, arguments on energy-saving windows, green energy technology and subsidy myths, reuse of old buildings, and building material recycling and reuse (Table 14.1).

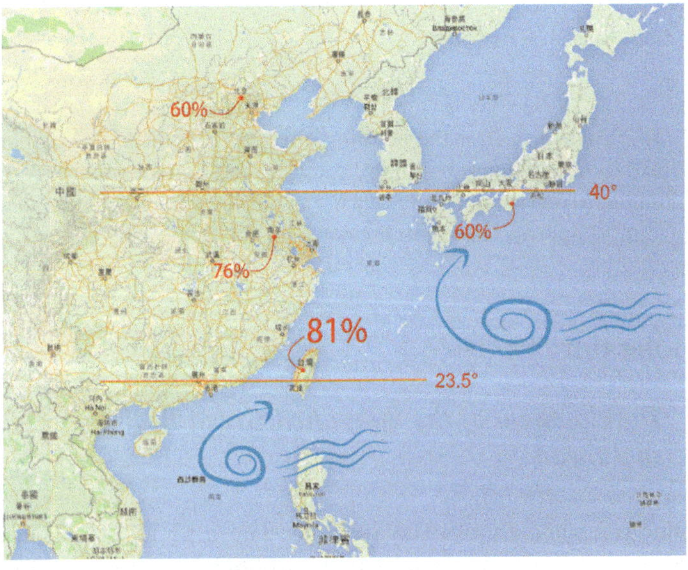


**Fig. 14.5** Taiwan's climate characteristic of high humidity

**Table 14.1** Evaluation methods and criteria for green buildings around the world

綠建築標章
GREEN BUILDING

Not only reduction but also efficient use of energy, which is also suitable for indoor environments and eliminating the sick building syndrome.

Now, around the world, green building assessment methods and classification systems both adopt energy performance and indoor environmental quality as evaluation criteria, as shown in the table.

| Assessment method | Evaluation Methods of Green Buildings around the world |
|---|---|
| USA LEED | Base for sustainable use, efficiency of water use, energy & atmosphere, materials & resources, indoor environmental quality and innovative design |
| Canada GB Tool | Resource consumption, economic load, indoor environmental quality,maintenance and service quality, economy, business management, transportation |
| UK BREEAM | Management, health and comfort, energy, water resource, materials, land usage, ecosystem and pollution |
| CHINA | Residential environment planning and design, energy and environment, indoor environmental quality, Residential water environment, materials and resources |
| JAPAN CABEE | Indoor environment, warm environment, temperature setting, exterior thermal load control, natural energy, high-efficiency equipment, monitoring and management system |
| TAIWAN EEWH | Biodiversity, green, water-based, daily energy saving, $CO_2$ reduction, waste reduction, water resources, indoor environmental quality and improvement of sewage and garbage |

#### 14.3.1.3 Simplified Green Environment

Behind the convenience provided by the environment is inevitable dependence. Excessive use of environmental control systems is contributed to by profitability and consumer behavior, hence the concept of the simplified green environment basically lies in avoiding excessive impact on the environment and also conserving the environment's coping potential. This section discusses issues such as valuing the residential living environment, water treatment in the living environment, and the environments of Dr. Tanaka's eco-environment pond versus the AG extreme green environment and extreme cultivation environment.

For example, the vertical greening fences adopted for construction sites contribute to better environmental quality, but it will be a waste of plants and a disaster to the development of sustainable environment if plants on vertical fences are not properly cared for and disposed of (Chien-Yuan Lin and Yin-Ling Huang 2010). A sophisticated and multidisciplinary green-energy, water-autonomous greenhouse system, by using water resources and solar energy in a rational way to aim at a higher level of well-being based on more products, more information, more services, and more experiences, could be an alternative technology approach towards a sustainable smart-green vertical greening in smart cities (Pai yao Hung and Kuang hui Peng 2017).

#### 14.3.1.4 Simplified Green Structure and Simplified Green Materials

A building structure contributes to carbon emissions during the building's life cycle and also affects the energy efficiency of the building; therefore, weight reduction and materials suited to the site are basic considerations. This section explores issues in market mechanisms and sustainable environments, building planning, locale incompatibility found under the global frugal style, wooden structure problems, and brick wall structure concerns. It also discusses the possible development of green structures and green materials for conventional buildings in Taiwan, encompassing use of water-permeable or impervious tiles, porous materials, roofing tiles, walls and tiles, leakage, thermal insulation, and stilt-style buildings.

#### 14.3.1.5 Simplified Green Living and Simplified Green Decoration and Finish

While enjoying a readily available comfortable living environment, we inevitable weaken the immunity to cope with environmental change. Among energy conservation strategies, the best ones are to encourage good daily living habits, to reduce dependence on equipment and facilities, and to made efficient use of resources. Since over 95% of building structures in Taiwan are constructed out of reinforced concrete, which is dull in appearance and problematic in terms of maintenance and functionality (e.g., sick buildings), interior decoration is typically regarded as a way of embellishing internal living space.

Simplified green living and green finishing and decoration concepts, on the other hand, emphasize health, environmental protection, and energy conservation. They aim to achieve a 20% energy saving from making good use of the physical environment of buildings, a 10% energy saving from daily power use concepts and patterns, and with regard to the remaining 70% of energy consumption, to offset 20% of it by using renewable or natural clean energy sources, with the remaining 50% to be reduced by using low-energy-consumption devices or equipment. Forty percent of material consumption can be saved in the building structure finishing, and the remaining 60% can be reduced with low-pollution healthy materials or recycled materials. This section presents ideals for ordinary space and living habits, designs inspired by Mother Nature, household sewage treatment concepts, humidity issues, intermediary space uses, applications of building infrastructure and geothermal air conditioning, residence healthiness inspection and improvements, sick housing, living audio systems, establishment of smart homes, and an analysis of the trends of i-Japan cities and housing planning.

### *14.3.2 Simplified Green Building Model for a Detached House on the Plains of Western Taiwan*

The normal pattern of climate on the plains of western Taiwan sees wet and cold northeasterly monsoon winds and mild southwesterly winds. Most regions have a wet and hot or wet and cold climate except for the hot summer in the southern regions. This section addresses issues such as sun shading in the southwestern direction, ventilation and convection, indoor humidity proofing, wall waterproofing, dew proofing, use of intermediary space and geothermal energy applications, outdoor extreme green environments, the scope of comfortable humidity levels and temperatures indoors, toxin-free finishing and decoration, and thermal window insulation treatment, etc., and finally creates a concept illustration of the "Simplified

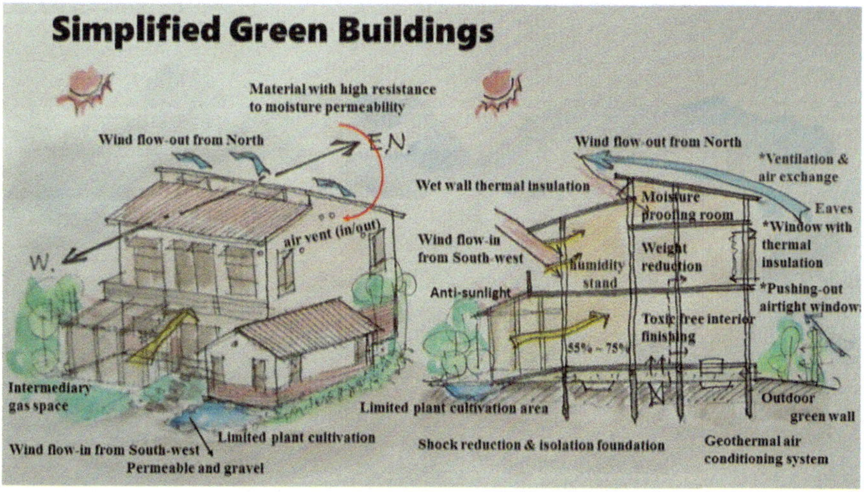


**Fig. 14.6** Concept of a simplified green building model for a detached house on the plains of western Taiwan

Green Building Model for a Detached House on the Plains in Western Taiwan" (Fig. 14.6).

Vegetation has a significant influence on microclimate directly by shading surfaces and channeling wind and indirectly by evapotranspiration of water. Proper house configuration, high albedo level, and strategic locations of surrounding landscape also can influence the thermal performance of the house by reducing air temperature and creating a comfortable environment for buildings (Misni et al. 2013).

### *14.3.3 Concept Illustration of a Comfortable Living Environment in the Northern Region of Taiwan*

In a typical year, the northern regions in Taiwan have a climate featuring sultry or wet and cold ambient air for 25% of the time. It is not advisable to let the air in, and therefore air conditioning systems are used with windows and doors being shut. The high humidity affects thermal conduction of windows and walls, heat convection, and the indoor heat gain rate. It is therefore important to address concerns about room structures absorbing moisture and being conducive to fungal growth, the reciprocating complementary features of demands for sun exposure of room windows in the summer and the winter, and heat exchange demands in closed rooms. The inferences on "The optimal temperature and humidity combination for room, composite structure building envelope and common windows wanting treatment with finishing" to enhance thermal insulation and reduce heat gain can be done by avoiding excessive indoor decoration and finishing, selecting an envelope material with higher humidity resistance, and devising a mechanism of proper intermediary

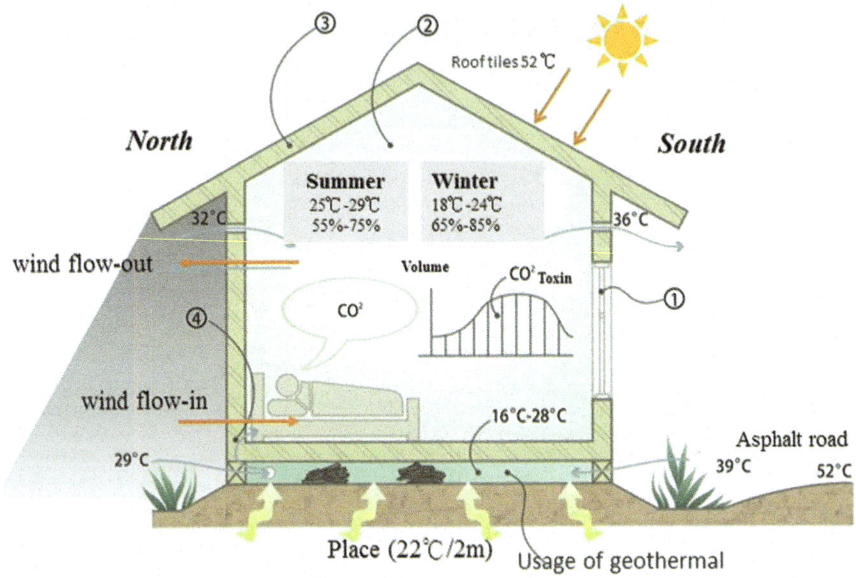


**Fig. 14.7** Concept of a model of a comfortable living environment in the northern regions of Taiwan
① Window opening
② Temperature and humidity
③ Adiabatic roof
④ Ventilation and air exchange

space for air exchange. In this fashion, with consequently little energy flow between the indoor and outdoor spaces, the air conditioning system is responsible for the loading of energy adjustments for the human body, lighting, and appliances indoors. This will be the optimal model of the green building environment for residential buildings in the climate conditions of the northern region in Taiwan. (Fig. 14.7).

### *14.3.4 Dialectic About the Efficiency of Energy Conservation Windows*

This study tries to demonstrate issues of green materials by considering the argument on the function of energy conservation windows as an example. Due to recent exaggeration of the function of energy conservation windows in Taiwan, Taiwan has seen booming construction of buildings with glass facade envelopes. A pair of contrasting models of the function or performance of thermal insulation glass was presented in a recent Green Building Materials Exhibition organized by Architecture & Building Research Institutes, Ministry of the Interior (Fig. 14.8). The models each

**Fig. 14.8** Contrasting models of the effects of energy conservation glass

came with a lighting bulb as an energy source, emitting light through energy conservation glass or non-energy-conservation glass to demonstrate the thermal resistance effect shown by the temperature difference felt by touching each glass sample. This was is used to prove that energy conservation glass is functional in lucratively reducing heat gain caused by solar radiation and in turn saving energy consumed by air conditioning systems. Two building models with glass facade envelopes were exposed under the emission of the light bulb. The temperature probe in the middle of the model with the non-energy-conservation glass displayed a temperature of 32.4 °C, whereas the model with energy conservation glass displayed a temperature of 28.1 °C. This experiment implied that the energy conservation glass led to a room temperature difference of 4.3 °C.

For further analysis of the argument, while energy conservation glass isolates the radiation energy of some wavelengths, the energy actually remains in the film of the glass and will be emitted to the indoor space through high moisture, therefore the only energy reduction happens partially outdoors, and the enthalpy outdoors is still transmitted to the indoor space through the glass and walls. As a result the energy conservation effectiveness is very limited. What misleads the public is the high tech image shown by the glass facade. In fact, in countries located at high latitudes, buildings need to maintain a higher temperature indoors with the sun's radiation energy during the daytime, plus features such as a comparatively lower cold radiation transmittance rate occur in low humidity, so the popularity of buildings with glass facade envelopes in high-latitude regions exists for a reason. However, this is not the case in Taiwan, due to her high humidity and low latitude. The misleading appeal of these glass products in fact adversely encourages architects and property owners to introduce glass facade envelopes to follow the trend. Unfortunately it is the wrong approach in terms of energy conservation.

## 14.3.5 Green Structure of AG Innovative Buildings

AG-LSRC is a construction method invented by Liu Chih-Peng during the period from 2000 to 2012. With this method, buildings are formed using a vibration reduction/isolation foundation, lightweight wall panels, and a steel cage. The structure combines vibration proofing, typhoon proofing, fire proofing, and ant proofing for a disaster-proof building that contributes to the development and advancement of green building concepts. Compared to the structure of reinforced concrete buildings common in Taiwan, the AG-LSRC structure features improved vibration proofing and reduced use of concrete achieved through weight reduction and adjustments in the foundation construction method. This, combined with improved heat insulation of the walls and windows, results in meaningful reduction of energy use and the carbon footprint.

The AG-LSRC light steel reinforced concrete construction method has been developed especially for a composite building structure to suit the natural environment in Taiwan, as a method that addresses cost effectiveness and ecological concerns, and falls into the following new edge process flows (Figs. 14.9 and 14.10):

### 14.3.5.1 Shock Reduction and Isolation Foundation

The absorption and transfer of earthquake energy is reduced by using an innovative foundation design and installing shock isolators. The energy is also released with a vibration isolating device, so as to dissipate impact energy, protect structural safety, and minimize consumption of structure materials (Fig. 14.11).

### 14.3.5.2 Cage Steel Frame

Building frameworks are built with combinations of hot rolled steel and cold rolled steel elements, with horizontal braces and wall panel assemblies to form a cage structure; hence the structure is lighter and tougher than a beam and column

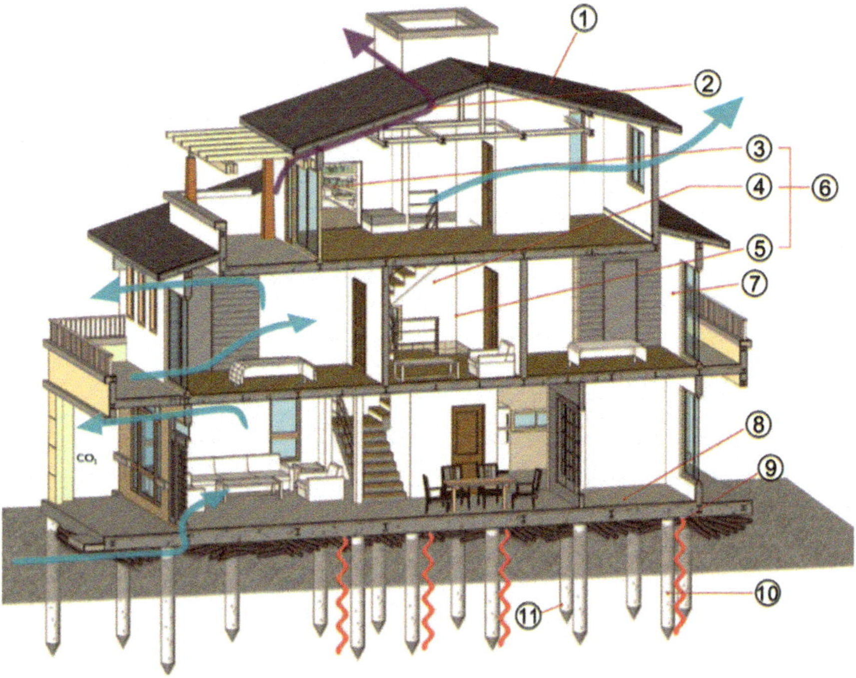


**Fig. 14.9** AG innovative composite structure building
① High-performance insulation
② Convection and radiation
③ Increased storage space
④ Without large beams and staircase reduction
⑤ Column angle reduction
⑥ Increased indoor space of 12%
⑦ Without dew formation and fireproof
⑧ Moisture proofing design on the floor
⑨ Isolation of foundation by 70%, shock reduction
⑩ Damping to two thirds
⑩ System of geothermal air conditioning providing 16–28 °C convection and air exchange

structure system, with comparable strength and construction work safety, speediness, precision, and cost effectiveness (Fig. 14.12).

#### 14.3.5.3 Lightweight Wall Panels

Wall panels built with a three-dimensional (3D) system of wall assembly and lightweight concrete have the advantages of weight reduction, high thermal insulation, condensation and dew proofing (Fig. 14.13).

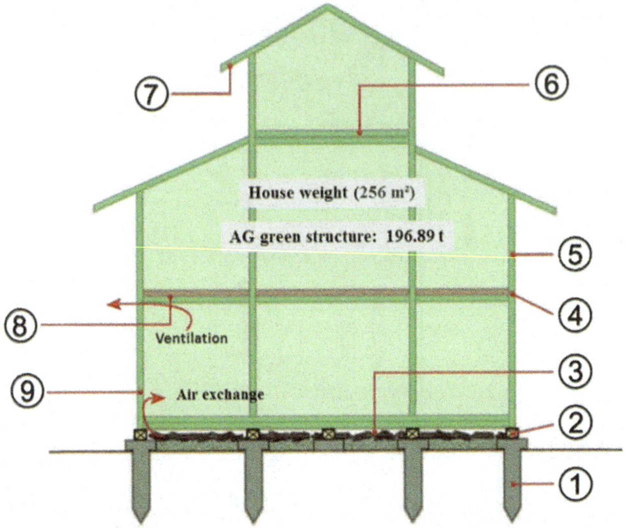


**Fig. 14.10** Structural system of the AG innovative composite structure
① Root-like damper base
② High-performance shock reduction and isolation foundation
③ Geothermal air conditioning system
④ Cage steel frame and building framework
⑤ Wet light steel reinforced concrete construction system
⑥ Dry preparation roof
⑦ Convection and radiation
⑧ Beam connection with columns
⑨ Column setting with smaller steel frames

### 14.3.6 Calculation of Related Carbon Emissions

The carbon footprint is defined as the $CO_2$ emitted directly or indirectly by the product during its production and its activities during its entire service life cycle. Currently there are no estimated guidelines or standards available for checking or assessment of $CO_2$ emissions during a building's service life cycle either domestically or overseas, and some building materials come without $CO_2$ emissions data. This study tries to define the life cycle of current general buildings' main structure and operation for assessing their carbon footprints, with eight stages identified: material production, material delivery, site operation, daily usage, daily repair and

**Fig. 14.11** Foundation in the AG innovative composite structure

**Fig. 14.12** Building the steel framework of the AG innovative composite structure

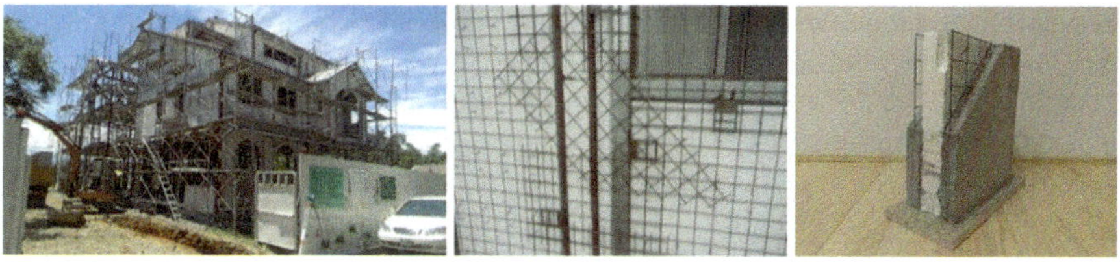

**Fig. 14.13** Building structure methods used in the AG innovative composite structure

maintenance, modification and renewal, demolition, and debris disposal. Each stage is defined and analyzed within the scope of the research and assumed conditions.

In the case analysis of the carbon footprint during the life cycle of a reinforced concrete building—the school building for the College of Engineering on the campus of National Ilan University—the case is reviewed in two separate situations: including daily use and excluding daily use. In the case of the building structure alone, it contributes to only 15% of the carbon footprint ratio, while the structure along with daily use contributes 85%. The report leads to two argument lines: firstly the carbon footprint lies mainly in the usage stage; secondly, a reinforced concrete structure is likely to lead to escalating energy consumption during its use in the future.

### *14.3.7 Influence of Wall Structure and Windows on the Indoor Environment*

According to recent Industrial Technology Research Institute (ITRI) research results, removal of the stored heat coming through the building's envelope accounted for 96% of the electricity use by household air conditioning, of which windows accounted for 57%, walls 17%, and the roof 22% (Chiou Jih-Jer 2009). This indicates that the building envelope and windows seriously influence the residence's temperature and humidity. Hence it is vital to review the types and combinations of building envelope structures and window insulation in order to understand their influence on the residence's temperature and humidity, and also carbon reduction.

Concerning the studies on the effect of the wall structure on the overall thermal transfer value (OTTV), the Technical Code for Energy Conservation Design of Buildings of Taiwan ENVLOAD (Envelope Load) is used to conduct an analysis of a building's heat gain (Lin, Hsien-Te, Yang, Kuan-Xiong 1997). In Hong Kong, the authority adapts the analysis results of a building's heat gain from the OTTV as the basis for estimating the air conditioning cooling load and for improvement, in view of the steady increase in electricity use by air conditioning year by year (Joseph et al. 2003).

ENVLOAD allows for taking into consideration the intrinsic factor of heat-generating objects and thus using different calculation patterns for specific locations, and also the overall heat of heat elements. Comparing the differences between OTTV and ENLOAD, OTTV takes into consideration the overall heat of the building envelope. Chou Chang-Hsien et al. (2013) adopt an approach of changing the building envelope structure, window glass material, and opening sizes as proper combinations, so as to conduct a comparison of the OTTV heat gains for the purpose of improving the indoor heat gain, saving construction costs and eventually achieving the goals of energy saving and carbon reduction.

# 14.4 Analysis of Materials and Methods

## *14.4.1 Basic Information on the AGS1 Experimental House*

"AGS1" stands for one of the typical three-story detached townhouses in the Tu-Shih Green Home Community developed by the AG-House group, located in Aspire Park, Longtan District, Taoyuan City, in northern Taiwan, where the community land area is about 3000 m$^2$, the site area of the experimental house is 232 m$^2$, and the area of all of the buildings is 256 m$^2$ (Fig. 14.14).

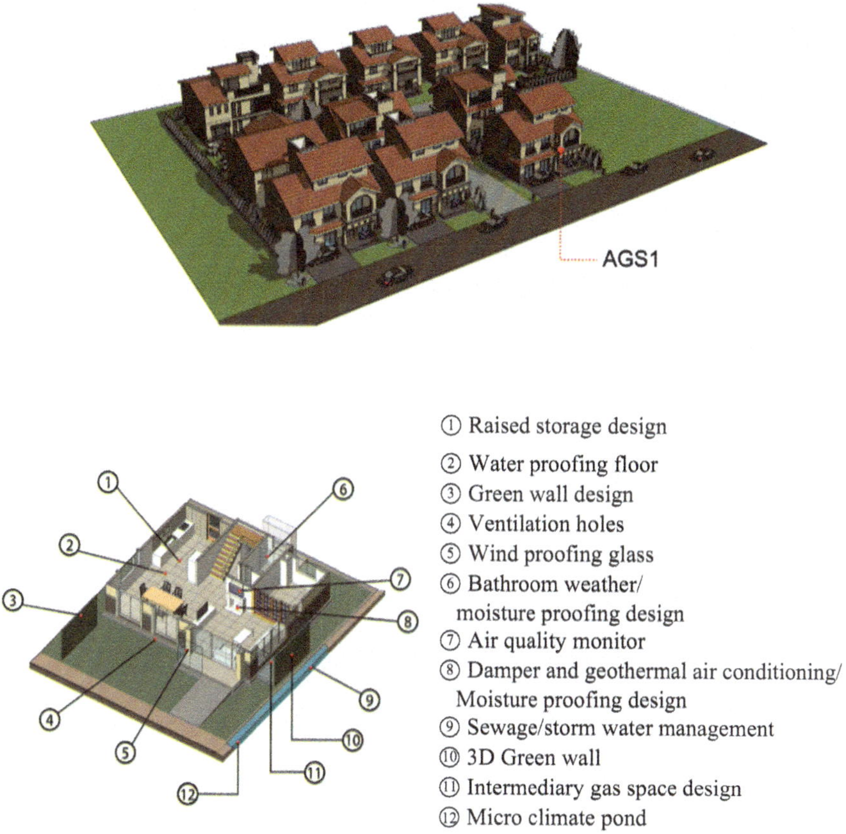


**Fig. 14.14** Layout of Tu-Shih Green Home Community and the AGS1 experimental house

The round floor design mainly consists of the foundation structure and geothermal air conditioning system, plus optimal use of space under the floor surface; outdoors are green walls, an intermediary gas space, and emergency or auxiliary attachments.

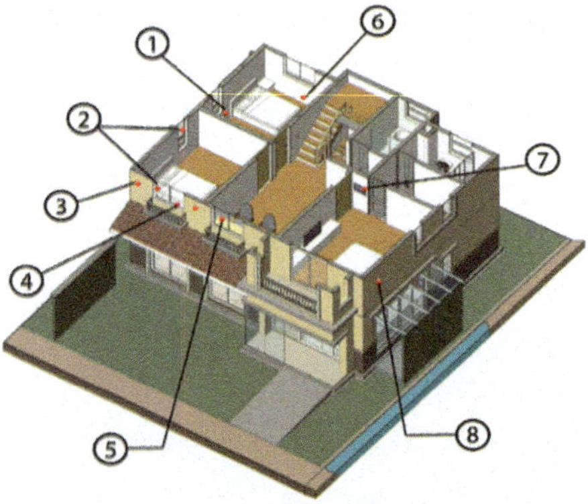


The second floor design consists of toxin-free interior finishing, an air quality monitor, windows with thermal insulation, energy conservation windows, air exchange ventilation, and wall thermal insulation.

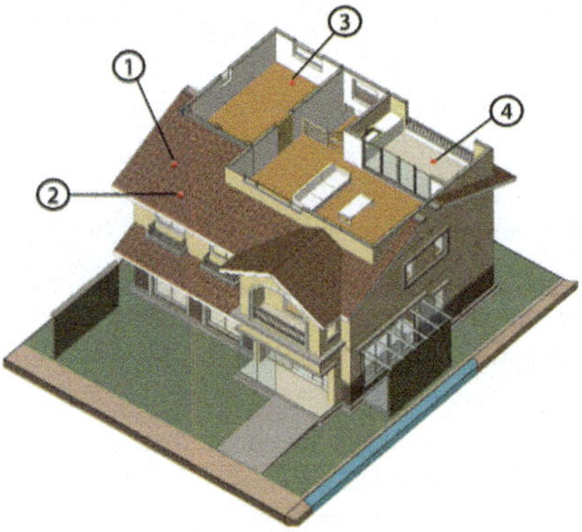


The third floor design consists of a roof with thermal insulation, moisture-proof room, balcony, water tank shelter, and dry preparation roof.

This study conducts calculations and analysis on data from the AGS1 experimental house and makes inferences on the aspects of carbon emissions and envelope thermal transfer values, taking those of a conventional reinforced concrete structure as the control group. Firstly this study establishes the basic data on the AGS1 experimental house related to analysis items and methods such as calculating carbon emissions of the reinforced concrete structure and composite structure, and calculating the thermal transmittance value (Ui) of the building envelope and windows, and then creates simulations of the possible combinations of heat proofing for the envelope and window openings.

### 14.4.2 Calculation of the Thermal Transmittance of the Building Envelope and Windows

This calculation targets the thermal transmittance of the building envelope and windows basically in accordance with the related methods for envelope thermal insulation calculation provided in the Green Building Design Technical Standards by the Construction and Planning Agency, Ministry of the Interior (CPA, MOI, 2O12). The calculation on the thermal conduction of the building envelope and windows is conducted on the basis of the structure's materials and combinations used; the calculation for the reinforced concrete building structure is done by the conventional methods for general buildings, while the calculation for the AGS1 experimental house structure is conducted from the related data and other data provided by the developers and those listed by the Green Building Design Technical Standards.

### 14.4.3 Assessment of Heat Gain for the Combination of the Building Envelope and Window Openings

In respect of assessing the heat gain of the combination of the building envelope and window openings, the assessment uses two methods, one taking the influence of window changes on heat gain into consideration, and the other considering the said influence along with thermal radiation.

### 14.4.4 Influence of Thermal Transmittance and Thermal Radiation on Various Combinations of Building Envelope and Window Openings

In view of the fact that a window is prone to the influence of heat gain in the outdoor environment accounted for by the sun's thermal radiation, this study tries to establish the heat gain by comparing the increase values with the Ui of the window. The calculation of the building envelope's thermal transmittance (Ui) uses the data provided in the Green Building Design Technical Standards (National Association of

Architecture 2012), while the heat gain at the window opening mainly results from the heat provided by the sun's thermal radiation in addition to the heat from the window's thermal transmittance. The research by Chou Chang-Hsien et al. (2013) utilizes an estimated calculation of the OTTV. The heat at the window opening resulting from the sun's radiation, based on climate data for the northern region of Taiwan during the period from 1980 to 2010, is estimated, and the building in question in this study was assessed during the air conditioning season between May and October. The OTTV results obtained from the calculation show the ratio between the total value of thermal transmittance and the main heat transfer amount, and with these results in mind, we can figure out the comparative ratios respectively from the wall, the envelope (Qw), the window openings (Qg), and windows subject to the sun's radiation (Qs) between the reinforced concrete building structure and the composite building structure.

By comparing the Qw, Qg, and Qs values calculated on the basis of the OTTV, the average Qw:Qg:Qs ratio of the reinforced concrete building structure was 19:1:3, while in contrast that of the composite building structure was 16:1:3. On the basis of the research results it is found that the envelope material is the key influence on the thermal transfer amount where the Qs value is three times that of Q. In other words, the average heat gain at a window equals three times the heat gain from thermal transmittance. In this study, the Ui value, i.e., the window's thermal transmittance, increased four times, taking a comprehensive view of the influence of thermal transmittance and thermal radiation on the heat gain of a building.

Group 1 is with the actual Ui—the heat gain change found at a 1 h interval.
Group 2 is with the Ui increased fourfold—the heat gain change found at a 1 h interval.

Based on the data analysis, this study provides a satisfactory analysis of the important difference between the reinforced concrete building structure and the composite building structure of the AGS1 experimental house, and also the conditions of heat gain changes in relation to window openings with different treatments under the parameter of raised heat gain influenced by thermal radiation.

## 14.5 Calculation and Analysis

### *14.5.1 Calculation of Weight and Carbon Emissions in a Reinforced Concrete Structure and the AGS1 Experimental House Structure*

The AGS1 experimental house in question has a composite structure weight that is only one third of that of the conventional reinforced concrete structure (196.89 t/622.51 t) (Fig. 14.15) due to firstly exclusion of a raft foundation, which

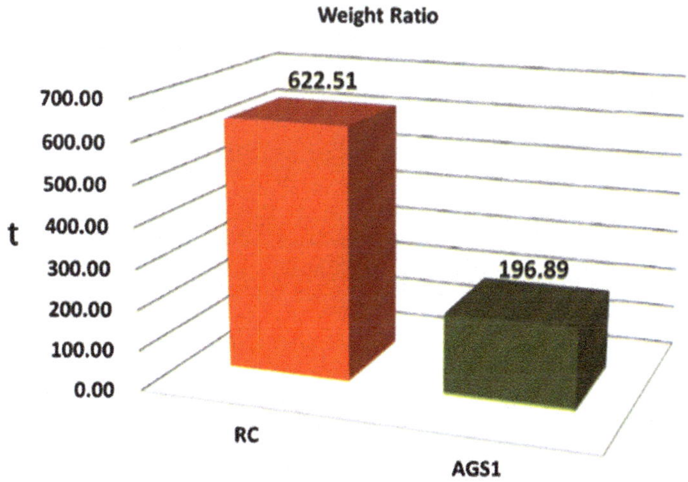


**Fig. 14.15** Weight comparison: AGS1 composite structure versus reinforced concrete structure

massively uses bars, cement and gravel; secondly the use of a structural steel system as the main framework; and thirdly use of 3D lightweight walls as well as light-weight aggregate concrete.

The AGS1 composite structure comes with structural steel, 3D lightweight walls, and lightweight aggregate concrete as its main components; excluding 40 years of daily use, the total carbon emissions of the reinforced concrete structure are 93.58 t, while those of the composite structure are 53.51 t; the latter represents a carbon emissions reduction of 43% in comparison with the reinforced concrete structure. When including 40 years of daily use, the total carbon emissions of the reinforced concrete structure are 374.31 t, while those of the composite structure are 259.37 t; the latter represents a carbon emissions reduction of 30.7% in comparison with the reinforced concrete structure (Fig. 14.16).

### *14.5.2 Calculation of the Building Envelope Thermal Transfer Rate in a Reinforced Concrete Structure and the AGS1 Experimental House Structure*

Figure 14.17 shows the thermal transfer rate (Ui) of each element in the components of the building envelope and windows in the reinforced concrete structure and those in the AGS1 experimental house's composite structure; the Ui readings for both the roof and walls of the AGS1 experimental house's composite structure proved lower than the Ui readings for those in the reinforced concrete structure; in terms of Ui,

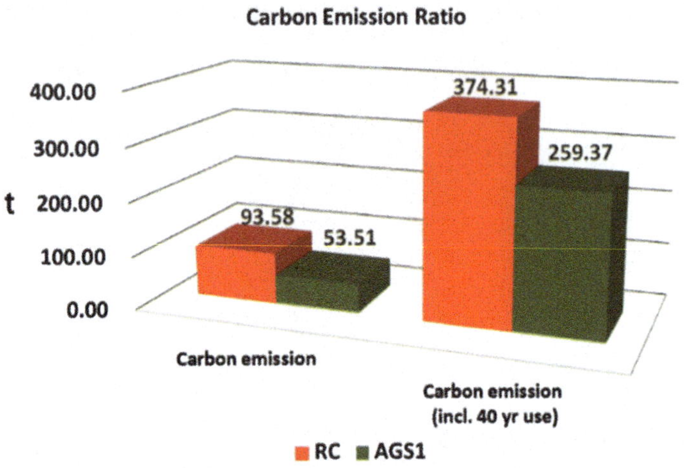


**Fig. 14.16** Comparison of carbon emissions: AGS1 composite structure versus reinforced concrete structure

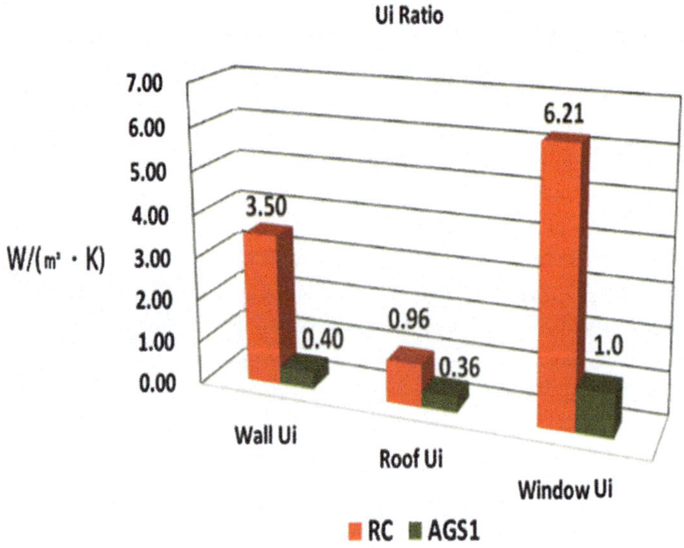


**Fig. 14.17** Ui ratio of each element between the AGS1 structure and the reinforced concrete structure

the readings for the walls in the reinforced concrete structure were 8.75 times higher than the readings for the walls in the AGS1 experimental house's composite structure. Figure 14.17 shows the data on the structures in question and the Ui readings of each component, where the ratio between the composite structure and the reinforced concrete structure is approximately 0.4:3.5 (11.4%) for the walls, while the ratio for the roof is approximately 0.36:0.96 (37.5%).

### *14.5.3 Assessment of Heat Gain with the Combination of the Building Envelope and Window Openings*

The AGS1 experimental house comes with a floor area of 230.8200 m$^2$, a pitched roof area of 140.74 m$^2$, a wall area of 302.52 m$^2$, a partition area of 118.45 m$^2$, and an opening area of 52.5400 m$^2$.

1. Influence of Thermal Conduction on the Heat Gain of Each Combination of Building Envelope and Window Opening

These are the heat gain findings on the specific window opening ratio in the walls of the AGS1 experimental house, in the comparison of the reinforced concrete structure with a thermal insulation roof and of the experimental house composite structure with the same. Compared to the ratio found in the average window opening in reinforced concrete walls, the heat gain in the AGS1 experimental house with envelope thermal insulation and window opening thermal insulation is only 14.8% (Fig. 14.18).

2. Influence of Thermal Conduction and Thermal Radiation on the Heat Gain of Each Combination of Building Envelope and Window Opening

In view of the fact that during the air conditioning season from May to October the average daily thermal radiation at the window location is taken into consideration only when the reading reaches three times the heat gain reading contributed by thermal conduction, this study directly expands the window thermal transfer rate

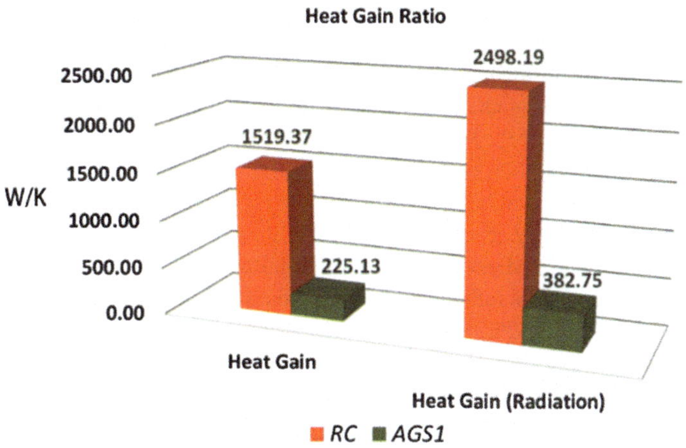


**Fig. 14.18** Heat gain in the AGS1 structure and the reinforced concrete structure

(Ui) to four times for taking an overall consideration over the fluency of thermal conduction and thermal radiation to the heat gain of the building structure. The result found in the reinforced concrete building structure with a thermally insulated roof reads 2498.19 W/K, while the resulting heat gain in the EPS panel in the AGS1 experimental house composite structure is 382.75 W/K—only 15.32% of that of the conventional roof and conventional window in the reinforced concrete building structure (Fig. 14.18).

## 14.6 Conclusion and Suggestions

### *14.6.1 Conclusion*

1. Through the literature review, this study first examines the regional climate characteristics of Taiwan, before discussing Taiwan's green buildings and the applications of green materials, elaborating the development model using a green environment, green structure, green materials, green life, and green decoration and furnishing under the "simplified" concept. Moreover, this study infers the proper models of green buildings suited to Taiwan's regional characteristics, including the correct combination of indoor temperature and humidity, reducing heat gain from the building envelope structure and windows, reducing unnecessary interior decoration, and selecting building envelope materials with high moisture resistance, a proper mechanism of air exchange in the intermediary space, and also a proper air conditioning model.
2. This study assesses and estimates the following by using the emissions coefficient method: the carbon footprint and overall carbon emissions at each stage of the AGS1 building in question, and the composite building structure compared with a reinforced concrete building, the results of which indicate reductions of 30.7% and 43% under conditions including or excluding daily use/occupancy for 40 years, respectively.
3. In estimating the influence of thermal transmittance and thermal radiation on heat gains of various combinations of building envelope and window openings by using OTTV on the basis of the heat gain results from the study on the AGS1 experimental house, it is found that when taking only thermal conduction into consideration, the AGS1 experimental house, with a composite structure plus PS panels, shows a Ui reading that is only about 14.8% of that of the reinforced concrete building with common windows; when taking both thermal conduction and thermal radiation into consideration, the influence is greater: the AGS1 experimental house shows a Ui of only about 15.32% of that of the reinforced concrete building.

### 14.6.2 Suggestions for Future Research

For the development of environmentally supportive green buildings, given Taiwan's various regional climate characteristics, it takes a more practical approach to tasks ranging from defining and establishing the scope of room comfort (temperature–humidity combinations for comfort under high humidity) to gaining indoor energy consumption data on rooms with windows closed under the various building conditions and various living styles of an average household over a whole year (the influence of humidity on air conditioning energy consumption). Then it takes effort to compile the related regulations and technical standards that are helpful to building development planning, building designs, and construction practices under the local laws, or to present better instructions for the design of typical residential buildings in Taiwan, so as to present residential building design guidelines that promote a green lifestyle, comfort, and health.

It is the practical information and knowledge acquired from the development model of simplified green buildings that demand more participation and implementation by staff from different agencies, and the research focus can be extended from the current residential building construction engineering to that of other functional buildings and their environments. Furthermore, it is advisable to promote future research on the performance tests and verifications of the AGS1 experimental house and the related short thesis papers, as they will facilitate positive development of the green building sector in Taiwan.

## References

Chih-Peng Liu (2015) Simplified green building. Elephant White Publisher Co., Ltd, Tai Chung City

Chiou Jih-Jer (2009) Good house. Yeren Publishing House, New Taipei City

Chou Chang-Hsien, Liu Chih-Peng, Yu Hsin (2013) The use of comparative assessment of the overall heat transfer combined in different building shell and windows. In: 21st conference on applications in civil and hydraulic engineering. National Ilan University, Ilan County

Hung P-y, Peng K h (2017) Green-energy, water-autonomous greenhouse system: an alternative-technology approach towards sustainable smart-green vertical greening in smart cities. Int Rev Spat Plann Sustain Dev 5(1):55–70

Joseph C, Lam JC, Tsang CL, Li DHW, Cheung SO (2003) Residential building envelope heat gain and cooling energy requirements. Building Energy Research Group, Department of Building and Construction, City University of Hong Kong, Tat Chee Avenue, Kowloon

Lin C-Y, Huang Y-L (2010) Planning review: application of ver landscape beautification in Taipei. Int Rev Spat Plan Sustain Dev 1(4):43–49. (2013)

Lin, Hsien-Te, Yang, Kuan-Hsiung (1997) Explanation and calculation for building energy conservation design of the building technology regulation in Taiwan. Ministry of the Interior, Construction and Planning Agency

Misni A, Baird G, Allan P (2013) The effect of landscaping on the thermal performance of housing. Int Rev Spat Plan Sustain Dev 1(1):29–48

National Association of Architecture (2012) The guide book for building energy-saving design and technical specifications. Mao-Long, Taipei

# Chapter 15
# Abu Dhabi's New Building Typologies: Beyond the Transit City

**Mohamed El Amrousi**

**Abstract** Abu Dhabi through its 2030 Vision is reshaping much of its urban identity, creating new forms of architecture that manifest sustainable design and development of new building typologies aimed at retaining its communities and improving the quality of life of its large expatriate community, which in the past has conceptualized Abu Dhabi as a transit city. This paradigm shift from its pragmatic urban fabric, planned in the 1970s, is illustrated by the new forms of architecture and urban spaces that are emerging in mega projects that recontextualize the concept of the brise-soleil and advocate environmental design principles. Masdar provides a prominent example of how to design better buildings and urban spaces to improve the quality of life. Many of its buildings are planned to generate energy and implement passive cooling effects, and act as life-scale icons of sustainability for community education, exhibiting diverse facade treatments and environmental design principles.

**Keywords** Abu Dhabi · Masdar · Sustainable architecture

## 15.1 Introduction

In the past decade, Abu Dhabi has begun to invest in mega projects to change much of its urban fabric from a city that was shaped by a grid street urban pattern and concrete block towers planned on modernism principles in the 1970s to a city that is reshaping its urban fabric with iconic projects aimed at developing a sustainable city, community, and spaces of belonging for many of its predominately expatriate community. Abu Dhabi's urban renewal projects run hand in hand with a rebranding of city image through forms of urbanism that include many iconic buildings such as Etihad Towers, Al-Dar Head Quarters, Ferrari World, Masdar City, the new Central Market, Al-Bahr Towers, and shorefront developments at Al-Raha Beach and on Al-Maryah, Al-Reem, and Al-Saadiyat Islands. Projects such as Abu Dhabi's Gate

---

M. El Amrousi (✉)
Department of Architecture and Design, Abu Dhabi University, Abu Dhabi, UAE
e-mail: Mohamed.amrousi@adu.ac.ae



Z. Shen et al. (eds.), *Green City Planning and Practices in Asian Cities*,
Strategies for Sustainability, https://doi.org/10.1007/978-3-319-70025-0_15

Tower as a symbol of global architecture that can be juxtaposed with more sustainable yet equally iconic projects such as the Abu Dhabi Louvre by Jean Nouvel, have acclaimed visibility manifesting the emergence of Abu Dhabi on the global map of modern cities (King 2004). In addition they manifest state interest in environmental sensibility and sustainability. Projects designed by star architects such as Norman Foster and Jean Nouvel represent—through their interpretations of tradition, site context, and the challenging environment—the broadness associated with the term *sustainable design*, which is no longer confined to regional expressions and adobe architecture of low profile. In contrast, these mega projects are indicative of the fact that public construction projects are now being used to manifest state interest in investing in sustainable design, which is likely to remain a key driver of the construction sector and the economy as a whole (Ponzini 2011).

In that context, Abu Dhabi's community is diverse, with many ethnic backgrounds formed by its expatriate and local community, representing a living example of an emerging global city. The term *sustainable architecture* is extremely broad, since it allows for multiple interpretations. It is also part of the concept of sustainable development, a term used in the Brundtland Report of 1987. From the Brundtland Report, sustainable development can be understood as a reaction to contemporary necessities of dwelling and urban development without sacrificing future development and resources. Urbanism is encouraged, with a reformation of the relationship toward the environment in order to avoid the social maladies of unplanned cities, which emerged as a result of the Industrial Revolution (Chust 2014). However, despite decades of effort, we are still in the midst of searching for a road map toward sustainability, reducing the environmental impact of rapid urbanism in challenging environments and new forms of cities that are postindustrial, within an increasingly connected and interrelated world (Li and Dang 2014). This chapter explores some of the new urban typologies that are emerging in Abu Dhabi. Within the framework of "Reinterpreting Sustainable Architecture—the Place of Technology," published by Guy and Farmer in 2001, six principles of sustainable architecture are outlined, as follows:

1. *Eco-technic logic—buildings and the global place*: This is based on the perception that science and technology can provide the solutions to environmental problems, through rational scientific methods that do not sacrifice the path of modernization.
2. *Eco-centric logic—building and the place of nature*: This principle is founded on a need for a radical reconfiguration of values and builds on the fact that the challenge of sustainable design is too complex to be dealt with as a technical problem.
3. *Eco-aesthetic logic—buildings and the new-age place*: Here the role of sustainable architecture is metaphorical and, as an iconic expression of societal values, acts to inspire and convey an increasing identification with nature.
4. *Eco-cultural logic—buildings and the authentic place*: Here, it is not the development of a new universal culture that is promoted but, rather, the preservation of a diversity of existing cultures. The issue here is how to define authenticity and the notions of locality and place.

5. *Eco-medical logic—buildings and the healthy place*: This shifts debates about sustainability from concerns about appropriate form and the wider cultural context of design toward social concerns for the sustaining of healthy environments.
6. *Eco-social logic—buildings and the community place*: This extends the social agenda of sustainability beyond a concern for the individual to encompass a political discourse and wider social factors such as common needs and goals (Guy and Farmer 2001).

While sustainable architecture is not necessarily about new technologies, but is about rethinking how to meet the need for growth, most of the emerging buildings in Abu Dhabi employ new technologies to reduce the challenging desert heat (Clark et al. 2009). The Abu Dhabi Urban Planning Council is taking serious measures directing Abu Dhabi toward sustainable architecture and development that enhances the quality of life, thus catering to creating a healthy city, with improved social, economic, and environmental conditions (Akadiri et al. 2012). This is supported by the Abu Dhabi 2030 Master Plan, which aims to regularize newly commissioned buildings along with the guidelines of Estidama. Communities residing in Abu Dhabi are beginning to replan their life from the perception of living in a transit city to becoming long-term residents. This change in social perception of the city from a temporary one to a permanent one offers the opportunity to create from Abu Dhabi an example of a sustainable city that caters to the needs of future investors. According to the Abu Dhabi 2030 Vision, this will be achieved through design of spaces where people want to live, work, and plan futures for their families, generating a catalyst for good quality of life (McDonald et al. 2009).

Masdar City exemplifies, on a small scale, a future for a sustainable healthy city through a master plan by Norman Foster and Partners, created to assume the role of the *eco-aesthetic logic*, which creates a new-age place. As outlined by Guy and Farmer, here the role of sustainable architecture is metaphorical and acts to inspire. Masdar's buildings act as an urban experience for sustainable design where technology and real-life buildings coevolve to attract significant amounts of attention and visitors as exhibitors of building typologies to be replicable in the living city (Guy and Farmer 2001). Since its initiation in 2006, it has become one of the important contributors to community education in relation to the environment and sustainability. It represents a real-live-scale urban model, exhibiting the latest technologies on its facade skins, which vary from energy-producing skins to perforated screens and louvers constructed with the latest technologies and materials. In this sense it reflects on Guy and Farmer's principle of *eco-technic logic* in which they highlight that buildings are designed to reflect the interconnectivity with the global world, which Abu Dhabi is becoming part of, and the perception that technology is the solution to challenging environments. Masdar also manifests a model of educating the community through an immersive experience, which links theory to practice as manifested by the facades of its buildings, many of which revisit Le Corbusier's designs through Norman Foster's designs. Foster's implementation of sun breakers, the use of pilotis to raise the building above the ground, and tightly woven clusters are successful as an icon for sustainability. The current part of Masdar that has been

**Fig. 15.1** Facades of Masdar City: stylistic diversity and materials

built is accessible to all and has investments in the area of renewable energy production, visible in its solar farm on-site, roofs covered by photovoltaic cells, and an offshore wind farm. These attributes contribute to communal awareness in the shift in Abu Dhabi's policy from the architecture of a transit city to one of sustainability and community development. As the capital of the United Arab Emirates, Abu Dhabi has set a 7% clean energy target, which will support investments in iconic environmentally based projects over the coming years, similar to Masdar. According to company officials, current investment in Masdar's renewable energy fields will benefit Abu Dhabi greatly in the medium and long terms. The Shams solar power plant is a good example in this direction (Oxford Business Group 2013: 119).

Masdar, despite its incompleteness, has provided a prominent example of how to design better buildings and urban spaces to improve the quality of life through sustainable design. Environmentally based architecture in Abu Dhabi depends on an investigation of its surrounding physical context, as it is difficult for architects and planners to balance, especially in oil-rich states, between the much-needed iconic architecture and sustainable design (Frobeen 2006). The range of variation in patterns displays a diversity of features and juxtaposition of contemporary architecture (Fig. 15.1). Masdar's designs attempt to reflect diversity, and validate modern technology and materials as essential contributors to sustainable design in cultural and environmental perceptions (Taylor 2009), In this respect, Masdar itself can be seen as a living laboratory—an experimental project that develops gradually and inspires the use of future technologies (Cugurullo 2013). Such paradigms cater to the multicultural community of Abu Dhabi in terms of the diversity of public and private spaces and facilities. Urban diversity was missing from the modernist downtown of Abu Dhabi, designed in the 1970s, and is being developed currently through shorefront developments at Al-Raha Beach and on Al-Maryah, Al-Reem, and Al-Saadiyat Islands, and a large urban renewal project of the downtown Corniche. The resulting enhancements in design, in turn, offer new opportunities for applied research and stimulate new investments as the urban environment once again becomes full of life and enterprise (Rapoport 2014). This is further cemented by emerging new spaces, such as the Yas Marina and other shorefront developments, exhibiting methods of sustainable architectural practices that shape spaces of gathering and express a collective consciousness of the presence of a multiethnic/expatriate community (Chiotinis 2006). In addition to its cultural developments and mega cultural projects in Al-Saadiyat, Abu Dhabi's interest in sustainability is visible via new projects that are emerging within the city. Masdar's effect as an example of a sustainable city has presented interesting solutions by using renewable sources of energy and a sustainable design approach used with frequency in order to justify projects with good levels of adaptation to the climate and real energy efficiency (Pantoja et al. 2012).

## 15.2 Revisiting the Mashrabiya Screen

In Abu Dhabi, many projects seek to incorporate strategies similar to those stated in its 2030 Master Plan and outlined by Estidama for achieving genuine architectural experiences designed according to environmental criteria. Projects that take into consideration Abu Dhabi's climatic and cultural environment play a major role in reshaping the living city, such as Al-Bahr Towers, with their mechanical shading devices, and the new Abu Dhabi Judicial Department on al Khaleej al Arabi Street (Fig. 15.2), a complex building that, despite its civic status, visually communicates the importance of environmental designs in public and private buildings. Abu Dhabi, through its 2030 Vision, represents an actual implementation—which has been felt by its residents over the past decade—of an example of urban development that promotes the identity of a modern Arab city, with sociocultural and environmental designs that have attributes of contemporaneity with an awareness of the need for a multiethnic identity (Mengusoglu and Boyacioglu 2013).

As outlined by the architects Dar al-Omran, the design of the new Abu Dhabi Judicial Department building represents an example of transformation of the relationship between classical forms of building typologies to a relationship based on physical perception related to the environment (Fig. 15.2). Here, an emphasis on the environmental efficiency of the development has stimulated a whole range of technological innovations in building fabric and servicing systems, such as solar

**Fig. 15.2** The new Abu Dhabi Judicial Department

shading, intelligent facades, double-skin walls, and shaded roofs. The design of overarching eaves and orientation of buildings reflects a sustainable design approach to achieve a passive cooling effect through their urban formations in addition to architectural elements that adhere to environmental design, such the use of "corridor facades," double glazing, and brise-soleils, all of which better familiarize architecture with its surrounding context. As a judicial building/courthouse it shifts debates about sustainability beyond the conventional form to the wider cultural context of design toward social concerns for healthy environments. The mixed-mode ventilation system created by the facades, which are unorthodox for a civic building, highlights the emergence of new building typologies in Abu Dhabi and the interest in sustainable design, thereby replacing the authoritative building typology associated with such buildings. This building can be understood as an example of Guy and Farmer's *eco-aesthetic logic*, where the building represents the possibility of designing civic buildings as healthy places (Guy and Farmer 2001).

Such examples revisit the juxtaposition between Regional and Modern, much debated in the designs of Hassan Fathy and Le Corbusier. Hassan Fathy tried to establish the concept of dwelling with an emphasis on revisiting a sense of place; in other words, he attempted to apply what Guy outlined as *eco-cultural logic*. Fathy attempted, through his designs of adobe architecture, to create dwellings that evolved from within the community in Egypt. Yet many of his buildings, such the New Gourna Village, failed to preserve the authenticity of place even though the right materials and building forms were used. Fathy's students, such as Abdel Wahed El Wakil and Rami El-Dahan, understood the importance of preservation of the

**Fig. 15.3** Contemporary screens, Masdar City

diversity of existing cultures, hence they applied the design elements of Hassan Fathy in different contexts, creating a unique sense of identity that reflects concern for the continuity of meaning between tradition and modernity (in terms of materials) with an emphasis on cultivation of cultural consciousness.

The diverse patterns of Masdar's screens highlight the idea of creating an Arab identity that reflects inclusivity of Abu Dhabi's multicultural community, which requires moving beyond formal links to tradition, creating designs of different materials and geometric vocabularies that transgress the constraints of regionalism (Fig. 15.3). Many of Masdar's buildings are raised above the ground to create air channels and cross-ventilation through formations of courtyards and stack ventilation (Pedrao 1996). Here, environmental design elements are conceived as separate modules and can be combined, regulated, and even articulated. Technically, the new facades exhibited in Masdar City can have dual functions, such as the ability to generate clean energy, regulate the amount of daylight, and avoid glare. Masdar's buildings represent a strategic and innovative concept project that reflects an increasingly connected and interrelated world, especially in rapidly developing economies, such as Abu Dhabi, which seeks to advocate itself as a modern Arab nation state with conciseness toward sustainability (Li and Dang 2014). The reinterpretation of traditional spaces such as shaded courtyards, narrow pedestrian streets, and buildings that are cantilevered or raised on columns offer a fundamentally different spatial context than the modernist designs of the downtown. Here is an attempt to exemplify the emergence of a new spatial–social ambience through the introduction of a new architectural vocabulary, which is elaborate and tectonically advanced.

## 15.3 Sustainable Architecture and Modernist Principles

Le Corbusier promoted forms, such as the brise-soleil and extended louvers, that aimed to regulate the environment surrounding the building envelope. Today new materials offer a revisiting of modernist design principles and can reduce the criticism of the architecture of the International Style as an architecture that failed to create distinctive regional identities. Norman Foster's design of Masdar and the new Abu Dhabi Central Market manifest an interest in exploring sustainable designs with diversity in patterns and materials. Abu Dhabi's new Central Market, designed by Foster + Partners, represents a project that connects heritage and modernity in an attempt to transform a vital part of the city center into a reinterpretation of an Arab souk, with cafes, a bazaar, and modern shops (Fig. 15.4). As one of the oldest sites in the city, the new project was intended to reinterpret the traditional marketplace and social space for Abu Dhabi. The project, opened in 2011, comprises a combination of high-end boutiques, cafes, and roof gardens, forming a new public space for

**Fig. 15.4** Abu Dhabi Central Market: shading devices and roof garden

**Fig. 15.5** Masdar: brise-soleil and pilotis creating shade and visual continuity

Abu Dhabi. In the Central Market, passive design takes into consideration building orientation and layout to reduce the absorption of heat from the sun, enhance inflow of daylight, and provide sun shading devices and wall greenery. The choice of materials was based on thermal resistance and their environmental and energy costs over the life cycle of the building (Aliagha et al. 2015). The Abu Dhabi Central Market screens included designs organized according to the rules of regular geometry and the relatively ambiguous term *arabesque*. Here the use of the brise-soleil produces an immediate visual impact. It follows the norms of the new architecture while at the same time forming a new vocabulary in the city landscape. Their patterns were organized according to rectilinear grids, in which derivatives play a prominent role (Al-Asad 1994). The covered Central Market explores the development of a variety of forms and acts as an interface between exterior and interior; instead of being an element of separation, the brise-soleil works as an element of transition.

Le Corbusier's designs offered a wide vocabulary of precedents for the brise-soleil; they all shared one aim, which was to regulate sunlight while at the same time allowing soft ventilation into the building. Today they are revisited by Norman Foster in Masdar (Fig. 15.5). Today the strong emphasis on environmentally based design is criticized for its limitations in terms of public acceptance because of inadequate interpretation of local lifestyles and standards of living, thereby ending up being impractical or inaccessible to the vast majority of the society (Rapoport 2014). In the case of the Central Market, the extensive use of the brise-soleil represents the inevitable necessity to adhere to global changes in the realm of architecture to produce new forms. In the Central Market, attempts to design for the environment cannot fully resolve the changes in the modern city resulting from globalization; however, they are attempts to adopt modern versions of traditional forms and patterns as a structure of ornament that can act as a vocabulary for sustainable in rapidly urbanizing Arab cities (Elsheshtawy, 2008). The interplay of light, shade, and shadow seen on the interior walls of the Central Market are part of the pattern formation of the building skin of the building. The new Central Market, on the other hand, locates several of its cafes either under the shade of its retreating colonnaded facade, in its multiple-height atrium, or on its landscaped roof. Here, community development toward sustainable design is done in an indirect manner, while the emphasis is on contemporary architectural translations of cultural meaning and sense into what, in many other places, has developed into a secular institution of the spectacle—the mall. However, its contemporary screens used extensively internally and externally have also been critiqued as interpretations of the past that have lost the proportions of ornaments developed over time, thereby resulting in buildings whose facades appear either too big or too small for their buildings (Schumacher 2010). Examples of *eco-medical logic* on an urban level are the new Abu Dhabi Central Park, which is still under construction and will have one million square feet of open space and associated recreational amenities, in addition to four kilometers of water canals to create new waterfront recreational facilities and improve the livability of the city and the public health of its community. It is also an example of transitional development of the city from a transit city in an oil-rich state toward a modern metropolis with more green spaces for socialization. Abu Dhabi's Corniche waterfront development plans, its Eastern Mangrove developments, and its new Central Park Project represent the interest in investing in various venues of sustainability—an attempt to experiment with sustainable living mediums to create a healthy city model that will reflect on its community, fostering responsibility toward the environment. Construction of such projects attempts to merge modernity and tradition via urban renewal and investment in green infrastructure in a desert environment. The change of environmental quality in the city center of Abu Dhabi will inevitably draw value and improve the urban environment, which consequently will affect the environmental quality inside and beyond the renewal area in an attempt to resolve urban stagnation and achieve economic, environmental, and social sustainability of the existing downtown area (Huang et al. 2016). Given the complexity and scale of these projects, here lived space is reconstructed as a means of establishing and consolidating place identity in a postmodern, multiethnic, and multicultural society. The combination of the spatial reproduction and environmentally oriented urbanism create a new image of conventional space in the city as going beyond the context of the transit city to emphasize the importance of sociocultural cohesion and sustainable communities.

## 15.4 Al-Bahr Towers, Building Envelopes, Identity, and Mechanization

Located on a highly visible site on Al-Salam Street, one of the main freeways leading to downtown Abu Dhabi, the Al-Bahr Towers manifest Guy and Farmer's principle of *eco-technic logic—buildings and the global place*. The building, through its implementation of mechanical shading devices on a massive vertical scale, represents an example of how science and technology can provide solutions to environmental problems (Guy and Farmer 2001). This is further manifested through use of iconic double-skin facades that respond dynamically to varying ambient conditions in Al-Bahr Towers; here use of screens on a massive scale as a form of responsive design to the daylight effect is manifested. The two cylindrical towers represent a breakthrough in sustainable design in Abu Dhabi in that they were designed to exemplify the possibility of constructing a neo-Islamic identity in the form of mechanized triangulated forms that re-enact the designs of Jean Nouvel in the Institut du Monde Arabe in Paris, designed in 1987. While the design of the Institute du Monde Arabe was challenged by mechanically failing to automatically respond to the environment, it still represented an innovation beyond the traditional *mashrabiya* screens in terms of materials and scale of application. In Al-Bahr Towers the same challenges are faced in that not all elements respond to sunrays due to the scale of the mechanized hexeract patterns constituting a massive vertical building skin, which covers two facades of the buildings (Fig. 15.6). Each triangle is coated with fiberglass and is programmed to respond to the movement of the sun as a way to reduce

**Fig. 15.6** Al-Bahr Towers, Abu Dhabi

solar gain and glare. Here the building skin is designed as a responsive elastic architectural skin assembled by a series of elastic tetrahedral modules. It contracts and expands without mechanical components such as motors or pistons. It is a kinetic, tent-like skin, which changes shape to meet various needs and environmental conditions, supported by a complicated mechanical system as well as manual control mechanisms (Khoo et al. 2011). The purpose of making such a complex system combines good attributes of the two principles (mechanical and natural) to ensure both good air quality and thermal comfort (Dahl 2010). Similar to the Institut du Monde Arabe, a new identity for high-rise towers in Abu Dhabi has been created—a design that reinforces state interest in advocating sustainable designs that are not frozen in time and constrained by attempts to authentically reproduce traditional forms. The triangulated patterns of the Al-Bahr Towers reflect the possibilities of application of the environmental designs stipulated by Estidama, in a living city and within contemporary time and space that manifests the dynamic nature of architectural design. The vertical application of the responsive hexeract elements that open and close are made visible by the location of the towers on Al-Salam Street, one of the main arteries from Abu Dhabi Island to its hinterland. The mere discussion as a result of the visibility of the towers triggers communal development and promotes contemporaneity of traditional ornaments that can be integrated onto modern buildings as a fragment, a whole, or a collage (Schumacher 2010).

## 15.5 Conclusion

The urban policies of the Emirate of Abu Dhabi are witnessing major urban developments that aim to rebrand the main Emirate of the oil-rich states as one that sustains cultural values and green infrastructures. In Abu Dhabi we are witnessing the ascendancy of a new genealogy of buildings that will change the image of the city—iconic structures built with sustainability as design criteria that aim to educate the community about the importance of environmental design. Guy and Farmer highlighted six principles through which they revisited sustainable architecture, some of which are more applicable than others in Abu Dhabi's emerging sustainable architecture. Masder has many applications of the principles of sustainable architecture revisited. Though still incomplete, it has set a prominent example of how sustainable design can be applied; its influence is being felt in the living city, especially in new projects emerging on the shorefronts of Abu Dhabi, such as Al-Maryah Island and the Al-Raha beach residence. In Abu Dhabi's new urban developments, Guy and Farmer's *eco-technic logic* and *eco-aesthetic logic* might be the most dominant because of its aspiration to become part to the global network of modern cities capable of attracting investors, similar to Singapore, Hong Kong, and Kuala Lumpur (King 2004).

## References


Akadiri PO, Chinyio EA, Olomolaiye PO (2012) Design of a sustainable building: a conceptual framework for implementing sustainability in the building sector. Buildings 2012(2):126–152

Al-Asad M (1994) Applications of geometry. In: Frishman, Khan (eds) The Mosque, history, architectural development and regional diversity. Thames and Hudson, London, pp 55–71

Aliagha G, Goh A, Abdullah M, Jaafar N, Eluwa S (2015) Investigating skill gaps in green building skills for energy efficiency. Adv Mater Res 1073-1076:1282–1287

Chiotinis N (2006) The request of sustainability and architecture as cultural paradigm. Manag Environ Qual Int J 17(5):593–598

Chust PP (2014) Towards a sustainable architecture. Arte y Sostenibilidad 10–11:227–250

Clark G, Kosoris J, Nguyen Hong L, Crul M (2009) Design for sustainability: current trends in sustainable product design and development. Sustainability 1:409–424

Cugurullo F (2013) How to build a sandcastle: an analysis of the genesis and development of Masdar City. J Urban Technol 20(1):23–37. Routledge

Dahl, T. (2010). Climate and architecture by Torben Dahl, Abingdon/New York, Routledge, 2010

Elsheshtawy Y (2008) Cities of sand and fog: Abu Dhabi's global ambitions. In: Elsheshtawy Y (ed) The evolving Arab city: tradition, modernity and urban development. Routledge, New York

Frobeen A (2006) Socially sustainable. Can Archit 51(4):62–64

Guy S, Farmer G (2001) Reinterpreting sustainable architecture: the place of technology. J Archit Educ 54(3):140–148

Huang KH, Pai JT, Liu JH (2016) Study of performance assessment for urban renewal project in Taipei city. Int Rev Spat Plan Sustain Dev 4(1):64–77

Khoo C, Salim F, Pallett J (2011) Designing architectural morphing skins with elastic modular systems. Int J Archit Comput 9(4):397–419

King DA (2004) Spaces of global cultures: architecture, urbanism, identity. Routledge, New York

Li Y, Dang A (2014) Sustainability indicators and analysis, introduction. Int Rev Spat Plan Sustain Dev 2(3):1–3

McDonald S, Malys N, Malienė V (2009) Urban regeneration for sustainable communities: a case study. Balt J Sustain 15(1):49–59

Mengusoglu N, Boyacioglu E (2013) Reuse of industrial built heritage for residential purposes in Manchester. METU J Fac Archit 30(1):117–138

Oxford Business Group (2013) Abu Dhabi 2013. Oxford Business Group, London

Pantoja M, Moreira P, Lomardo L (2012) Towards the performative architecture through the tectonic vision. Int Sci J Archit Eng 2012

Pedrao A (1996) The use of the brise-soleil in Brazilian modern architecture. Docomomo Journal 1996 conference proceedings—Fourth international Docomomo conference, 183

Ponzini D (2011) Large scale development projects and star architecture in the absence of democratic politics: the case of Abu Dhabi, UAE. Cities 28:251–259

Rapoport E (2014) Utopian visions and real estate dreams: the eco-city past, present and future. Geogr Compass 8(2):137–149. Routledge

Schumacher L (2010) "Facadism" returns, or the advent of the "duck-orated shed". J Archit Educ 63(2):128–137

Taylor J (2009) Shiny, slick and sustainable: Abu Dhabi's Masdar City is an experiment in sustainability built on oil money. But will it work. Altern J 35(5):18

# Chapter 16
# Green Energy Water-Autonomous Greenhouse System: An Alternative Technology Approach Toward Sustainable Smart–Green Vertical Greening in a Smart City

**Paiyao Hung and KuangHui Peng**

**Abstract** By means of "going greener," "getting smarter," and "converging smart–green," an innovation-driven smart city could take steps toward greater sustainability and aim at greater human well-being. Vertical greening means a vertical triumph of greenery in a high-density urban area; it is well suited to displaying the level of smartness and greenness in a city. But conventional vertical greening is used in an open-field way, unprotected, threatened by climate disasters such as high wind speed and heavy rainfall, and with lack of control of climate conditions and plant-response-based circumstances. Then there are the challenges of energy saving, reduced $CO_2$ emissions, and reductions in water use and in pesticide use. A greenhouse system could instead solve different facets of the problems of conventional vertical greening because a greenhouse system could be developed to achieve an optimal balance between efficient environmental control and efficient plant use of available resources. This appears to be more intellectually justifiable, adaptable, and innovative, and appears to make it much easier to be smart–green and sustainable in a smart city. The purposes of this chapter are to summarize the major concepts and trends in smart city, vertical greening, and new greenhouse technologies and approaches by reviewing relevant subjects of research, and to present a novel prototype, discussing its innovations and advantages to reveal that the proposed green energy water-autonomous greenhouse system (GEWA system), being a sophisticated and multidisciplinary system by using water resources and solar energy in a rational way, could be fit for an alternative technology approach toward sustainable smart–green vertical greening in a smart city. Aimed at improving responsiveness,

P. Hung
Doctoral Program in Design, College of Design, National Taipei University of Technology, Taipei, Taiwan

K. Peng (✉)
Institute of Design, Taipei University of Technology, Taipei City, Taipei, Taiwan
e-mail: khpeng@ntut.edu.tw



Z. Shen et al. (eds.), *Green City Planning and Practices in Asian Cities*, Strategies for Sustainability, https://doi.org/10.1007/978-3-319-70025-0_16

efficiency, and performance for environmental sustainability, resource sustainability, and material and technological sustainability, and also aimed at greater well-being.

**Keywords** Green energy water-autonomous greenhouse system · Sustainability · Smart-green · Vertical greening · Smart city

## 16.1 Introduction

### *16.1.1 Background*

The twenty-first century has been described as the "urban century." Cities around the world are faced with complex social and ecological challenges caused by population growth, urbanization, and climate change. To envisage these challenges, cities are increasingly concerned with providing innovation-driven, more environmentally responsive, resource-efficient, and material- and technology-performance-based answers to strengthen the future viability of cities and to improve the quality of life for citizens in future urban communities. Accordingly, the use of the terms "green city" and "smart city" has sharply increased in recent years.

Both "green" and "smart" are umbrella terms. According to Attmann, "green" involves a combination of values—environmental, social, political, and technological—and thus seeks to reduce negative environmental impacts by increasing efficiency and moderation in the utilization of materials, energy, and development space, so green is an abstract concept that requires inclusion of the terms "sustainability," "ecology," and "performance" (Attmann 2010). According to Hatzelhoffer, considering the term from an Anglo-American perspective, "smart" can have a whole array of meanings. It can be used in the sense of brisk, elegant, competent, or fashionable, as well as meaning clever or intelligent (Hatzelhoffer et al. 2012). An international scientific team led by Caragliu (2009) drew up six criteria and formulated the following definition of a smart city: "We believe a city to be smart when investments in human and social capital and traditional (transport) and modern (ICT) communication infrastructure fuel sustainable economic growth and a high quality of life, with a wise management of natural resources through participatory governance" (Hatzelhoffer et al. 2012). According to Posada, "sustainability" is broader in its reach than "green," addressing the long-term impacts of the built environment on future generations and demanding an examination of the relationship between ecology, economics, and social well-being (Kwok and Grondzik 2011).

To sum up the above, then, under the overall umbrella concepts, both "green" and "smart" could be converged into "smart–green" on the basis of sustainable materials and technologies, resources, and the environment. By means of "going greener," "getting smarter," and "converging smart–green," an innovation-driven smart city could take steps toward greater sustainability and aim at greater human well-being.

Vertical greening, according to KuangHui Peng et al., can increase greening amounts, reduce the urban heat island effect, improve the quality of outdoor and indoor air, beautify urban landscapes, lower indoor temperatures, increase energy efficiency, protect building structures, and reduce noise. Expanding the use of vertical greening is a good way to rehabilitate high-rise congregated house building facades and sustain the green wall system to improve the sustainable environment (Peng et al. 2015). Moreover, beyond the environmental function of providing horticultural cultivars for the urban landscape, vertical greening can also be a type of agricultural cropping for food production in the city. Actually, vertical greening means a vertical triumph of greenery in a high-density urban area and its surface environment; it is well suited to displaying the level of smartness and greenness in a city.

However, conventional vertical greening is used in an open-field way, unprotected, threatened by climate disasters such as high wind speed and heavy rainfall, and with lack of control of climate conditions and plant-response-based circumstances. Then there are the challenges of energy saving, reduced $CO_2$ emissions, and reductions in water use and in pesticide use. Water scarcity and the large demand for primary energy are still serious handicaps for the sustainability of the actual production system.

On one hand, progressive technological solutions developed from traditional to modern and from conventional to unconventional will help vertical greening itself to be more smart–green and sustainable. But on the other hand, a greenhouse system could instead solve different facets of the problems of conventional vertical greening because a greenhouse system that minimizes the use of water and energy could be developed by an increasingly sophisticated and multidisciplinary approach to achieve an optimal balance between efficient environmental control and efficient plant use of available resources. This appears to be much more intellectually justifiable, adaptable, and innovative, and makes it much easier to be smart–green and sustainable in a smart city.

### *16.1.2 Motivation and Objectives*

The purposes of this chapter are as follows:

1. To summarize the major concepts and trends of the smart city, vertical greening, and new greenhouse technologies and approaches by reviewing relevant subjects of research that focus on factors of water saving, green energy management, natural ventilation, and system integration, which make it possible for the progressive greenhouse, instead of vertical greening, to become more sustainable and smart–green.

2. To present a novel prototype of a green energy water-autonomous greenhouse system (GEWA system), which is mainly operated by using just solar energy, with a considerable reduction of water use due to closed water recycling inside the greenhouse with recovery by evapotranspiration–condensation, and by integrating networking solutions, converging smart–green technology.
3. By discussing the innovations and advantages, to reveal that the proposed GEWA system, which was developed by an increasingly sophisticated and multidisciplinary approach to become more responsive, efficient, and functional, is fit for an alternative technology approach toward sustainable smart–green vertical greening in a smart city.

## 16.2 Literature Review

### *16.2.1 New-Generation Greenhouses*

For overcoming drought, the Cycler Support project (Buchholz et al. 2008) aimed to investigate the potential for growing food on a base of unconventional water sources and to describe a long-term scenario including a new generation of water-efficient greenhouses. A group of new greenhouse technologies allows collection of condensed water from air water vapor within greenhouses. By using this kind of much less conventional water together with harvested rainwater, it is possible to reach a water-autonomous situation of irrigation water supply in many regions of the world. A major point is that the natural water cycle can be circumvented and water efficiency can be drastically improved by the new greenhouse technologies. Even a greenhouse could provide condensed water regained after being evaporated by plants with recycling rates of up to 80% and water consumption reduced by 95% compared with open-field intensive production. In the Cycler Support project, five new-generation greenhouse model research areas are proposed, including: (1) a closed greenhouse for food; (2) a closed greenhouse for nonfood crops, including greenhouse integrated solid-state fermentation; (3) an open greenhouse with natural convection, built on mountain slopes, using saline water from the sea for evaporative cooling, with integrated aquafarming for fish and algae production using wastewater and solid waste from the fish farming; (4) a model urban area for wastewater preselection in urban areas with use of gray water in greenhouse projects; and (5) a concentrated solar power project with cooling water recycling in closed greenhouses (Buchholz et al. 2008).

The Naples event of the GreenSys2007 Symposium (Giacomelli 2007) demonstrated that there is room for improvement in greenhouse cropping. Regarding innovation in greenhouse engineering, Giacomelli pointed out in his address that (1) greenhouse components and design directly impact crop growth; (2) correct assessment of the importance of crop–greenhouse interactions is needed; (3) real-time measurement of the maximum number of parameters is necessary; (4) other than the

Watergy project, engineers have to find ways and means to substantially reduce energy and water use and to ensure that an acceptable return on investment can be achieved when high enough yields are produced (Giacomelli 2007). In the opinion of Prof. Stefania De Pascale, convenor of GreenSys2007, a greenhouse system is in some respects already an energy-saving system compared with open-field agriculture and is an excellent environment to achieve an optimal balance between efficient environmental control and efficient plant use of the available resources. In Naples, the main focus was on a famous prototype of a closed greenhouse named Watergy (Giacomelli 2007).

### *16.2.2 The Watergy Project*

The research in the Watergy project is funded by the European Union's 5th Framework Program, promoting energy, the environment, and sustainable development (Zaragoza et al. 2007).

The Watergy project proposes two prototypes for application of a novel humid air solar collector. The first is constructed in Almeria (Spain), and it is a closed greenhouse for solar thermal energy capture, water recycling, water desalination, and advanced horticultural use. The second is constructed in Berlin (Germany), and it is a greenhouse with an autonomous supply of heat and also of clear water; it is connected to the building and purifies its residual gray water. In the context of sustainable architecture, the Watergy system means that this concept of zero energy is complemented with that of water autarchy. The autonomous and local way to treat urban residual water means that, on one hand, decentralization of the water supply can be contemplated with self-sufficient systems able to close their water cycle locally; on the other hand, intensive agriculture can be freed from its enormous water consumption. The concept of solar collectors is that, on one hand, humid air allows storage of more thermal energy at a given temperature and the same amount of energy can be transported by much lower air volume flow sustained by natural buoyancy; on the other hand, the evaporation and condensation processes increase the efficiency of the heat transfer (Zaragoza et al. 2007).

The Watergy project was widely discussed as part of many subjects of research, including the following:

1. Thermal control for optimized food production and gray water recycling by a new solar humid air collector system (Buchholz et al. 2004)
2. Functioning of the system for solar thermal energy collection, water treatment, and advanced horticulture (Zaragoza et al. 2005)
3. Use of the simulation environment Smile to simulate thermal and fluid dynamic processes including water interactions between plants and air (Jochum and Buchholz 2005)
4. Description of a passive cooling and dehumidification strategy (Buchholz et al. 2006)

5. Critical discussion following the results of EI Ejido in Almeria, Spain (Zaragoza and Buchholz 2008)
6. Exploration of a suitable method to provide the required automatic adaptation to an adaptive model for greenhouse control (Speetjens et al. 2009)
7. Development of a physics-based model, based on enthalpy and mass balance, to predict system behavior (Speetjens et al. 2010)

### 16.2.3 Water Saving

Water-saving concepts were also discussed as part of many subjects of research, including the following:

1. Novel high-technology solutions in greenhouse production can lead the way to highly efficient water use production techniques and can alleviate the water shortage problem (Van Kooten et al. 2008).
2. The technical aspects and results of a trial using a fully closed greenhouse showed advantages in reductions of energy, water, and chemical crop protection (Opdam et al. 2005).
3. Compared with irrigated crops outdoors, the seasonal evapotranspiration (ET) of greenhouse horticultural crops is relatively low in comparison with irrigated crops outdoors, due to the lower evaporative demand inside the greenhouse (Orgaz et al. 2005).

### 16.2.4 Energy Management

Energy management concepts were also discussed as part of many subjects of research, including the following:

1. The final energy efficiency is determined by improvements in energy conversion, reductions in energy use for environmental control, and the efficiency of crop production; the developments range from new modified covering materials to innovative and energy-conservative climate control equipment, a plant response–based control system, and integrated energy-efficient greenhouse design (Bakker et al. 2008).
2. The greenhouse system is treated as a solar collector with an absorber plate (the greenhouse soil) and a cover system consisting of three semitransparent parallel layers (the greenhouse cover, the humid air, and the plants); there are some general relations for estimating the amounts of solar energy absorbed by the greenhouse components and lost to outside the greenhouse (Abdel-Ghany and Al-Helal 2011).
3. An energy management concept is to maximize the utilization of solar energy through seasonal storage by removing excess sensible and latent heat due to the

lack of ventilation in a fully closed greenhouse; although a greater amount of solar energy can be harvested in a fully closed greenhouse, in reality a semi-closed greenhouse concept may be more applicable (Vadiee and Martin 2012).
4. Many thermal energy storage systems such as TES, UTES, SCW, PCM, and BETS could be considered as seasonal storage or short-term storage; a theoretical model could be developed to carry out the energy analysis (Vadiee and Martin 2013).
5. Fluorescent solar concentrators, photoselective materials, and other materials could be considered as solar radiation manipulations in greenhouse claddings to provide advantages for plants (Lamnatou and Chemisana 2013).

### *16.2.5 Natural Ventilation*

Natural ventilation concepts were also discussed as part of many subjects of research, including the following:

1. Relative to a naturally ventilated greenhouse, the wind direction significantly affected the flow patterns both inside the greenhouse and at the roof openings, and also affected the ventilation rate and the air and crop temperature distributions (Teitel et al. 2008).
2. Excessively high internal temperatures have negative effects on the yield and quality of almost all greenhouse crops because that ventilation is generally insufficient. The reduced $CO_2$ levels and the creation of high humidity adversely affects the inside air composition, and condensation on the cover also reduces the transmission of solar radiation (Perez Parra et al. 2004).
3. To maximize ventilation when wind speeds are low, buoyancy-driven combined roof and sidewall ventilation should be used (Baeza et al. 2009).
4. There is a unique relationship between water use efficiency and the coupling of the greenhouse environment to the outside air; increasing the capacity of the cooling system could reduce the ventilation needs of semiclosed greenhouses and increase water use efficiency (Katsoulas et al. 2015).

### *16.2.6 New Technological Approaches*

The new approaches of greenhouse technology were also discussed as part of many subjects of research, including the following:

1. All three geographic areas share the need to have optimized climate control based on the crop response to the greenhouse environment. For more efficient greenhouses, progress in Northern Asia is being made in using a solar collector and developing new heating strategies. Important subjects being addressed in the Netherlands are energy conservation and increasing mechanization. In the

Mediterranean there is growing interest in semiclosed greenhouses with $CO_2$ enrichment and control of excessive humidity (Montero et al. 2011).
2. To achieve a sustainable greenhouse that is energy neutral, consumes only the essential amount of water, and has minimal negative environmental impact, recent years have witnessed the development of photovoltaic cells for power generation, insect-proof screens, and the use of computational fluid dynamics (CFD) simulation tools to investigate the effects of the structure shape, ventilator size, and arrangement on the microclimate (Teitel et al. 2012).
3. The low-energy concept could be refined by combining energy-saving methods with improved control of the greenhouse microclimate and also by improving the cropping system and using new cultivars. The closed greenhouse has been developed and propagated as an energy-producing greenhouse; the greenhouse should be operated semiclosed to improve the use of solar energy for heating (Tantau et al. 2011).
4. The passive greenhouse uses only renewable energy sources such as geothermal energy, wind, and sun, by means of cool water heat pumps, wind turbines, and photovoltaic panels. Thereby they are fully free of any energetic infrastructure and can be installed in remote areas, so they offer a fundamental sustainable agricultural resource and a global ecological reconstruction opportunity (Balas 2014).

### *16.2.7 Vertical Greening Technology*

According to Peng, Kuo, et al., vertical greening of building-related items with an impact factor were pointed out as follows: for the item of plants, the factors are category, growth nature, eco-nature, planting method, maintenance method, growth period, and sense; for the item of vertical greening technology, the factors are climbing, hanging, and module; for the item of the building external environment, the factors are building type, greening position, impact factor, additional substance of the facade, community environment, material of the facade, and disaster. With regard to the influencing factors mentioned above, the wall microclimate in a windy environment, illumination with sunshine, and temperature changes need to be carefully handled, and it is especially necessary to find suitable treatments for rising wind, descending wind, and vortexes, and to prevent the vertical greening side from being stripped by the wind (Peng et al. 2015).

According to Hemming, Speetjens, et al., in Taiwan, open-field vegetable production is threatened by subtropical climatic disasters, such as high wind speeds and heavy rainfall, which can cause the destruction of whole crops; also, vegetable production is threatened by pests and diseases, resulting in a high need for pesticides. Greenhouse production systems are able to provide protection for crops (Hemming et al. 2014).

## 16.3 Green Energy Water-Autonomous Greenhouse System

### *16.3.1 Novel Prototype of a Green Energy Water-Autonomous Greenhouse System*

The presented novel prototype of a green energy water-autonomous greenhouse system has been patented as an invention in Taiwan for a period from March 21, 2015 to August 24, 2031. The patent number is I-477230.

#### 16.3.1.1 Summary of the Prototype

The objective of the present prototype is to provide a green energy water-autonomous greenhouse system, which can be installed in suspension outside a window in order to automatically supply electric power for its internal operations by means of the solar power generation device configured therein, such that a plant ecological environment system similar to a greenhouse can be maintained within the window box.

The system capable of achieving the aforementioned objective comprises a frame body, with a plantation trough installed at the bottom; a solar power generation device, installed at the top of the frame body; a thermoelectric cooling chip board, including a plurality of thermoelectric cooling chips; and a window body structure, installed on the frame body, wherein the front, rear, left, right, and upper frames of the frame body all allow the installation of a window body structure, the thermoelectric cooling chip board is set up within the frame body, the solar power generation device is installed on the top of the frame body, and the solar power generation device is connected to the thermoelectric cooling chip board by way of a battery so as to provide electric power for the operations of the thermoelectric cooling chip board.

Therefore, the solar power generation device converts absorbed light energy into electric energy to drive the thermoelectric cooling chip board to operate, and the operations of the thermoelectric cooling chip board can reduce the surrounding temperature, thereby generating condensed water to drop-irrigate the plants cultivated in the plantation trough. In addition, the sunlight illumination further causes the temperature in the system to rise, such that the water molecules within the system may evaporate, and wet warm air is continuously heated and humidified and ascends by natural buoyancy and with assistance from the air circulation enhanced by a built-in ventilation device; as such, a plant ecological environment system can be achieved by way of this continuous circulation process.

The frame body includes a front frame, a rear frame, a left frame, a right frame, and an upper frame, and the window body structure is installed on the front frame, the rear frame, the left frame, the right frame, or the upper frame of the frame body. Moreover, the window body structure in the front frame, left frame, or right frame of the frame body may include a prism sheet splitter glass or a dimmer glass.

The window body structure may be a manual pushed-out window, an electric projected-out window, or a fixed window. In addition, a solar photovoltaic chip can be additionally installed within the window body structure, thereby acting as a splitter or dimmer glass to regulate and control the power supply, and the solar photovoltaic chip may include CIGS solar cells.

Condensing media may be placed under the thermoelectric cooling chip board, such that the cooling terminal of the thermoelectric cooling chip board can reduce the temperature and generate condensed water on the condensing media.

The solar power generation device on the top of the frame body may include a monocrystalline silicon solar photovoltaic panel and a battery, and the battery is connected to the monocrystalline silicon solar photovoltaic panel and the thermoelectric cooling chip board, wherein the interior of the monocrystalline silicon solar photovoltaic panel includes a solar photovoltaic chip, which can be monocrystalline silicon solar cells.

At least two water storage containers can be added to the inside of the frame body, wherein one of the water storage containers can be placed under the thermoelectric cooling chip board and the other one can be placed beneath the vegetation media of the plantation trough, and two such water storage containers are connected by way of a water pipeline.

The plantation trough includes the vegetation media and at least one plant, wherein the plant may be an ornamental plant or an edible plant, the plantation trough may be a light-porous moisture-retentive vegetation media trough, and the vegetation media may be soil or porous moisture-retentive media. Also, a siphon can be connected between the vegetation media in the plantation trough and the water storage containers.

A recycle pipeline and an inlet can be additionally installed in the system, wherein the recycle pipeline can be used to recycle condensed water or rainfall, and an end of the inlet is connected to the water storage container under the vegetation media in the plantation trough.

An electronic supervisory system can be additionally connected to the system, thereby allowing control over the window body structure (regarding the open and closure operations, as well as illumination adjustments of the window body structure) and the thermoelectric cooling chip board. Moreover, a ventilation device capable of ventilation amount adjustments, a thermohygrometer, or a barometer can be additionally installed in the system, and the electronic supervisory system controls the ventilation device, the hygrometer, the thermometer, or the barometer.

A heat recycle dehumidification ventilation device can be additionally installed in the system, thereby providing the features of inward air dehumidification, heat exchange, and ventilation, wherein the heat recycle dehumidification ventilation device may be set up between the bottom of the solar power generation device and the frame body or otherwise between the frame body and the top of the thermoelectric cooling chip board.

#### 16.3.1.2 Detailed Descriptions of the Preferred Embodiments

The aforementioned and other technical contents, aspects, and effects in relation to the present prototype can be clearly appreciated through the detailed descriptions concerning the preferred embodiments of the present prototype in conjunction with the appended drawings.

Refer initially to Fig. 16.1 (panels 1, 2A, 2B, 3A, and 3B), wherein a stereo combinatorial structure view, a front stereo view, a rear stereo view, and a lateral view of a green energy water-autonomous greenhouse system according to the present invention are respectively shown. From this figure, it can be seen that the system (1) comprises a frame body (11), having a plantation trough (16) installed at the bottom; a solar power generation device, installed at the top of the frame body (11) (and including a monocrystalline silicon solar photovoltaic panel (121) and a battery (122)); a thermoelectric cooling chip board (13) (which includes a plurality of thermoelectric cooling chips (131) and condensing media (132)); and a window body structure (14) installed on the frame body, wherein the front, rear, left, right, and upper frames of the frame body (11) all allow the installation of a window body structure (14). In addition, the thermoelectric cooling chip board (13) is installed on the upper side within the frame body (11), the solar power generation device is installed on the top of the frame body (11), and the solar power generation device is connected to the thermoelectric cooling chip board (13) by way of the battery (122) so as to provide electric power for the operations of the thermoelectric cooling chip board (13).

It can be appreciated from Fig. 16.1 (panels 1, 2A, and 2B) that the frame body (11) includes a front frame, a rear frame, a left frame, a right frame, and an upper frame, and the window body structure (14) is installed on the front frame, the rear frame, the left frame, the right frame, or the upper frame of the frame body (11). Herein the window body structure (11) may be a manual pushed-out window, an electric projected-out window, or a fixed window, so that in case the window body structure (11) is a manual pushed-out window or an electric projected-out window, as shown in Fig. 16.1 (panels 2A, 2B, and 3B), it can be pushed outward to open in order to perform air circulation for the interior of the system (1). Furthermore, a solar photovoltaic chip can be additionally installed within the window body structure (14), thereby acting as a splitter or dimmer glass to regulate and control the power supply, and the solar photovoltaic chip may be CIGS solar cells.

Also, the window body structure (14) in the front frame, left frame, or right frame of the frame body (11) may include a prism sheet splitter glass or a dimmer glass.

Moreover, condensing media (132) may be placed under the thermoelectric cooling chip board (13), such that the cooling terminal of the thermoelectric cooling chip board (13) can reduce the temperature and generate condensed water on the condensing media (132).

Additionally, the solar power generation device may include a monocrystalline silicon solar photovoltaic panel (121) and a battery (122), and the battery (122) is connected to the monocrystalline silicon solar photovoltaic panel (121) and the thermoelectric cooling chip board (13), wherein the interior of the monocrystalline silicon solar photovoltaic panel (121) includes a solar photovoltaic chip, which can be monocrystalline silicon solar cells.

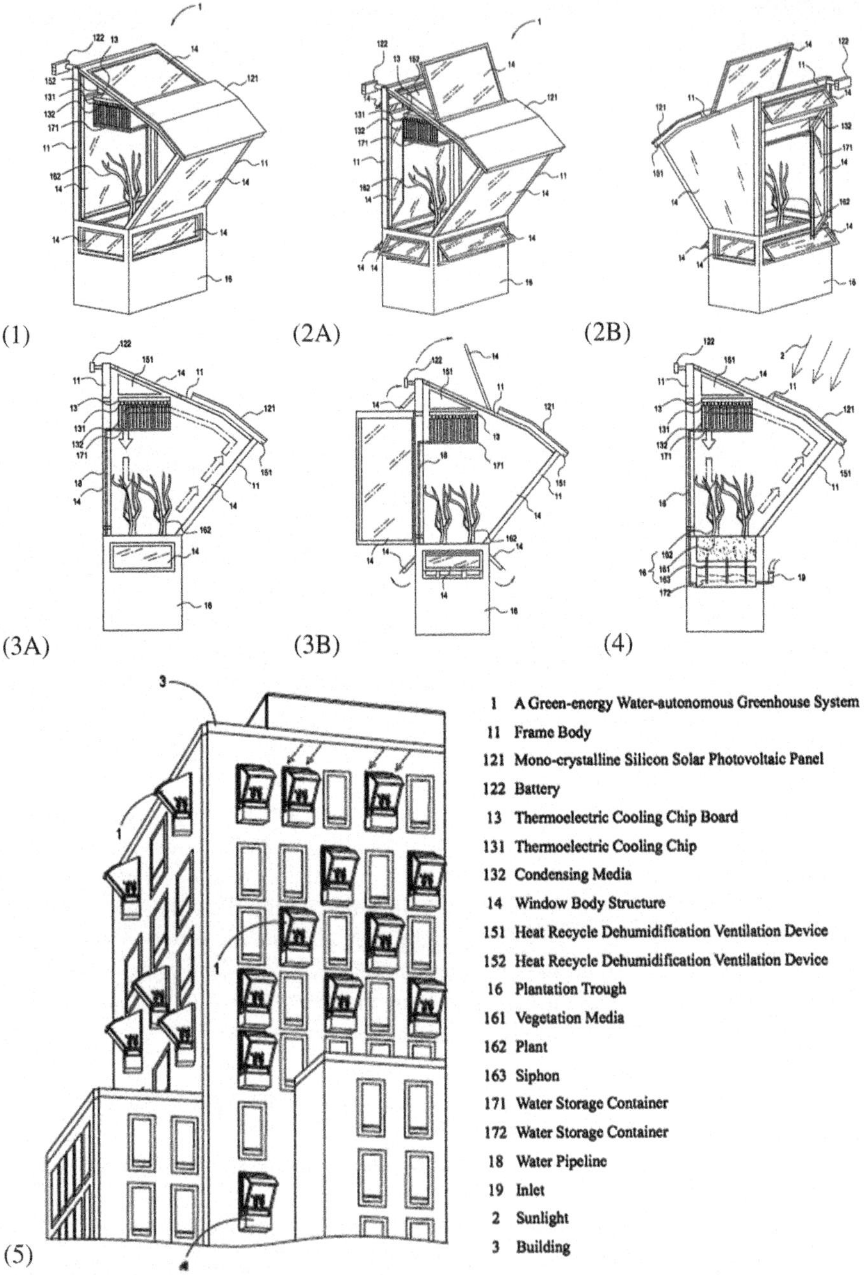


**Fig. 16.1** Drawings and major component symbol description (please see Sect. 16.3.1.2 for more details)

Also, heat recycle dehumidification ventilation devices (151, 152) can be additionally installed in the system (1) so these heat recycle dehumidification ventilation devices (151, 152) are capable of providing the features of inward air dehumidification, heat exchange, and ventilation, wherein one heat recycle dehumidification ventilation device (151) may be set up between the monocrystalline silicon solar photovoltaic panel (121) of the solar power generation device and the frame body (11), and the other heat recycle dehumidification ventilation device (152) may be installed between the frame body (11) and the thermoelectric cooling chip board (13).

Furthermore, as shown in Fig. 16.1 (panel 4), the plantation trough (16) can include the vegetation media (161) and at least one plant (162), wherein the plant (162) may be an ornamental plant or an edible plant, the plantation trough (16) may be a light-porous moisture-retentive vegetation media trough, and the vegetation media (161) may be soil or porous moisture-retentive media. Also, a siphon (163) can be connected between the vegetation media (161) in the plantation trough (16) and the water storage container (172).

In addition, it can be observed from Fig. 16.1 (panel 4) that at least two water storage containers (171, 172) can be added to the inside of the frame body (11), wherein a water storage container (171) can be placed under the thermoelectric cooling chip board (13) and the other one can be placed beneath the vegetation media (161) of the plantation trough (16), in which these two water storage containers (171, 172) are connected by way of a water pipeline (18), so as to transfer water held in one water storage container (171) into the other water storage container (172).

Next, it can be seen from Fig. 16.1 (panel 4) that a recycle pipeline (not shown), as well as an inlet (19), can be additionally installed in the system (1), wherein the recycle pipeline facilitates the recycle usage of condensed water or rainfall, and an end of the inlet (19) is connected to the water storage container (172) under the vegetation media (161) in the plantation trough (16), such that the user can also irrigate manually via the inlet (19).

In addition to the abovementioned implementations, other devices can also be configured in the system (1) according to the present prototype. For example, an electronic supervisory system (not shown) may be alternatively connected into the system (1), such that the electronic supervisory system can control the window body structure (14) and the thermoelectric cooling chip board (13). Furthermore, a ventilation device capable of ventilation amount adjustments (not shown), a thermohygrometer capable of detecting the temperature and humidity in the system (1) (not shown), or a barometer capable of detecting air pressure values in the system (1) (not shown) can be additionally installed in the system (1).

The electronic supervisory system can further control the ventilation device, the thermohygrometer, or the barometer, such that the user can remotely monitor the environmental variations within the system (1) and control the ventilation device so that the ventilation device can automatically adjust the states of air speed, air withdrawal, and air exhaustion in order to adjust the temperature in the system (1). What is more, the electronic supervisory system also allows the user to remotely monitor the environmental variations in the system, and the electronic supervisory system can control the window body structure to open or close automatically and manipulate the thermoelectric cooling chip board (13) to adjust the internal temperature.

Consequently, as shown in Fig. 16.1 (panel 4), when the sunlight (2) illuminates the monocrystalline silicon solar photovoltaic panel (121), the solar power generation device converts absorbed light energy into electric energy and transfers electric power to the thermoelectric cooling chip board (13), as well as the ventilation device (not shown); therefore, when the thermoelectric cooling chip board (13) operates, the cooling end of the thermoelectric cooling chip (131) in the thermoelectric cooling chip board (13) can reduce the temperature, and condensed water created by means of the condensing media (132) can drop down due to gravity and irrigate the plant (162) in the plantation trough (16). Meanwhile, because the sunlight (2) illuminates the system (1), the internal temperature may increase and the potential energy difference between the plant (162) and the vegetation media (161) (e.g., soil) in the plantation trough (16) causes water molecules to evaporate and rise up to the top, and the air circulation can be enhanced with the ventilation device, such that the wet warm air can be continuously heated and humidified in order to ascend, thereby achieving a plant ecological environment system capable of autonomous circulations by means of this continuous circulation process.

To further describe the contents of the present invention in detail, a diagram for an embodiment according to the present invention is herein proposed. As shown in Fig. 16.1 (panel 5), the system (1) can be directly installed in suspension outside a building (3), such that the system (1) can serve for the purposes of heat insulation, sound insulation, air filtering of the building (3), landscape effects, or an illumination box.

#### 16.3.1.3 Brief Description of the Drawings

Figure 16.1 shows the drawings and major component symbol descriptions, including a stereo combinatorial structure view, a front stereo view, a rear stereo view, two lateral views, an operation diagram, and an embodiment diagram for a green energy water-autonomous greenhouse system according to the present invention.

## 16.4 Discussion

### *16.4.1 Innovations*

Figure 16.2 shows the system function and illustrations of the present invention.

Innovations in the presented prototype of a green energy water-autonomous greenhouse system are pointed out as follows:

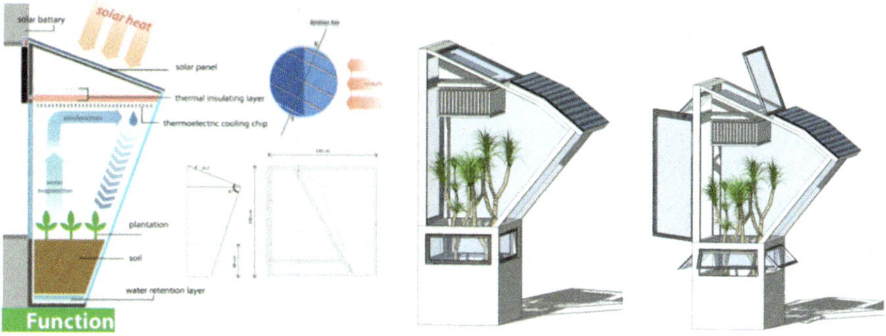


**Fig. 16.2** Innovations in the system

1. A system is a set of things interconnected in such a way that they produce their own pattern of behavior over time. The system may be buffeted, constricted, triggered, or driven by outside forces. But the system's response to these forces is characteristic of itself, and that response is seldom simple in the real world. A system is more than the sum of its parts; it may exhibit adaptive, dynamic, goal-seeking, self-preserving, and sometimes evolutionary behavior (Meadows 2008). Just like the system defined above, the green energy water-autonomous greenhouse system (named GEWA), interconnected by a set of elements, as shown in Figs. 16.1 and 16.2, is indeed a coherently organized system that achieves an adaptive eco-environment.
2. The system's characteristic key functioning starts with the use of a thermoelectric cooling chip board on the basis of the Peltier effect, which make it possible that the evaporation and condensation processes for a small water cycle could be completed inside the greenhouse, and an autonomous water supply could be accomplished.
3. GEWA enhances sophisticated and multidisciplinary approaches and technologies mentioned in most reviewed research subjects and is an attempt to achieve an adaptive semiclosed climate-responsive passive greenhouse system with the following points of invention:

   (1) There is automatically supplied electric power for the internal operations by means of a solar power generation device installed at the top of the frame body. The solar power generation device is connected to a thermoelectric cooling chip board by way of a battery to provide electric power to drive it to reduce the surrounding temperature, thereby generating condensed water.
   (2) The condensed water is phase-changed by evapotranspiration into water vapor ascending with sunlight-illuminated wet warm air by natural buoyancy.
   (3) An electronic supervisory system can be connected, thereby allowing control over the operations of the window's opening–closing and illumination adjustments, the thermoelectric cooling chip board, the ventilation device,

the heat recycle dehumidifier, the hygrometer, the thermometer, the barometer, and other measuring devices.

4. The special integrated mechanisms applied in the system include:
   (1) The photovoltaic effect of the solar power system
   (2) Purification of water
   (3) Photocatalysis on the surfaces of the structure, materials, equipment, and devices
   (4) A fuzzy control strategy for the climatic sensor system
   (5) The photoelectrochromic effect of the window glass
   (6) Prism sheet splitter glass
   (7) Other intelligent automations
5. GEWA could be developed as a cyber-physical system.

### 16.4.2 Advantages of the General System

Compared with other conventional application technologies, GEWA further offers the following advantages:

1. A plant ecological environment system can be maintained autonomously by the system.
2. The window of the present prototype facing indoors can be opened to supply the generated oxygen into the room, thereby attaining the purposes of a green environment and beautification effect, and the people within the room can also enjoy the delightful landscape.
3. The present prototype can be installed extending outward from a window, does not occupy indoor space, and is very convenient for arrangements. Meanwhile, such an outside-window installation can be helpful for adding suitable ornamental plants or alternatively cultivating edible plant vegetation to act as a small-sized vegetation farm.
4. Environmental variations within the system can be monitored remotely by an externally connected electronic supervisory system, and the environmental changing factors can be controlled and suitably regulated to maintain the optimal growth environment for the plants.

### 16.4.3 Advantages: Interaction Across Building Systems

GEWA, as an eco-environment, could have many interactions across building systems. It could be an experimental platform, a climate station, or a cloud point, and could indeed be a third environment in between the outside and inside building environments. Being a third environment, GEWA could be more adaptive than conventional vertical greening by an additional function of mitigating the environmental impact of a building. It could become a cyber-physical system.

An example is described by a CFD simulation to investigate the influence of GEWA on the wind flow from the outside to the inside of the building. As shown in Fig. 16.3, wind flow with average wind speed of 2.43 m/s with a southwest wind direction was simulated. Figure 16.3a–c shows a CFD model and a grid pattern with 2,753,520 computing grids; Fig. 16.3d shows a wind flow section in a range of 0–3 m/s; Fig. 16.3e shows wind speed distribution at 1.5 m high in a range of 0–3 m/s; and Fig. 16.3f shows the wind speed equalizer surface at a wind speed of 1 m/s. From the results of the analysis, the wind flow pattern can be realized and the effects on the environment can be evaluated.

Other environmental factors and parameters such as temperature and humidity could also be simulated.

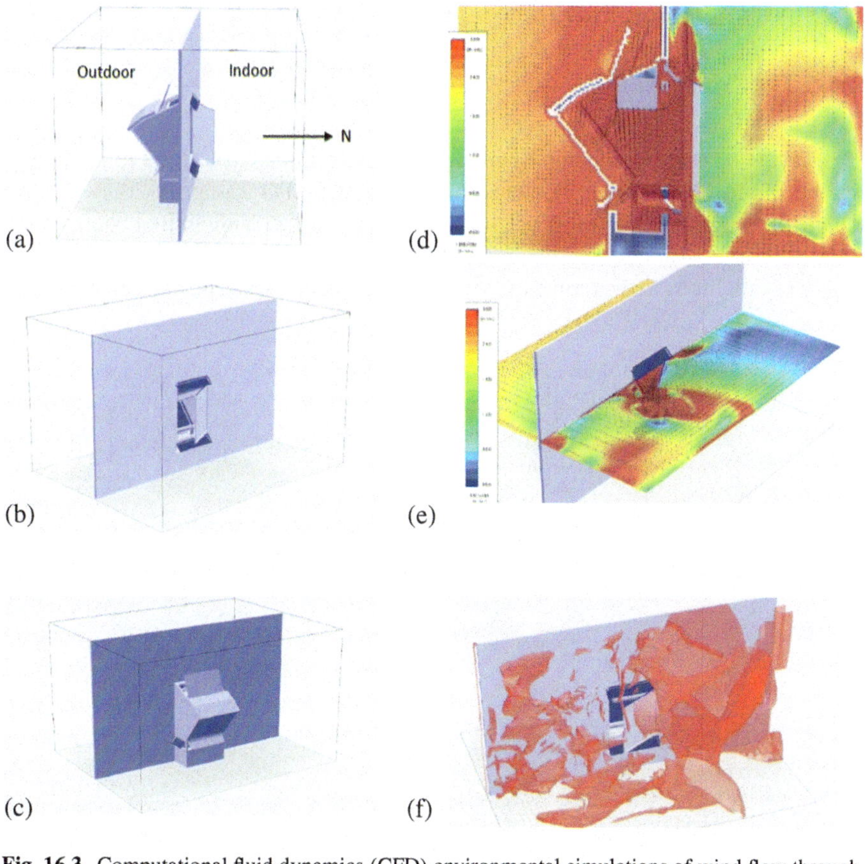


**Fig. 16.3** Computational fluid dynamics (CFD) environmental simulations of wind flow through a greenhouse (simulated by Windperfect DX). (**a–c**) CFD model and grid pattern with 2,753,520 computing grids. (**d**) Wind flow section in a range of 0–3 m/s. (**e**) Wind speed distribution at 1.5 m high in a range of 0–3 m/s. (**f**) Wind speed equalizer surface at a wind speed of 1 m/s

#### 16.4.3.1 Advantages: Incorporating Greenery into Buildings

Figure 16.4 shows an example of the incorporation of GEWA into the Mega Holdings Building, Taipei, using a real "before" photograph and a simulated "after" photograph.

GEWA could incorporate a variety of greenery into buildings on the walls and openings; it could be a window garden, a sky garden, or a vertical farm. It could develop a vertical greenery networking system in a smart city.

It is expected that GEWA will be used to incorporate greenery into buildings as an alternative way of vertical greening to open up a very interesting possibility for increasing the attractiveness of the cityscape and to enhance progressive urban revitalization in a smart city.

## 16.5 Conclusion and Recommendations

### *16.5.1 Conclusion*

The saying by Parakramabahu the Great, King of Sri Lanka from 1153 to 1186, used as the Motto of the Water Tribune Conference at EXPO Zaragoza 2008—"Not a single drop of rain-water must be allowed to go to the sea without first being of service to people"—announced the greatest importance of efficient usage of water and water saving for us to overcome the growing drought and water scarcity in the world.

**Fig. 16.4** Incorporating greenery into a building: before and after

Green technologies of efficient natural solar energy usage, including energy generation and energy retention, are also most important for humankind in "going greener" and "getting smarter" naturally and wisely.

A sophisticated and multidisciplinary green energy water-autonomous greenhouse system—by using water resources and solar energy in a rational way to aim at greater well-being based on more products, more information, more services, and more experience—could be an alternative technology approach toward sustainable smart–green vertical greening in a smart city.

There are also good sayings for considering innovations:

> "Foresight should take the place of experience."—Shimon Peres, former President of Israel (July 15, 2007 to July 24, 2014)

> "Simplicity is the ultimate sophistication."—Leonardo da Vinci

> "There are three fundamental principles of evolution: mutation, selection and cooperation. Mutation leads to diversity upon which selection can act. Cooperation allows the construction of higher levels of organization."—Martin Andres Nowak

Aimed at improving responsiveness, conservation, efficiency, and performance for environmental sustainability, resource sustainability, and material and technological sustainability, and also aimed at greater well-being.

### *16.5.2 Recommendations*

It is recommended that follow-up studies be conducted in the following areas:

1. A real model of the prototype should be established to conduct a relevant experimental study and to contribute to building big data for research.
2. Research studies should be continued to search for a more progressive GEWA by enriching qualities of the design from variety, leaps of imagination in materials and form, sheer elegance of craft and composition, and inspirational ideas for increasing the attractiveness of the cityscape and to enhance progressive urban revitalization in smart cities.
3. For a more progressive GEWA, aspects such as energy efficiency, water saving, construction methods, performance of materials and smart devices, system mechanisms, environmental benefits, and economics must be further examined.

## References

Abdel-Ghany AM, Al-Helal IM (2011) Solar energy utilization by a greenhouse: general relation. Renew Energy 36:189–196

Attmann O (2010) Green architecture: advanced technologies and materials. McGraw-Hill Education, New York

Baeza EJ, Perez Parra J et al (2009) Analysis of the role of sidewall vents on buoyancy-driven natural ventilation in parral-type greenhouses with and without insect screens using computational fluid dynamics. Biosyst Eng 104:86–96
Bakker JC, Adams SR et al (2008) Innovative technologies for an efficient use of energy. Acta Hortic 801(1):49–62
Balas MM (2014) Seven passive greenhouse synergies. Acta Polytechnica 11:199–210
Buchholz M, Zaragoza G et al (2004) Thermal control of an isolated greenhouse for optimized food production and greywater recycling: the Watergy project, International Astronautical Federation—55th International Astronautical Congress 2004, vol 3, pp 1681–1688
Buchholz M, Buchholz R et al (2006) Temperature and humidity control in the watergy greenhouse. Acta Hortic 719:401–407
Buchholz M, Schmidt M et al (2008) Overcoming drought. Implementation guide of the "Cycler Support" project, RTD FP 6-INCO. Ref. Nr. 031697
Caragliu A, Del Bo C, Nijkamp P (2009) Smart Cities in Europe, Series Res
Giacomelli G (2007) Innovation in greenhouse engineering for GreenSys2007, New AG International, GreenSys pages at: www.newaginternational.com
Hatzelhoffer L, Humboldt K et al (2012) Smart city in practice. Jovis Verlag GmbH, Berlin
Hemming S, Speetjens SL et al (2014) Greenhouse design for vegetable production in subtropical climate in Taiwan. Acta Hortic 1037:65–74
Jochum P, Buchholz M (2005) How to simulate thermal and fluid dynamical processes in closed greenhouses including water interactions between plants and air. Acta Hortic 691:553–560
Katsoulas N, Sapounas A et al (2015) Reducing ventilation requirements in semi-closed greenhouses increases water use efficiency. Agric Water Manag 156:90–99
Kwok AG, Grondzik WT (2011) The green studio handbook. Taylor & Francis, New York
Lamnatou C, Chemisana D (2013) Solar radiation manipulations and their role in greenhouse claddings: fluorescent solar concentrators, photoselective and other materials. Renew Sust Energ Rev 27:175–190
Meadows DH (2008) Thinking in systems—a primer. Chelsea Green Publishing, White River Junction
Montero JI, Van Henten EJ et al (2011) Greenhouse engineering: new technologies and approaches. Acta Hortic 893:51–64
Opdam JJG, Schoonderbeek GG et al (2005) Closed greenhouse: a starting point for sustainable entrepreneurship in horticulture. Acta Hortic 691:517–524
Orgaz F, Fernandez MD et al (2005) Evapotranspiration of horticultural crops in an unheated plastic greenhouse. Agric Water Manag 72:81–96
Peng KH, Kuo YC et al (2015) The use of vertical greening in urban rehabilitation to improve sustainability of the environment in Taiwan. Int Rev Spat Plan Sustain Dev 3(1):5–16
Perez Parra J, Baeza E et al (2004) Natural ventilation of parral greenhouses. Biosyst Eng 87(3):355–366
Speetjens SL, Stigter JD et al (2009) Towards an adaptive model for greenhouse control. Comput Electron Agric 67:1–8
Speetjens SL, Stigter JD et al (2010) Physics-based model for a water-saving greenhouse. Biosyst Eng 105:149–159
Tantau HJ, Schmidt U et al (2011) Low energy greenhouse—a system approach. Acta Hortic 893:75–84
Teitel M, Ziskind G et al (2008) Effect of wind direction on greenhouse ventilation rate, airflow patterns and temperature distributions. Biosyst Eng 101:351–369
Teitel M, Baeza EJ et al (2012) Greenhouse design: concepts and trends. Acta Hortic 952:605–620
Vadiee A, Martin V (2012) Energy management in horticulture applications through the closed greenhouse concept, state of the art. Renew Sust Energ Rev 16:5087–5100
Vadiee A, Martin V (2013) Thermal energy storage strategies for effective closed greenhouse design. Appl Energy 109:337–343
Van Kooten O, Heuvelink E et al (2008) New developments in greenhouse technology can mitigate the water shortage problem of the 21st century. Acta Hortic 767:45–52

Zaragoza G, Buchholz M (2008) Closed greenhouses for semi-arid climates: critical discussion following the results of the Watergy prototype. Acta Hortic 797:37–42

Zaragoza G, Baeza E et al (2005) The Watergy greenhouse: a closed system for solar thermal energy collection, water treatment and advanced horticulture. In: SAE international conference on environmental systems, paper #: 2005-01-2919

Zaragoza G, Buchholz M et al (2007) Watergy project: towards a rational use of water in greenhouse agriculture and sustainable architecture. Desalination 211:296–303

The manufacturer's authorised representative in the EU is Springer
Nature Customer Service Centre GmbH, Europaplatz 3, 69115 Heidelberg,
Germany. If you have any concerns regarding our products, please
contact ProductSafety@springernature.com

Printed and bound by CPI Group (UK) Ltd, Croydon, CR0 4YY
30/09/2026
02225587-0006